Diagnostic Pathology of the Intestinal Mucosa

William O. Dobbins III

Diagnostic Pathology of the Intestinal Mucosa

An Atlas and Review of Biopsy Interpretation

With 124 Illustrations in 173 Parts

Springer-Verlag
New York Berlin Heidelberg
London Paris Tokyo Hong Kong

William O. Dobbins III, M.D., Professor of Internal Medicine, Department of Internal Medicine, University of Michigan School of Medicine, Ann Arbor, MI 48109; Associate Chief of Staff for Research, Veterans Administration Medical Center, Ann Arbor, MI 48105, USA

Library of Congress Cataloging in Publication Data
Dobbins, William O., 1932-
Diagnostic pathology of the intestinal mucosa : an atlas and review of biopsy interpretation / William O. Dobbins.
p. cm.
Includes bibliographical references.
ISBN-13:978-1-4612-7946-4
1. Intestinal mucosa—Biopsy—Atlases. 2. Intestinal mucosa—Histopathology—Atlases. I. Title.
[DNLM: 1. Intestinal Diseases—diagnosis—atlases. 2. Intestinal Mucosa—pathology—atlases. WI 17 D632d]
RC804.B5D63 1990
616.3'407'58—dc20
DNLM/DLC
for Library of Congress 89-21937
CIP

Softcover reprint of the hardcover 1st edition 1990

Typeset by Caliber Design Planning, Inc., New York, New York.

9 8 7 6 5 4 3 2 1

ISBN-13:978-1-4612-7946-4 e-ISBN-13:978-1-4612-3276-6
DOI: 10.1007/978-1-4612-3276-6

Contents

Acknowledgments

The author is grateful to the following individuals who generously supplied illustrations for this text: Elizabeth D. Hay, James L. Madara, Marian Neutra, and Jerry S. Trier at Harvard Medical School; C.P. Leblond at McGill University; John C. Partin at SUNY, Stony Brook; L. Andrew Staehelin at the University of Colorado; Wilfred M. Weinstein at UCLA; Robert L. Owen at UCSF; John Heuser and Robert J. Rothbaum at Washington University; and Cyrus E. Rubin at the University of Washington. Indeed, greatest credit goes to Cy Rubin who taught me all that I know about intestinal morphology and who continues to teach me.

The following publishing companies permitted me to reproduce the illustrations supplied by the individuals listed above. The company and the journal or text (in parentheses) are: Academic Press, Inc. (*Pathobiology of Cell Membranes*, Vol. I, 1975, BF Trump, ed.); Alan R. Liss, Inc. (*American Journal of Anatomy*); Churchill Livingstone, Inc. (*Pathology of the Colon, Small Intestine, and Anus*, 1983, HT Norris, ed.); Elsevier Science Publishing Co., Inc. (*Gastroenterology*); John Wiley and Sons, Inc. (*Diagnostic Electron Microscopy*, 1978, BF Trump and RL Jones, eds.); Raven Press (*Physiology of the Gastrointestinal Tract*, 1987, L.R. Johnson, ed.); Rockefeller University Press (*Journal of Cell Biology*); and Williams and Wilkins Co. (*Gastroenterology*).

Introduction

This book is designed to provide a thorough overview of intestinal mucosal structure defined in peroral or endoscopic biopsy specimens obtained in normal or disease states. It is intended to serve students, clinical and investigative gastroenterologists, and research and clinical pathologists. All are invited to pick and choose those portions that best suit their needs. Routine methods of processing tissues for light microscopy, electron microscopy, histochemistry, and light and electron microscopic immunoperoxidase techniques are provided so that the novice in intestinal structure may have an easy reference for setting up a reasonably sophisticated morphologic laboratory. The ultimate goal of this text is to provide a thorough review of mucosal biopsy pathology of the small intestine, while providing sufficient morphological and functional or pathophysiological correlations so that the reader comes away with an intellectually satisfying understanding of intestinal structure and function as well as with a practical grasp of the value of mucosal biopsy of the intestine in the diagnosis of disease. In some measure this text is an amalgam of three of my publications dealing with small bowel biopsy at light microscopy (1) and at electron microscopy (2), and of the review of the human intestinal epithelium as a biological membrane (3). However, much more is contained in this text and I have borrowed liberally from other authorities in intestinal mucosal structure and pathology.

There are a number of reviews of intestinal ultrastructure (4–6) and of biopsy pathology of the intestine (7–10) but none gives in-depth reviews of all aspects of mucosal structure and none correlates the diverse variety of available light and electron microscopic information in a unified manner so as to promote an in-depth understanding of the pathology of intestinal mucosal disease. I have avoided extensive discussions of nonspecific and minor pathological changes seen in biopsy specimens. Such discussions have no more value than do the nonspecific changes have in making a diagnosis. Noting their presence is sufficient.

Emphasis is placed upon disease states in which peroral biopsy (either capsule or endoscopic) is helpful in making diagnosis. Other disease states in which peroral biopsy is not likely to be helpful, e.g., ischemic enteritis, are not discussed. I have attempted to use simple straightforward descriptive terms and to avoid such terminology as "villous atrophy" and "crypt hyperplastic villous atrophy" and have preferred to follow C.E. Rubins's example of describing architectural changes in villi as being mild, moderate, and severe, with or without crypt lengthening.

Discussions that speculate about the etiology of disease causation, e.g., celiac sprue, Crohn's disease, are avoided. There is ample (often mindless) material in this regard in the references cited in the bibliographies. Intestinal neoplasms and tumors, other than lymphoma, are not discussed because they are rarely diagnosed by peroral biopsy.

References

1. Dobbins WO III: Small bowel biopsy in malabsorptive states. In *Contemporary Issues in Surgical Pathology, Vol. 4. Pathology of the Colon, Small Intestine and Anus.* HT Norris (Ed.), New York, Churchill Livingstone Inc., 1983, pp 121–165.
2. Dobbins WO III: Diagnostic pathology of the intestine and colon. In *Diagnostic Electron Microscopy, Vol. I.* BF Trump and RT Jones (Eds.), New York, John Wiley, 1978, pp 253–339.
3. Dobbins WO III: Human intestinal epithelium as a biological membrane. In *Pathobiology of Cell Membranes,* BF Trump (Ed.), New York, Academic Press, 1975, pp 429–469.
4. Pfeiffer CJ, Rowden G, Weibel J: *Gastrointestinal Ultrastructure–An Atlas of Scanning and Transmission Electron Micrographs,* Tokyo, Igaku Shoin, 1974.
5. Toner PG, Carr KE, Wyburn GM: *The Digestive System–An Ultrastructural Atlas and Review,* London, Butterworth & Co., 1971.
6. Shiner M: *Ultrastructure of the Small Intestinal Mucosa,* Berlin, Springer-Verlag, 1983.
7. Lee FD, Toner PG: *Biopsy Pathology of the Small Intestine,* Philadelphia, J.B. Lippincott Company, 1980.
8. Rotterdam H, Sommers SC: *Biopsy Diagnosis of the Digestive Tract–Biopsy Interpretation Series,* New York, Raven Press, 1981.
9. Whitehead R: *Mucosal Biopsy of the Gastrointestinal Tract, Vol. 3, Major Problems in Pathology,* JL Bennington (Ed.), Philadelphia, W.B. Saunders, 1985.
10. Day DW: *Biopsy Pathology of the Oesophagus, Stomach and Duodenum,* New York, John Wiley, 1986.

1
Processing of Biopsy Specimens for Light and Electron Microscopy

Methods for the processing of biopsy specimens and tissues for light and electron microscopy are quite routine and have been discussed in detail elsewhere.[1,5] The methods outlined briefly here should be helpful to pathologists and gastroenterologists who have a special interest in the histopathology of gastrointestinal mucosal lesions.

Rubin and colleagues[1] have pointed out five main sources of misinterpretation of small-bowel biopsy specimens:

1. Failure of the clinician to orient and handle the biopsy optimally
2. Failure of the technician to process and section the biopsy adequately
3. Failure of the clinician to provide adequate information to the pathologist
4. Inadequate appreciation by the pathologist of the spectrum of normal intestinal histology
5. Failure of the pathologist to appreciate the clinical significance of the interpretation

A sixth important factor is inadequate tissue sampling. This chapter will briefly discuss sampling and processing of biopsy specimens for both light and electron microscopy. Formulas for the solutions involved in processing are given at the end of each section. Chapter 2 will discuss interpretation in detail.

Sampling

Biopsy specimens may be obtained by suction or by endoscope. With the suction method, specimens should be obtained routinely near the easily identified fluoroscopic landmark the ligament of Treitz, located at the duodenojejunal junction. Using this as the standard biopsy location facilitates comparisons among patients and assessment of therapeutic response in the same patient. Rubin and colleagues[1] have discussed the suction biopsy technique in detail, and interested readers are referred to this paper.

Because suction biopsy necessitates fluoroscopic placing of the biopsy tube, I recommend that specimens be obtained endoscopically under direct vision. This permits sampling of those areas that show the greatest gross changes and takes into account the fact that many mucosal lesions are patchy in distribution. By using the large ("jumbo") biopsy forceps one can obtain specimens even larger than those obtained with the suction biopsy technique.

With endoscopic biopsy, specimens must be obtained more proximally than those obtained with suction biopsy. Thus, unless focal lesions are observed proximally, the specimens should be obtained as far distally as possible. If one keeps in mind that confusing histological artifacts are more likely to be present in the proximal than in the distal duodenum, the proximal duodenum accurately reflects the distal duodenum.[6]

Because proximal small-intestine lesions may be patchy, four or more biopsy specimens should be obtained during each sampling. This permits processing of three specimens by routine light microscopy and of one specimen by electron microscopy or biochemical analysis.

In the case of a patient with malabsorption, before a biopsy specimen is taken the prothrombin time should be verified as normal and a careful history should be obtained to discern evidence of bleeding phenomena.[5] If such evidence is found, a complete clotting workup is probably indicated. All patients should avoid taking aspirin for 72 hours before the biopsy procedure.

Orienting the Sample

The physician who performs the biopsy should be knowledgeable about the method of orientation of the tissue and should be careful not to traumatize it. The specimens should be flattened carefully and mounted on a monofilament plastic mesh or on a thin plastic disk so that there will be no doubt about its proper orientation at the time of sectioning by the laboratory technician. The value of the monofilament mesh is that it leaves no material adherent when the mesh is removed before sectioning. This protects the cutting edge of the microtome knife from needless injury.

Processing Specimens for Light Microscopy

Processing for light microscopy consists of the following steps:

- Fixation
- Dehydration
- Embedding
- Sectioning
- Staining

Fixation

After the specimen is oriented on mesh or disk it is immediately fixed in Bouin's fixative (my preference) or in Hollande's fixative for at least two hours but no longer than three days. Hollande's-fixed tissue is easier to section than formalin-fixed tissue. More important, there is less shrinkage in tissue fixed in Bouin's or Hollande's fixative and therefore few artifactual gaps in the tissue sections. (The formula for Bouin's solution is given at the end of this section.)

Dehydration

Dehydration can be done manually or mechanically (Autotechnicon). The procedures are as follows.

Manual Dehydration

Agitate solutions occasionally. Carry through in 1-oz jars.

Soak in 70% alcohol.	30–40 min
Soak in 95% alcohol.	30 min
Soak in absolute alcohol, four times.	15–30 min each time
Soak in chloroform or xylene, three times.	15 min each time
Place in Paraplast* paraffin (56–60°C temp range) (in small beakers in 60°C oven).	2 hr

Autotechnicon Dehydration

Hold in 70% alcohol.	
Soak in 70% alcohol.	40 min
Soak in 95% alcohol.	30 min
Soak in absolute alcohol, two times.	30 min each time
Soak in chloroform (clearing agent–mixes with both alcohol and paraffin; causes less distortion than xylene).	30 min
Soak in chloroform.	40 min
Place in Paraplast* paraffin (56–60°C temp range).	2 hr

Embedding

The specimen should be embedded as quickly as possible after dehydration. Because serial sectioning of the specimen is important, the use of Paraplast paraffin (56–58°C temperature range) for embedding is preferable. The specimen is removed gently from the monofilament plastic mesh by bending the mesh and exposing one side of the specimen. The specimen is picked up from paraffin in small beaker at 60° C with warmed forceps and placed in the embedding mold (Peel-A-Way†) filled with paraffin. Following Autotechnicon dehydration, the use of Tissue Tek stainless-steel base molds (no. 4121) for embedding facilitates the ease with which the tissue can be oriented for appropriate sectioning. An identification tag containing the name of the patient, the date of biopsy, and a letter for the particular specimen is placed at the top of the specimen. (The cut surface of the specimen is easier to cut than the intact surface; thus it should be considered the bottom so that it hits the knife first.)

Sectioning

The paraffin block is mounted on a "chuck," which is attached to a rotary microtome, and the block is trimmed down fairly close to the specimen so that two to

---

*Monoject Scientific Division of Sherwood Medical, St. Louis, MO 63103, USA.
†Peel-A-Way Scientific, South El Monte, CA 91733, USA

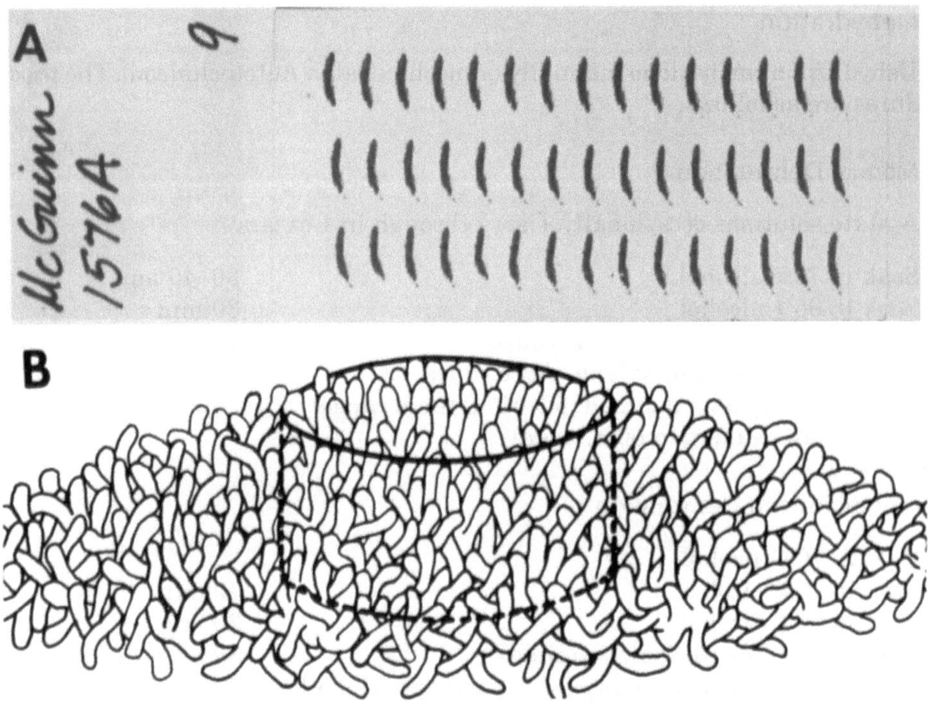

Figure 1-1. Serial sectioning. (A) Photograph of a typical microscopy slide containing three ribbons of 15 sections each. (B) Diagram of an oriented biopsy specimen to illustrate the central core which is most likely to contain sections with the best oriented villi. (A and B reproduced from Dobbins,[5] by permission of Churchill Livingstone Inc., New York.)

three ribbons per slide containing approximately 30 serial sections per ribbon can be obtained (Figure 1-1A). The specimens are then sectioned at 4 μm through the limits of the central core (Figure 1-1B). (If the knife is kept clean with xylene it will not need frequent sharpening.) The cut sections are floated in a 55°C water bath containing 50 to 100 mg of gelatin. The sections are then placed on clean slides and are dried in a 60°C oven, or on a warming plate at 60°C, for about an hour just prior to staining. Three to four slides should be prepared from each sampling. The slides are allowed to cool down somewhat, which prevents the specimens from falling off the slides during staining.

Serial sectioning of small-intestine biopsies is not essential, though it is very helpful in ascertaining the presence or absence of four normal villi in a row in a normal specimen. Step sectioning of rectal biopsies is essential to detect focal lesions and granulomas in inflammatory bowel disease.[7]

Staining

Rubin and colleagues[1] have recommended hematoxylin and eosin (H & E) staining with Alcian blue and saffron. The saffron stains collagen and differentiates edema from collagenous scarring. I prefer to use H & E on one slide and periodic acid Schiff (PAS)-hematoxylin on a separate slide. It is also helpful to leave several slides unstained for those occasions when special stains are found to be needed. The staining procedures are as follows.

Hematoxylin and Eosin—Routine Stain Using Harris' hematoxylin

Removal of paraffin–decerating (steps 1 and 2).

1. Soak in xylene.	3 min
2. Soak in xylene.	3 min

Hydration (to use aqueous stain) (steps 3–7).

3. Soak in absolute alcohol.	2 min
4. Soak in absolute alcohol.	2 min
5. Soak in 95% alcohol.	2 min
6. Soak in 80% alcohol.	2 min
7. Rinse in tap water.	2 min
8. Soak in Harris' hematoxylin, pH 2.95 (filter daily).	1½ min (approx.)
9. Wash well with tap water.	Briefly
10. Dip in 0.5% concentrated HCl in 70% alcohol (make fresh each time you stain).	5–10 dips
11. Wash with tap water (may need distilled).	Briefly
12. Dip in saturated lithium carbonate.	2–5 dips
13. Wash well in tap water.	3 min (minimum)
14. Examine under microscope to check color: Cytoplasm should be gray, nuclei a healthy blue.	
15. Dip in 0.5% buffered eosin, pH 5.3.	10–20 dips (approx.)
16. Rinse in buffered water.	4–5 dips

Dehydration (to use organic mounting medium) (steps 17–19).

17. Dip in 70% alcohol.	4–5 dips
18. Dip in 95% alcohol.	8 dips
19. Soak in 100% alcohol (differentiate with this	Briefly
under microscope).	Briefly
20. Wash with toluene (clearing agent) or xylene.	
21. Mount in Histoclad or Permount.	

Other Stains Frequently Useful for Interpreting Intestinal Biopsy Specimens

Alcian Blue, H & E

1. Deparaffinize sections to water.	
2. Rinse in 3% acetic acid.	2 min
3. Stain in 0.1% Alcian blue in 3% acetic acid.	10 min
4. Wash in running water.	2 min
5. Rinse in distilled water.	Briefly
6. Soak in alkaline alcohol (sodium borate 0.5% in 1% ethanol).	2 hr
7. Wash in running water.	2 min
8. Follow H & E procedure.	

Congo Red

1. Hydrate to water.	
2. Soak in 0.5% congo red in 50% alcohol.	5 min
3. Soak in 0.2% potassium hydroxide in 80% alcohol.	1–3 min
4. Rinse in tap water.	Briefly
5. Follow H & E procedure (results: amyloid, red; nuclei, blue).	

Iron Hematoxylin

1. Hydrate to water.	
2. Soak in 4% iron alum.	30 min
3. Wash in running water.	10 min
4. Stain in 0.5% hematoxylin.	30 min
5. Wash in running water.	10 min
6. Destain in saturated aqueous picric acid; check by washing in tap water and then dipping in ammonia water: background very pale, nuclei bright blue.	
7. Wash in running water.	20 min (minimum)
8. Counterstain in 0.5% aqueous eosin.	2 min
9. Dehydrate, clear, and mount.	

Periodic Acid-Schiff (PAS)

1. Deparaffinize sections to water.	
2. Soak in 0.5% periodic acid.	5 min
3. Rinse in running water.	1 min
4. Soak in Schiff's reagent.	15 min
5. Rinse in sulfurous acid, three times	2 min each time
6. Wash in running water.	10 min
7. Soak in hematoxylin.	1½ min
8. Decolorize and blue as for H & E.	
9. Wash in graded alcohols to xylene.	

Masson

1. Deparaffinize sections to water.	
2. Heat Harris' hematoxylin to 58°C; stain sections in oven.	5 min
3. Rinse in 95% alcohol.	Briefly
4. Differentiate in picric alcohol.	10 min
5. Wash in running tap water.	10 min
6. Stain in Ponceau de xylidine (red).	5–8 min
7. Differentiate in 1% aqueous phosphomolybdic acid.	5 min
8. Without rinsing, place in aniline blue (ca. 5%; saturated solution in 2.5% acetic acid in water).	2 min
9. Rinse in tap water.	Briefly
10. Place in fresh 1% aqueous phosphomolybdic acid.	5 min
11. Place in 1% acetic acid in distilled water.	

12. Rinse or dip in distilled water. Briefly
13. Dehydrate. Quickly
14. Clear and mount. Results: muscle, red; fibrin, orange; cytoplasm, red; nuclei, dark to pink; red blood cells, orange; connective tissue, blue; nerves, light pink.

Stock Solutions

Alcian Blue (50 cc of 1% solution)

3.3% acetic acid	45 cc
1% aqueous Alcian blue	5 cc

Mix. (Make 3.3% acetic acid by adding 6.6 cc glacial acetic acid to 194.4 cc distilled water.)

Ammonia Water for Iron Hematoxylin Stain

Ammonium hydroxide (approx. 29%)	6–10 drops
Distilled water	100 cc

Mix.

Bouin's Solution for Fixation

Saturated aqueous picric acid	750 cc
38% formaldehyde	250 cc
Glacial acetic acid	50 cc

Mix and store. The solution is stable indefinitely. (Prepare saturated aqueous picric acid by adding 9 g picric acid to 750 cc distilled water. Prepare 38% formaldehyde by adding 95 g paraformaldehyde to 250 cc distilled water at 60°C and slowly adding approximately 8 cc 1N sodium hydroxide until all paraformaldehyde is depolymerized and solubilized as formaldehyde.)

Buffered Eosin, pH 5.3, for Routine H & E Stain

95% alcohol	105 cc
Distilled water	1000 cc
Eosin Y	5 g

Add the alcohol to the distilled water; add Eosin Y. To 200-cc aliquots of this stock solution add 0.20 M acetic acid 2.7 cc and 0.20 M sodium acetate 3.0 cc.

Buffered Water, pH 5.3, for Routine H & E Stain

0.20 M acetic acid	1.2 cc
0.20 M sodium acetate	2.4 cc
Distilled water	200 cc

Mix.

Harris' Hematoxylin, for Routine H & E Stain

Hematoxylin crystals	5 g
Alcohol, 95% or absolute	50 cc
Ammonium or potassium alum	100 g
Distilled water	1000 cc
Mercuric oxide	2.5 g

Dissolve the hematoxylin ($C_{16}H_4O_6$; from heart wood of *Hematoxylon campechianum*, grown in West Indies) in the alcohol; dissolve the alum in water by means of heat; mix the two solutions together. Bring mixture to a boil as rapidly as possible. Remove from heat and add mercuric oxide; bring to a boil again. (Use an extra large flask and watch for splash when adding mercuric oxide.) As soon as the solution becomes dark purple (immediate reaction), cool by plunging at once into a basin of cold water. The solution is ready to use as soon as it is cool. This solution keeps for years in a tightly stoppered bottle. Be sure to filter before using, at least daily.

Hematoxylin for Iron Hematoxylin Stain

Hematoxylin	0.5 g
95% alcohol	10 cc
Distilled water	100 cc

Dissolve hematoxylin in the alcohol and add water. Place in a bottle with a cotton plug and allow to ripen for four to five weeks. For use, dilute with equal parts of distilled water. The solution may be used repeatedly.

Iron Alum for Iron Hematoxylin Stain

Ferric ammonium sulfate	4 g (pick only violet crystals)
Distilled water	100 cc

Mix.

Picric Alcohol for Masson Stain

95% alcohol saturated with picric acid (about 7%)	1 part
95% alcohol	1 part

Mix.

Ponceau de Xylidine (Red) for Congo Red Stain

Ponceau de xylidine	1 g
Acetic acid	1 cc
Distilled water	99 cc

Mix.

Schiff's Reagent for PAS Stain

Distilled water	400 ml
Basic fuchsin	2 g

1N hydrochloric acid	40 ml
Sodium or potassium metabisulfite	2 g
Activated charcoal	0.5 g

Bring water to boil; add fuchsin. Cool to 50 °C, filter, add HCl. Cool to 25 °C, add metabisulfite. Stir, store overnight in dark. Add activated charcoal, shake thoroughly and filter. Store at 0–5 °C.

Sulfurous Acid Rinse for PAS Stain

10% sodium or potassium metabisulfite	6 ml
1N hydrochloric acid	5 ml
Distilled water	100 ml

Mix just before using.

Processing Specimens for Electron Microscopy

The method that we find most suitable for fixation, en bloc staining, and embedding of peroral biopsies of the small intestine, of rectal biopsies obtained at the time of sigmoidoscopy or colonoscopy, and of biopsy specimens obtained at the time of surgery is as follows:

1. Place whole tissues as soon as possible in fixative at 4 °C or at room temperature. Fixative: 2.5% glutaraldehyde; 3% paraformaldehyde; 0.075% $CaCl_2$; 0.1 M sodium cacodylate buffer; pH 7.4 and osmolality 1500.
2. After five minutes mince tissue into 0.5-mm sections, being careful to cut across the biopsy specimen parallel to villi and/or crypts of Lieberkühn.
3. Allow to remain in fixative for 2 to 24 hours.
4. Wash at room temperature in 0.1 M cacodylate buffer for three changes of solution, five minutes each wash, or leave overnight in the buffer with cap off the container so that excess glutaraldehyde can evaporate.
5. Postfix at 4 °C for one to two hours in solution (initial fixation may be at this step) consisting of 2% osmium tetroxide; 0.1M sodium cacodylate buffer; pH 7.4 and osmolality 300.
6. Wash, three changes for five minutes each, with cold sodium hydrogen maleate buffer, 0.05M, pH 5.2
7. Stain en bloc in 2% uranyl magnesium acetate in 0.05M sodium hydrogen maleate buffer, pH 4.2, in dark at 4 °C for 2 to 24 hours.[8]

En bloc staining considerably enhances contrast of membranes[8,9] and increases ease of focusing when obtaining electron micrographs.

8. Wash (three changes) in maleate buffer 0.05M, pH 5.2.
9. Dehydrate and embed by method of Luft[10]:
   a. Soak in:

35% ethanol	15 min
70% ethanol	15 min
95% ethanol	15 min
100% ethanol	15 min
Propylene oxide	15 min

Propylene oxide 15 min
Pour off second change of propylene oxide and add 1 to 2 ml of fresh propylene oxide.

b. Add 1 to 2 ml of freshly prepared epoxy (A + B + tris-dimethylaminomethyl phenol [DMP-30*]) resin and mix according to proportions below.

A (ml)	B (ml)	Total (ml)	DMP-30 (ml)
6	4	10	0.15–0.20
9	6	15	0.25
15	10	25	0.38–0.40

Prepare A and B solutions as follows:

Mixture A		Mixture B	
Epon 812*	100 ml	Epon 812	100 ml
DDSA	145 ml	NMA	78.5 ml

c. Wait one hour and add equal volume (2–4 ml) of epoxy resin mixture.
d. Allow 4 to 12 hours for infiltration and then embed in the epoxy resin mixture in flat polyethylene chambers. To obtain sections of intestinal mucosa always oriented in the same fashion, be sure that each minced portion of tissue settles to the bottom of the chamber with villi and crypts parallel to the surface of the chamber.
e. Polymerize:
   At 35°C 12–24 hr
   At 45°C 12–24 hr
   At 60°C 12–24 hr

   Polymerization may be shortened by placing epon-embedded specimens in the oven at 60°C for 18 to 24 hours. After polymerization, great care is required to mount the trimmed blocks so that thick sections (1.0–1.5 μm) will be parallel to villi and crypts (Figure 5-1, Chapter 5).

Surgical and autopsy specimens, to be suitable for electron microscopy, should be routinely fixed in buffered 4% paraformaldehyde free of methanol,[21] in phosphate-buffered 4% glutaraldehyde,[13] or most simply in phosphate-buffered 4% formalin, technical grade, containing 10% to 15% methanol.[14] The latter fixative obviates the need to predetermine the desirability of electron microscopy study. However, I recommend that the interested reader review the references carefully before proceeding with routine electron microscopy of surgical or autopsy specimens.[15]

* Available from Ernest F. Fullam, Inc., Schenectady, NY. The epoxide equivalent (WPE) (see page 12) of the epon is supplied with each order, and each order includes directions for preparing stock mixtures A and B so that the appropriate anhydride/epoxide molar ratio of 0.7 is obtained on mixing.[11] Epon 812, lot #3HHG1, epoxide equivalent 157, is mixed with DDSA (Mixture A) and with NMA (Mixture B) as above in order to give an A:E ratio of 0.7. DDSA–dodecenyl succinic anhydride; NMA–nadic methyl anhydride. Also available from E.F. Fullam, Inc.

Stock Solutions

Osmium Fixative

Mixture A	
4% osmium tetroxide	2 g
Distilled water	50 ml
Mixture B	
0.2M Na cacodylate, pH 7.4	50 ml

Dissolve osmium tetroxide in distilled water. Mix 1 ml of this solution with 1 ml Na cacodylate in a weighing bottle. Result: 2 ml 2% osmium in 0.1M cacodylate, pH 7.4, osmolality 290 to 310. Prepare mixture just prior to use; keep 4% osmium tetroxide solution on ice, in a tightly stoppered bottle.

Glutaraldehyde Fixative

25% glutaraldehyde	20 ml
Calcium chloride	150 mg
Distilled water	80 ml
Paraformaldehyde	6 g
Sodium cacodylate, 0.2 M, pH 7.4	100 ml

We prepare our own fixative by dissolving 6 g of paraformaldehyde in 80 ml of distilled water (see section on light microscopy processing for details). We add 20 ml of 25% glutaraldehyde,* 150 mg of CaCl, and 100 ml of 0.2M sodium cacodylate buffer. The final solution is 200 ml 2.5% glutaraldehyde, 3% formaldehyde, 0.075% $CaCl_2$, and 0.1M cacodylate, pH 7.4, osmolality about 1500. The pH of the final mixture is checked and adjusted with sodium hydroxide or hydrochloric acid if necessary.

Cacodylate Buffer

Mix 100 ml solution A and 8.4 ml solution B to get 0.2M cacodylate at pH 7.4

Solution A		Solution B
16 g Na cacodylate or 21.4 g Na cacodylate • 3 H_2O	in 500 ml H_2O	16.5 ml conc. HCl in 1 liter H_2O.

NaH-Maleate NaOH Buffer (0.05M)

Stock solutions:

0.2M NaH-maleate
Maleic acid, 23.2 g
NaOH, 8 g
Distilled water, q.s. 1000 ml

0.2M NaOH
NaOH, 8 g
Distilled water, q.s. 1000 ml

*Electron Microscopy Sciences, Fort Washington, PA 19034.

Staining solutions:

pH at 25°C	0.2M NaH-Maleate	0.2M NaOH
5.2	50 ml	7.2 ml
6.0	50 ml	26.9 ml

Dilute each solution to 200 ml to obtain a molarity of 0.05.

Formulation of Epoxy Resin Mixtures

The cutting properties of epoxy resins depend on the anhydride/epoxide (A/E) molar ratio of the final mixture. Since different lots of epoxy resins show considerable variation, the proportions of the mixture should be changed when a new lot of epoxy resin is used.

Luft[16] studied the cutting qualities of epoxy resins at various A/E ratios. An A/E ratio of 0.90 to 1.0 produces blocks that are very highly cross-linked but very difficult to cut. At A/E ratios of about 0.60 to 0.65 there is a marked deterioration of the cutting properties. An A/E ratio of 0.70 was selected as optimal.

Table of Properties

Epoxy resins	Molecular weight	Specific gravity	WPE no.* (g/eq)	Viscosity (cps, at 25°)
Epon 812	306†	1.23	140–165	900–1500
Anhydrides				
DDSA	266	1.002		
NMA	178	1.236		

*The WPE (weight per epoxide) number is the number of grams of resin that contain one gram-equivalent of epoxide. It therefore corresponds to the equivalent weight and is sometimes called the epoxide equivalent. The WPE number is different in different lots of resin. The manufacturer titrates each lot of resin and obtains an accurate WPE number.
†This is an average molecular weight.

To calculate the proper proportions of a resin mixture, use the formula

$$\frac{\text{(Volume of anhydride)(sp gr anhydride)}}{\text{Mol weight anhydride}} = 0.70 \times \frac{\text{(vol epoxy resin)(sp gr epoxy resin)}}{\text{WPE no.}}$$

For Epon 812 and DDSA, this becomes

$$\frac{\text{(x ml DDSA)(1.002 g/ml)}}{\text{266 g/eq}} = 0.70 \text{ eq/eq} \times \frac{\text{(100 ml Epon)(1.23 g/ml)}}{\text{WPE no. g/eq}}$$

or

$$\text{ml DDSA:100 ml Epon} = 22{,}900/\text{WPE no.}$$

For WPE no. = 157, proportions are 146 ml DDSA:100 ml Epon. Similarly,

$$\text{ml NMA:100 ml Epon} = 12{,}400/\text{WPE no.} = 79 \text{ ml NMA:100 ml Epon}$$

A convenient table[11] lists properly calculated proportions of resin mixtures for all commercially available epoxide equivalents of Epon 812.

Staining

Thick sections for light microscopy are obtained first and are routinely stained with the alkaline methylene blue-azure II[17] method or with methylene blue, azure II, and basic fuchsin.[18] The latter method gives good differentiation of collagen, smooth muscle, elastic tissue, cytoplasm, and nuclei. Using the thick section to determine the precise area designated for electron microscopy, the plastic block is trimmed for thin sectioning and subsequent electron microscopy.

Thin sections are routinely stained with uranyl magnesium acetate for 15 minutes followed by lead citrate for 5 minutes.[19] Uranyl magnesium acetate is much more soluble in water than uranyl acetate; does not form white precipitates, as does uranyl acetate; and has remained stable in solution for over eight years in our laboratory. We find it useful to filter the uranyl solution through a 0.45-μm millipore filter (using a 1-ml tuberculin syringe and attached plastic holder for the 13-mm diameter filter) just prior to staining. We keep the solution in a light-tight box, since uranium is sensitive to light.

The uranyl stain should be used first, because it apparently acts as a mordant for lead. Uranyl and lead salts have some specificity:

	Lead	Uranyl
Polysaccharide	+++	+
Protein	++	++
Nucleoprotein	+	+++

More detailed discussions of staining of thin sections for electron microscopy are in the paper by Zobel and Beer.[20]

We have found Venable and Coggeshall's[21] method of using commercially prepared lead citrate to be very convenient, and we now routinely use this in our laboratory. The lead citrate solution is prepared as follows:

Place 0.025 g lead citrate in a 10-ml screw-top centrifuge tube.
Add 10 ml of distilled water and 0.1 ml of 10N NaOH.
Shake well until dissolved.

This solution is stable for about one month.

We have determined that pure lead citrate* disintegrates at 280 to 285 °C; it has no clear-cut melting point. This disintegration accounts for the fine speckling seen along membranes of specimens that have been too intensely beamed in the electron microscope. This problem can be avoided by using a small condenser aperture (hole of about 200 μm), by reducing beam intensity to 100 μamp, by avoiding crossover, or by using a modern transmission electron microscope.

References

1. Perera DR, Weinstein WM, Rubin CE: Symposium on pathology of the gastrointestinal tract. Part II. Small intestine biopsy. *Hum Pathol* **6**:157–217, 1975.

* Obtained from K&K Laboratories, 121 Express Street, Plainview, NY 11803.

2. Goldman H, Antonioli DA: Progress in pathology: Mucosal biopsy of the esophagus, stomach, and proximal duodenum. *Hum Pathol* **13**:423–448, 1982.
3. Whitehead R: *Mucosal Biopsy of the Gastrointestinal Tract* (3rd ed). Philadelphia, WB Saunders, 1985.
4. Lee FD, Toner PG: *Biopsy Pathology of the Small Intestine.* Philadelphia, JB Lippincott, 1980.
5. Dobbins WO III: Small bowel biopsy in malabsorptive states. In HT Norris (Ed), *Contemporary Issues in Surgical Pathology.* Vol 4. *Pathology of the Colon, Small Intestine and Anus.* New York, Churchill Livingstone, 1983, pp 121–165.
6. Gear EV Jr, Dobbins WO III: The histologic spectrum of proximal duodenal biopsy in man. *Am J Med Sci* **257**:90–99, 1969.
7. Surawicz CM: Serial sectioning of a portion of a rectal biopsy detects more focal abnormalities in inflammatory bowel disease. *Dig Dis Sci* **27**:434–436, 1982.
8. Karnovsky MJ: The ultrastructural basis of capillary permeability studied with peroxidase as a tracer. *J Cell Biol* **35**:213–236, 1967.
9. Dobbins WO III: Morphologic aspects of lipid absorption. *Am J Clin Nutr* **22**:257–265, 1969.
10. Boom A, Deams W Th, Luft JH: On the fixation of intestinal absorptive cells. *J Ultrastruct Res* **48**:350–360, 1974.
11. Burke CN, Geiselman CW: Exact anhydride epoxy percentages for electron microscopy embedding (Epon). *J Ultrastruct Res* **36**:119–126, 1971.
12. Lynn JA, Martin JH, Race GJ: Recent improvements of histologic technics for the combined light and electron microscopic examination of surgical specimens. *Am J Clin Pathol* **45**:704–713, 1966.
13. Chambers RW, Bowling MC, Grimley PM: Glutaraldehyde fixation in routine histopathology. *Arch Pathol* **85**:18–30, 1968.
14. Carson FL, Martin JH, Lynn JA: Formalin fixation for electron microscopy: A re-evaluation. *Am J Clin Pathol* **59**:365–373, 1973.
15. Dobbins WO III: Diagnostic pathology of the intestine and colon. In BF Trump, RT Jones (Eds), *Diagnostic Electron Microscopy*, Vol. I. New York, John Wiley 1978, pp 253–339.
16. Luft JH: Improvements in epoxy resin embedding methods. *J Biophys Biochem Cytol* **9**:409–414, 1961.
17. Richardson KC, Jarrett L, Fincke EH: Embedding in epoxy resins for ultrathin sectioning in electron microscopy. *Stain Technol* **35**:313–323, 1960.
18. Huber JD, Parker F, Odland GF: A basic fuchsin and alkalinized methylene blue rapid stain for epoxy-embedded tissue. *Stain Technol* **43**:83–87, 1968.
19. Frasca JM, Parks VR: A routine technique for double staining ultrathin sections using uranyl and lead salts. *J Cell Biol* **25**:157–161, 1965.
20. Zobel CR, Beer M: The use of heavy metal salts as electron stains. *Int Rev Cytol* **18**:363–400, 1965.
21. Venable JH, Coggeshall R: A simplified lead citrate stain for use in electron microscopy. *J Cell Biol* **25**:407, 1965.

2
Biopsy Interpretation—Light Microscopy

Because the interpretation of gastrointestinal mucosal biopsy specimens requires considerable skill, it merits the same specialization within pathology that has been so useful in the interpretation of needle biopsy specimens of the liver, lung, and kidney.[1-3] There is no substitute for experience. Thus it is desirable to have a select group of pathologists with a major interest in gastrointestinal histopathology interpret the biopsies.

The villous architecture is the first structure to be assessed. Other key features include the surface epithelium, the lamina propria, and the structures of the submucosa.

Normal Villous Architecture

Villous shape and length vary considerably, a fact that must be appreciated to avoid interpretive errors. A variety of methods are available for the quantitative assessment of villous architecture, but these are rarely required for routine diagnostic assessment of small-bowel biopsies. Villi are not necessarily tall fingerlike structures standing in a row perpendicular to the muscularis mucosae. Rather, many villi bend in various directions and vary in shape, from slender, fingerlike structures to leaf-shaped structures. It is reasonable to follow Rubin and colleagues' dictum that if there are four normal fingerlike villi in a row in *any* serial section of a biopsy specimen, the biopsy can be considered normal[1] (Figure 2-1). At times, however, especially when the biopsy specimens are small, one may have to modify this dictum.

Tangential sectioning of villi is the most common source of misinterpretation of normal biopsy specimens. The tangential sectioning artifact (Figure 2-2) can be avoided completely by attention to appropriate orientation of the specimen prior to fixation, during embedding, and during attachment of the block to the microtome chuck. It is also avoided by obtaining the sections within the central core of the biopsy specimen (see Figure 1-1B). Tangential sectioning is easily recognized by the presence of multilayered cross sections of crypts, or by villi that appear to be broad and short and have an apparently multilayered surface epithelium.

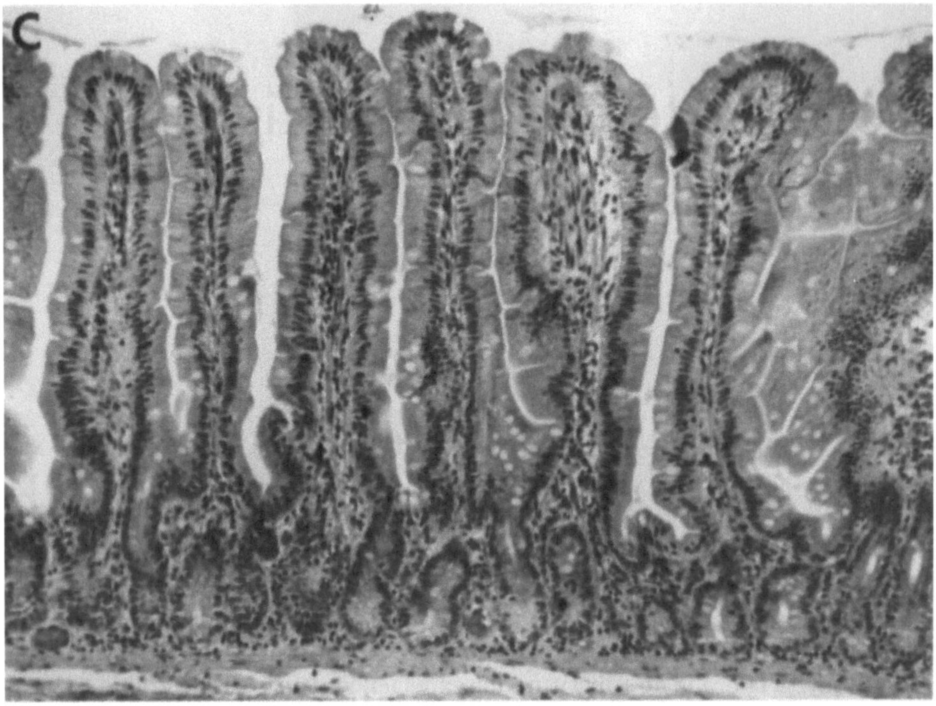

Figure 2-1. Normal section of proximal jejunum from the central core of an oriented biopsy specimen demonstrating six normal villi in a row. H & E stain, × 132

If the biopsy specimens must be obtained from the proximal duodenum rather than from the duodenojejunal junction, the Brunner gland artifact may be present.[1,4] Brunner glands in the first and second portions of the duodenum may distort the overlying villi so that they are broadened and shortened (Figure 2-3). Distorted villi may be present even when Brunner glands are not included within the biopsy specimen (Figure 2-4). Occasional small-bowel biopsy specimens appear to consist of gastric antral mucosa. In this case it is likely that the suction biopsy tube was inadvertently placed in the stomach rather than in the small bowel. Such a problem should not occur during endoscopic biopsy. Occasional nests of gastric mucosa may be found in the proximal duodenum and are often biopsied by the endoscopist who can't resist sampling all isolated focal nodular changes seen in the duodenum.[5]

In shallow biopsy specimens that do not include the muscularis mucosae the villi often appear shortened and broadened because they lack the underlying muscle bundle to hold them together. Biopsy trauma may be very subtle. Generally, separation of the epithelium from the underlying lamina propria and gaps in epithelial continuity without inflammatory changes in the underlying lamina propria are due to trauma. Lymphoid follicles are occasionally found within the normal mucosa and may distort adjacent villi so that they appear to be abnormal (Figure 2-5).

Most of the artifacts just described are easily recognized and should not cause interpretive error. Even unrecognized, such artifacts should not be confused with severe mucosal lesions such as those seen in celiac sprue. However, if patchy mucosal changes are likely clinically, additional specimens may be necessary to distinguish artifacts from such patchy changes.

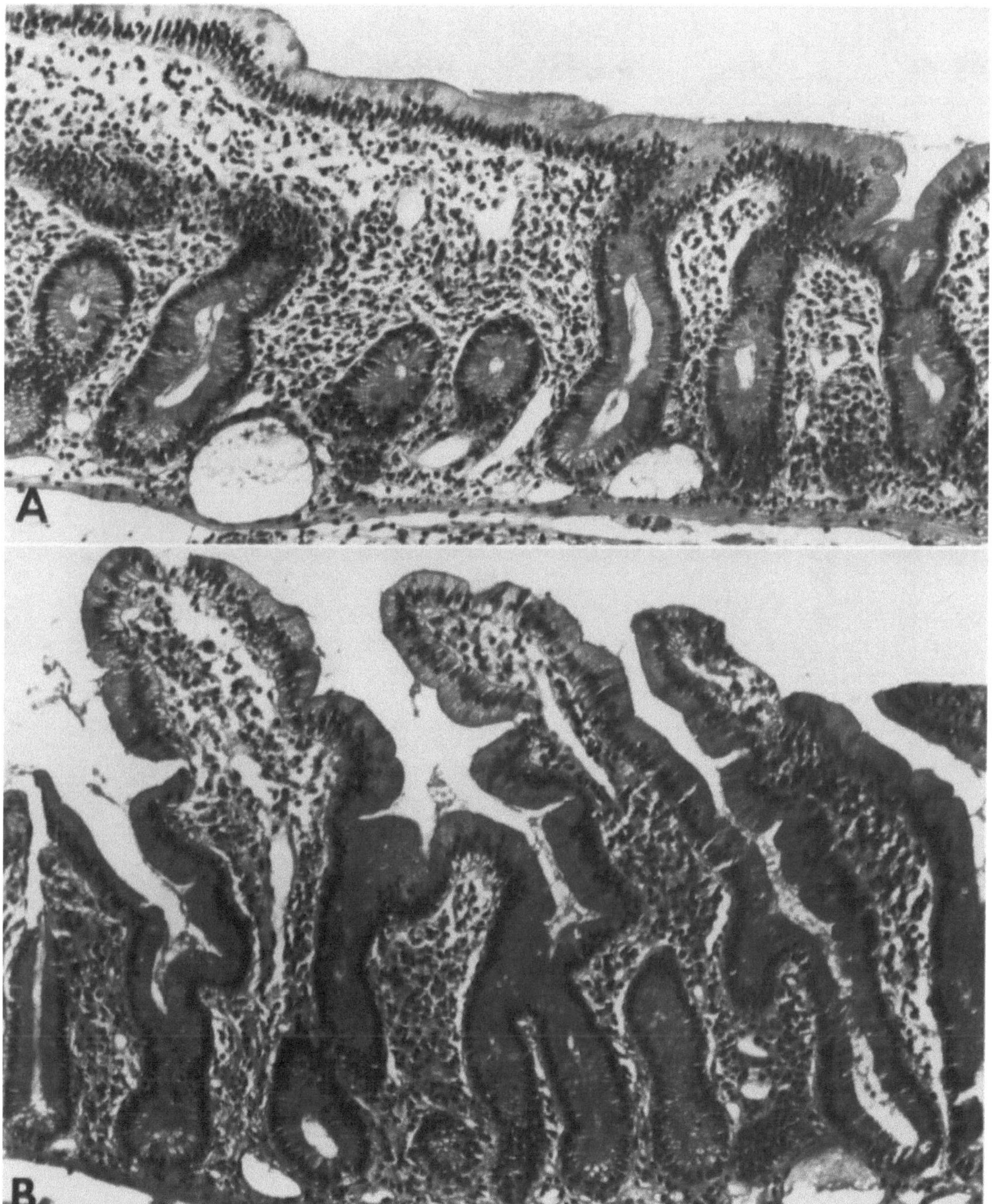

Figure 2-2. Tangential sectioning artifact. (A) Normal biopsy specimen showing a "flat" surface, an appearance often confused with the severe lesion of celiac sprue. H & E stain, × 150. (B) Normal section of the biopsy specimen illustrated in (A). H & E stain, × 150

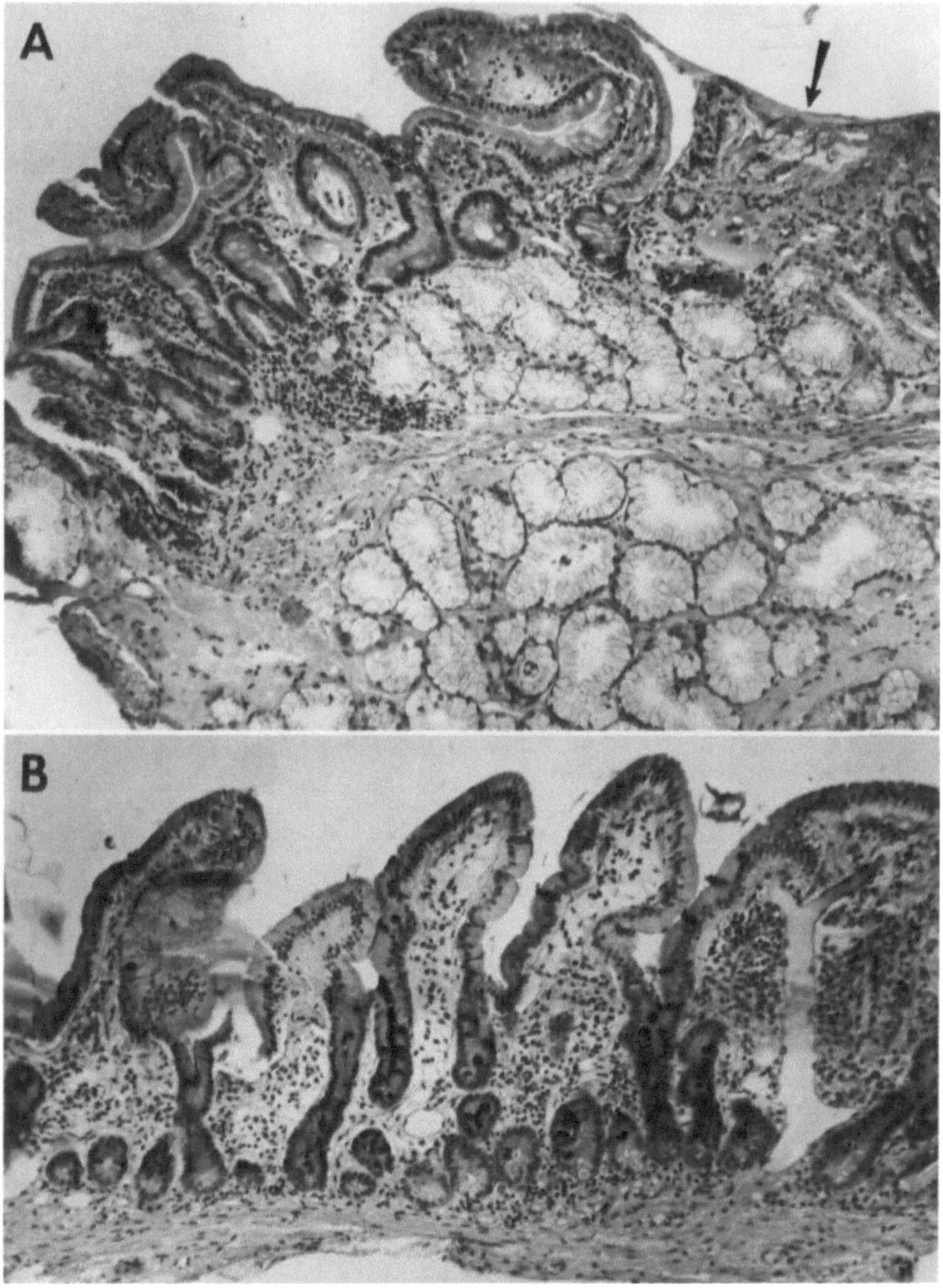

Figure 2-3A,B.

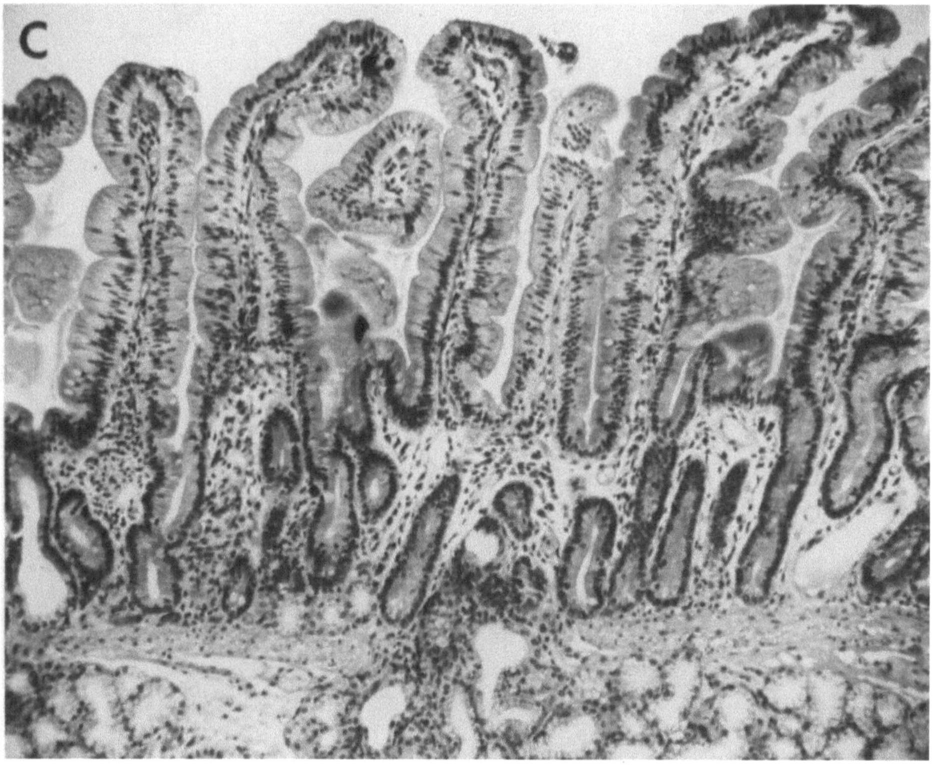

Figure 2-3C. Artifacts in the proximal duodenum. (A) Section of duodenal bulb biopsy specimen showing severe Brunner gland artifact. The glands are present both above and below the muscularis mucosae. The villi are shortened and distorted. Note severe traumatic artifact (arrow). Such an artifact is frequently seen when Brunner glands make cutting the specimen difficult. H & E stain, × 132. (B) Normal section of duodenal bulb biopsy specimen obtained only a few millimeters away from the specimen in (A). There are four normal villi in a row, but these are shortened and broadened and the lamina propria contains an increased cellular content. Note the leaf-shaped villus on the right, which contains a prominent central lacteal. The pallisading of the nuclei of the epithelium at the apex of several villi is an artifact of tangential sectioning. H & E stain, × 132. (C) Section of biopsy specimen obtained from duodenal bulb showing that villous architecture may be normal even in the presence of Brunner glands. Usually, proximal duodenal biopsy specimens accurately reflect the jejunal and distal duodenum in both health and disease (Gear and Dobbins,[4] Chapter 5). H & E stain, × 132. (A, B, and C reproduced from Dobbins[3] by permission of Churchill Livingstone, Inc., New York.)

Normal Epithelium

Tall columnar absorptive cells predominate in the epithelium of villi of biopsy specimens obtained at the duodenojejunal junction (Figure 2-1). Occasional goblet cells are seen between the absorptive cells, and approximately one intraepithelial lymphocyte is usually present for every five epithelial cells.[6,7] Mitoses are found only in the crypt region, and Paneth cells are located at the bases of crypts. Enteroendocrine cells can occasionally be discerned in H & E stained sections, and undifferentiated crypt cells (best defined by electron microscopy) round out the epithelial cell population.

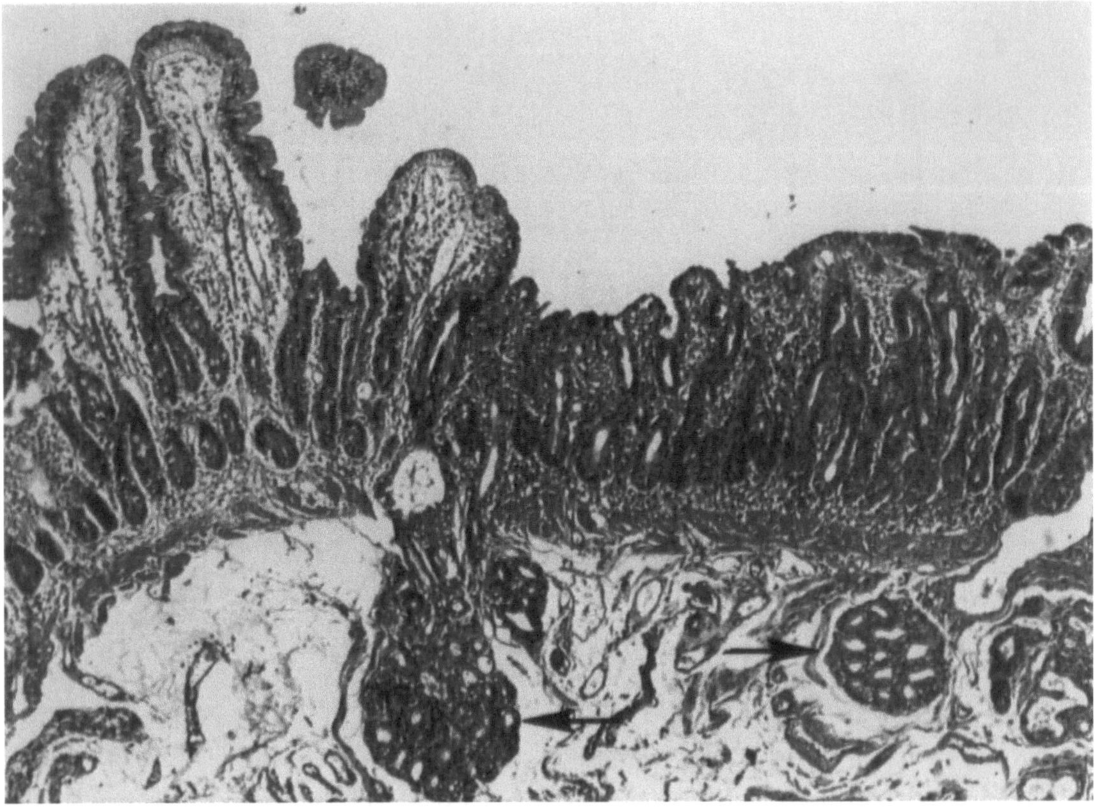

Figure 2-4. Light microscopy photograph showing unusual Brunner gland artifact in which an area of mucosa between two portions (arrows) of Brunner glands is flattened while villous structure to the left and right (not shown) of the artifact is normal. H & E stain, × 80

Normal Lamina Propria

Suction used in obtaining the biopsy specimen often results in mild hemorrhage and edema within the lamina propria. The presence of these changes must be interpreted with great caution. Plasma cells tend to be the most common cells found within the lamina propria; lymphocytes and macrophages are less common. Fibroblasts align themselves parallel to the basal lamina. Lymphoid nodules are occasionally present in mucosal biopsy specimens of normal individuals (Figure 2-5). Macrophages, often containing prominent inclusions, are sometimes found at the tips of villi. Eosinophilic leukocytes, when looked for, are frequently observed within the lamina propria, but only rarely are polymorphonuclear leukocytes seen outside of blood vessels in normal specimens. Leukocytes within areas of artifactually induced hemorrhage obviously have no pathological significance.

It is important to note that lymphoid cells are normally found within the lamina propria, and their numbers vary widely depending on the country of origin of the patient–the intestinal mucosa of individuals who live in tropical

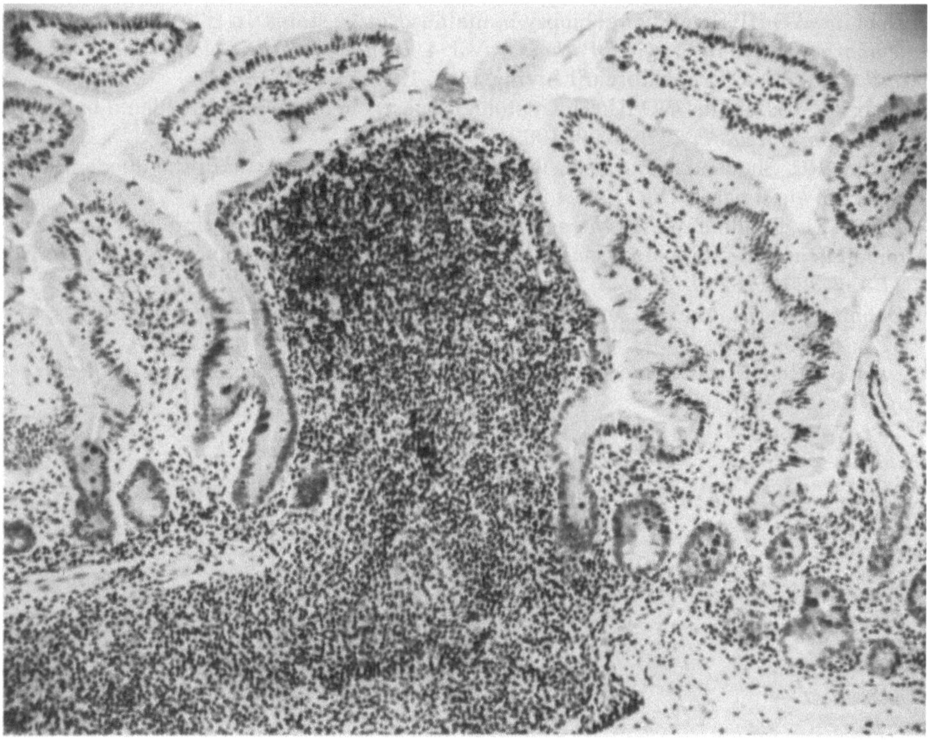

Figure 2-5. Section of proximal duodenal biopsy specimen in which a lymphoid follicle distorts a single villus in an otherwise normal specimen. H & E stain, × 132. (Reproduced from Dobbins,[3] by permission of Churchill Livingstone, Inc., New York.)

areas tends to contain increased numbers of round cells. When this is the only finding, it is better to note that the round cell content of the lamina propria appears to be increased rather than that chronic inflammation is present.

Normal Muscularis Mucosae

The intestinal mucosa is separated from the submucosa by a thin layer of smooth muscle called the muscularis mucosae. This is the third component of the intestinal mucosa. This muscle layer is generally poorly developed, but may on occasion be prominent, consisting of an inner layer of circular fibers and an outer layer of longitudinally arranged fibers. This thin layer of muscle lies just beneath the intestinal glands that open into the crypts of Lieberkühn. Occasional muscle cells extend from this thin muscle layer into the villus core.[8]

References

1. Perea DR, Weinstein WM, Rubin CE: Small intestinal biopsy. *Hum Pathol* **6**:157–217, 1975.
2. Rubin CE, Dobbins WO III: Peroral biopsy of the small intestine. A review of its diagnostic usefulness. *Gastroenterology* **49**:676–697, 1965.

3. Dobbins WO III: Small bowel biopsy in malabsorptive states. In HT Norris (Ed), *Contemporary Issues in Surgical Pathology.* Vol. 4. *Pathology of the Colon, Small Intestine and Anus.* New York, Churchill Livingstone, 1983, pp 121–165.
4. Gear EV Jr, Dobbins WO III: The histologic spectrum of proximal duodenal biopsy in man. *Am J Med Sci* **257**:90–99, 1969.
5. Spiller RC, Shousha S, Barrison IG: Heterotopic gastric tissue in the duodenum: A report of eight cases. *Dig Dis Sci* **27**:880–883, 1982.
6. Austin LL, Dobbins WO III: Intraepithelial leukocytes of the intestinal mucosa in normal man and in Whipple's disease: A light and electron microscopic study. *Dig Dis Sci* **27**:311–320, 1982.
7. Dobbins WO III: Progress report: Human intestinal intraepithelial lymphocytes. *Gut* **27**:972–985, 1985.
8. Lane BP, Rhodin JAG: Fine structure of the lamina muscularis mucosae. *J Ultrastruct Res* **10**:489–497, 1964.

3
Biopsy Interpretation—Electron Microscopy

Intestinal Epithelial Cells

Mammalian intestinal epithelial cells include:

Absorptive cells
Goblet cells
Undifferentiated crypt cells
Paneth cells
Enteroendocrine cells
Tuft cells (fibrovesicular, caveolated)
Cup cells
M cells

The morphology of absorptive cells will be described in the greatest detail; the structural aspects of the other cells will be described only when they differ from those of the absorptive cell.

Absorptive Cells

Intestinal epithelial cells form the interface between the external environment (the intestinal lumen) and the internal environment (the lamina propria, which contains an extensive vascular, lymphatic, and nerve supply). Intestinal absorptive cells are tall, about 22 to 26 μm in height, and columnar. The luminal surface is specialized to form a striated or brush border (Figure 3-1). Beneath the brush border is the terminal web, a clear area free of organelles and inclusions except for occasional vesicles. These vesicles are involved in transport of a variety of substances into the cell interior;[1] they provide membranes for expansion and elongation of microvilli, and they mediate the final step of intracellular movement (transfer to the brush border membrane) of sucrase-isomaltase.[2] Adjacent absorptive cells are attached by a junctional complex, characteristic of epithelial cells in general, consisting of a tight junction (zonula occludens), intermediate junction (zonula adherens), and desmosomes (macula adherens) (Figure 3-2). Also occasionally present (in flounder intestine) is a fourth component of the

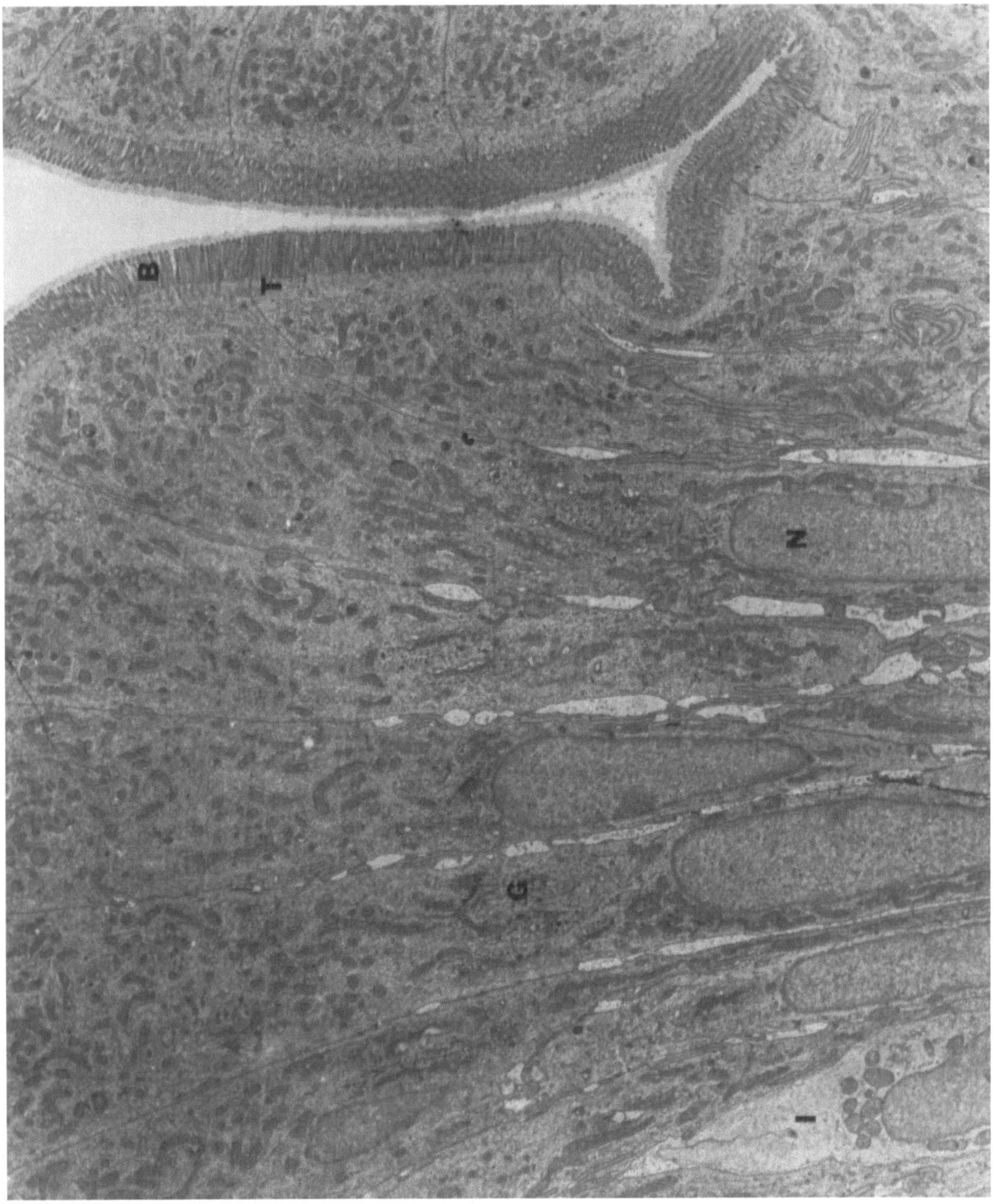

Figure 3-1. Low-magnification electron micrograph showing usual appearance of epithelial cells over the apical one third of the villus. Note tall columnar epithelial cells with basally to centrally located nuclei (N), obvious brush border (B), and terminal web (T) generally free of organelles except for occasional transport vehicles. Note prominent supranuclear Golgi complex (G). Intraepithelial lymphocyte (I). × 10,000

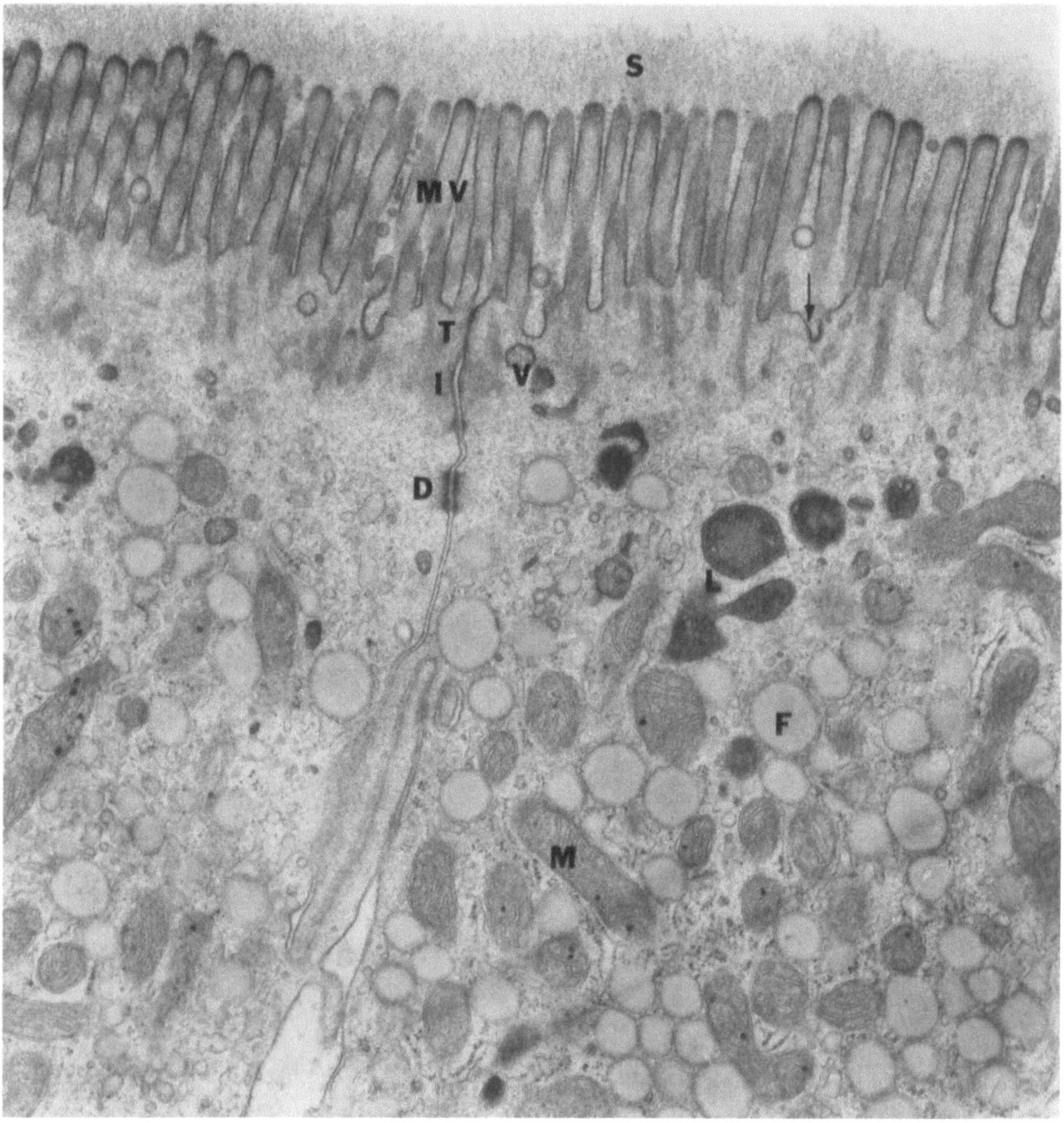

Figure 3-2. Intermediate-magnification electron micrograph showing the apical portion of two adjacent epithelial cells during fat absorption. The junctional complex consists of a tight junction (T), an intermediate junction (I), and a desmosome (D). Note prominent surface coat (S) or glycocalyx; microvilli (MV); secretory vesicles (V); lysosomes (L); mitochondria (M); and absorbed lipid (F) within profiles of the smooth endoplasmic reticulum. Membrane invaginations at the base of microvilli, called apical pits (arrow), take up ferritin but not fat (Cardell RR, Badenhausen S, Porter KR. *J Cell Biol* **34**:123, 1967). × 25,000. (Reproduced from Dobbins,[1] by permission of Academic Press, Inc., New York.)

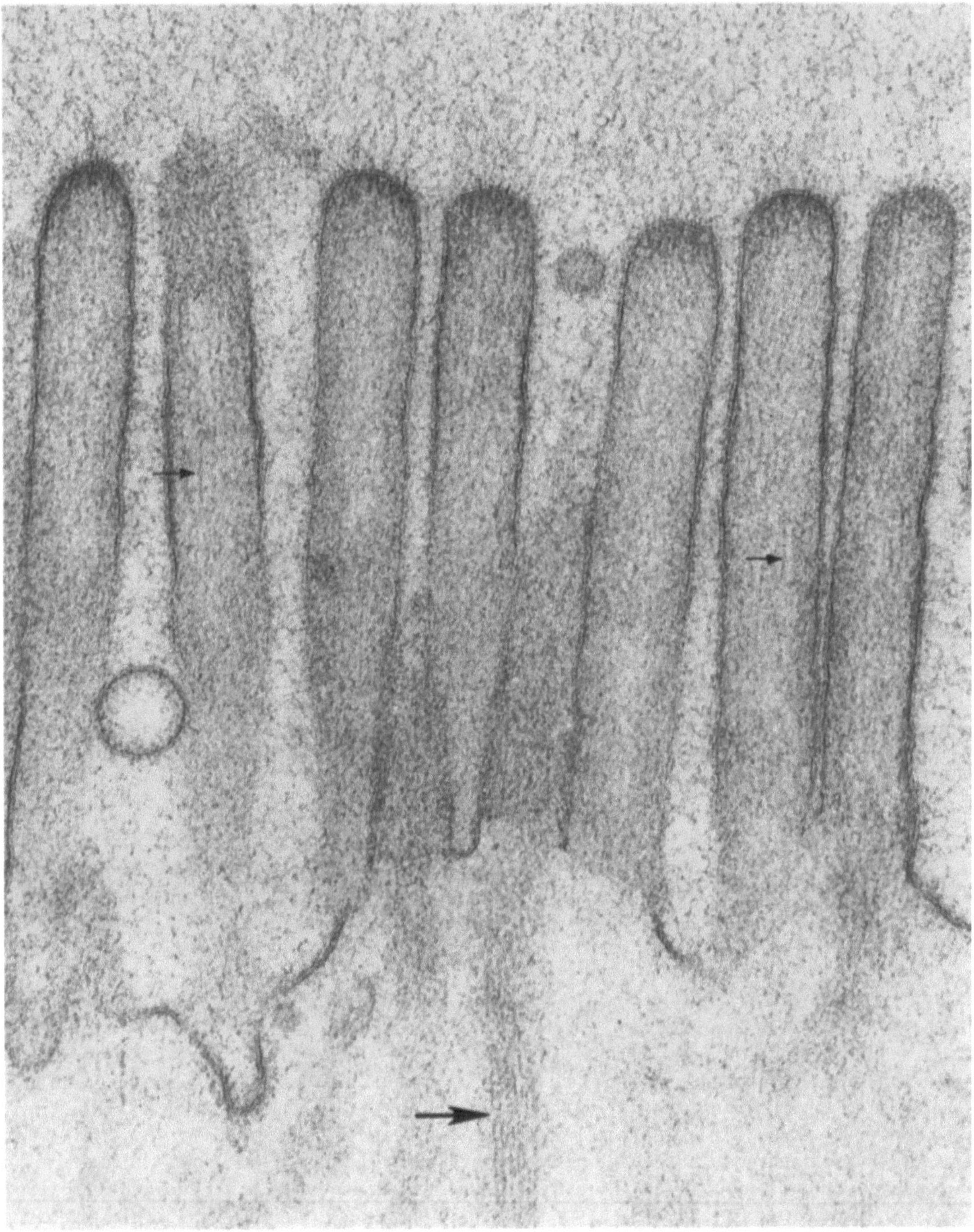

Figure 3-3. High-magnification electron micrograph of microvilli in which actin filaments are just visible within the microvilli (small arrows) and extending into the terminal web (large arrow). Note prominent glycocalyx radiating from the outer leaflet of the plasma membrane. Fixed with osmium tetroxide alone, × 100,000. (Reproduced from Dobbins,[1] by permission of Academic Press, Inc., New York.)

junctional complex, the gap junction (macula communicans). Elongated mitochondria are abundant in the apical cytoplasm (beneath the terminal web), and a prominent Golgi apparatus is present above the basally located nucleus. The absorptive cells rest on a thin basal lamina.

The *brush border* of the intestinal epithelial cell consists of an array of tightly packed microvilli (Figure 3-3). Within each microvillus is a bundle of 20 to 30 actin filaments.[3-7] The fact that the actin filaments have the same polarity (all are arranged longitudinally) clearly suggests that microvillar contraction takes place. However, such contraction has not been convincingly demonstrated.[5] When the brush borders of mouse intestinal epithelial cells are examined by the quick-freeze, deep-etch, and rotary-replication method (Figure 3-4), the actin filaments are found to extend into the terminal web as rootlets in compact bundles, and they remain compact within the terminal web.[5] Even though myosin is a major cross-linker in the terminal web, its arrangement vis-à-vis the actin filaments (Figure 3-5) of rootlets suggests that the myosin plays a purely structural role between the rootlets.[5] The actin rootlets in the terminal web sit on a dense network of intermediate filaments (tonofilaments containing keratin) and are linked by many cross connections of fine filaments (see below).

Actin filaments within microvilli are cross-linked by bridges that are attached to the actin-filament bundle at periodic intervals of 33 nm along its length and appear to be arranged as twin spirals around the bundle.[7] The bridges range from 15 to 30 nm in length and 2 to 7 nm in width.[7] The actual attachment of the bridges to the plasma membrane may be through an integral membrane protein called vinculin. The bridges themselves consist of two proteins, villin and fimbrin.[6] The integral membrane protein (vinculin?) cannot be visualized on the P face of the microvillus membrane using freeze-fracture techniques.[4] The mechanism of end-on attachment of the actin filaments to the apex of microvilli is not clear. Membrane particles are also not visible on the P face at the apex of microvilli. Each filament bundle is embedded in a dense matrix, composed in part of an α-actininlike protein, at the tip of microvilli. The lateral bridges are composed of a 110-kilodalton (kd) protein with four associated calmodulin molecules.[7]

The *terminal web* region is organized in a trilayered fashion, with each layer composed of different arrangements and types of filaments. Each layer correlates with the three levels of the junctional complex (tight, intermediate, and desmosome)[6,8] (Figure 3-6). The *apical* region is adjacent to the tight junction and contains predominant rootlets. The *intermediate* region consists of two zones, an apical and a basal zone. The apical zone contains rootlets and a band of thin, 6- to 7-nm filaments that run parallel to the cell membrane. The basal zone contains rootlets and thin filaments and thicker 10-nm tonofilaments associated with the rootlets. The *desmosome* region has few rootlets but contains numerous 10-nm tonofilaments (Figures 3-2, 3-5, 3-6).

A notable feature of the intermediate junction region is the presence of a circumferential ring of thin filaments that is composed of actin, myosin, and α-actinin. Burgess[6] has shown in vitro the presence of contractions in the terminal web region causing the microvilli to be fanned apart in response to adenosine triphosphate (ATP) and to δ-S-ATP. Because myosin, α-actinin, tropomyosin, and actin have been localized to this region of the terminal web (immediately adjacent to the intermediate junction), Burgess proposed that intestinal

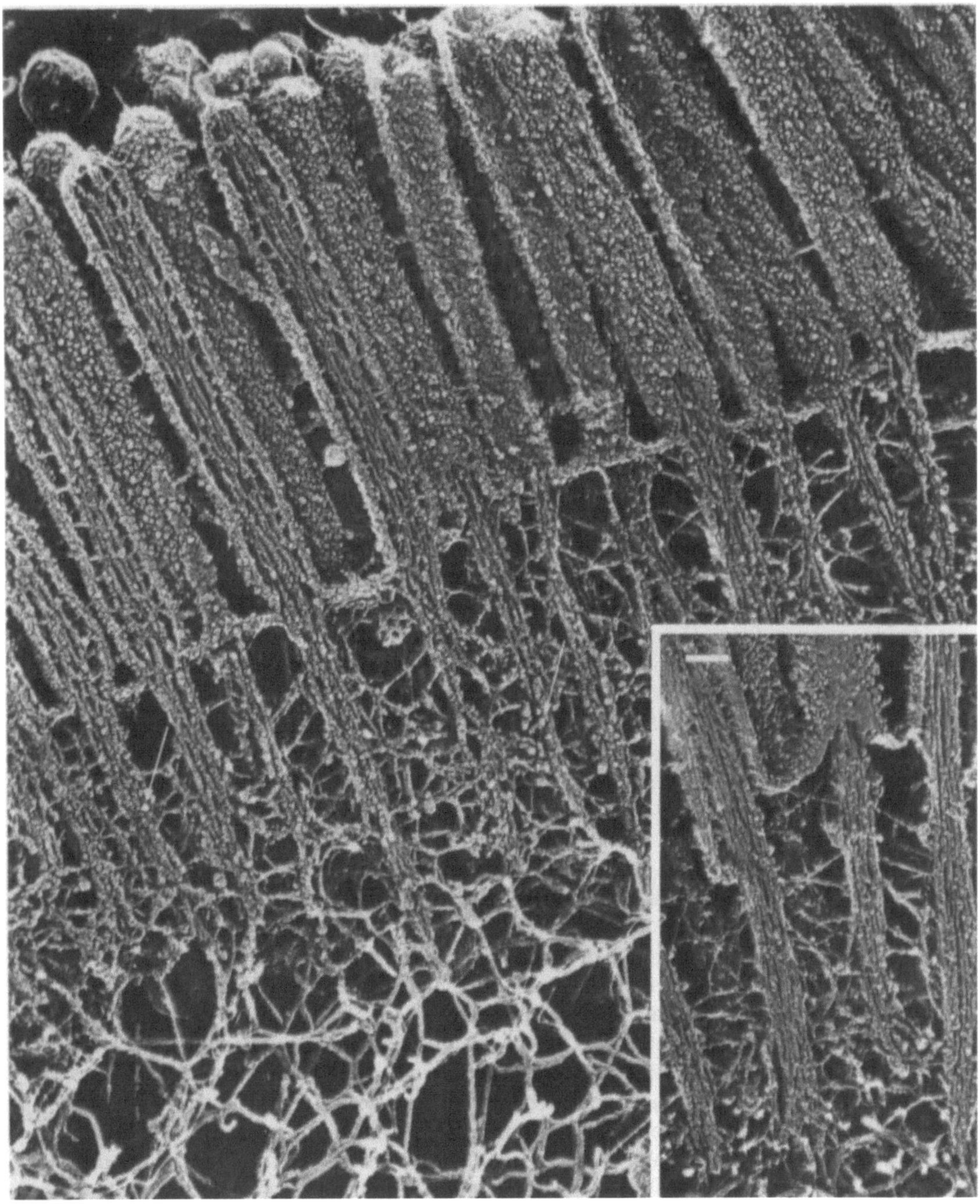

Figure 3-4. Scanning electron micrograph of intestinal epithelial cell following quick freeze, deep etch, and rotary replication, which illustrates actin filaments extending from microvilli into the terminal web and remaining compact "as rootlets." Note more delicate fibers that appear to serve as cross-links between the rootlets. The small dots (arrows) on the rootlets may be cross-fractured fine fibrils. × 77,000 (inset × 97,000). Hirokawa N, Tilney LG, Fujiwara K, Heuser JE. Reproduced from *The Journal of Cell Biology* **94**:425–443, 1982, by copyright permission of The Rockefeller University Press. Micrograph produced by Dr. John E. Heuser of Washington University School of Medicine, St. Louis, MO.)

Figure 3-5. Diagrammatic illustration of the structural features of the apical portion of intestinal epithelial cells to include microvilli, the terminal web, and the junctional complex. (Reproduced from Madara and Trier,[9] by permission of Raven Press, New York.)

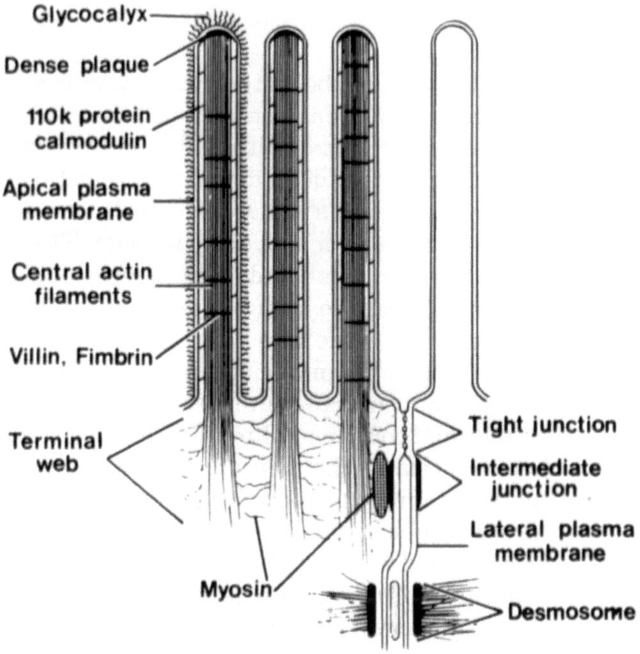

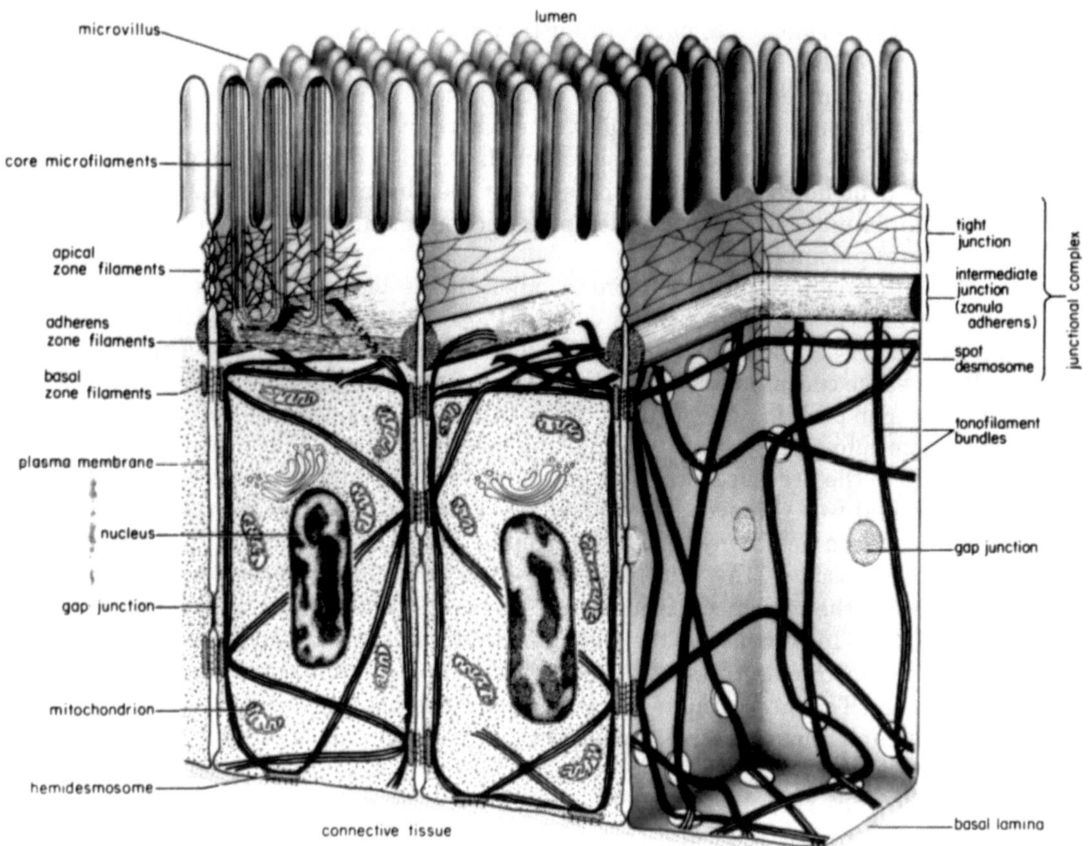

Figure 3-6. Diagrammatic illustration of organization of cytoplasmic filaments in columnar epithelial cells with emphasis on the association of filaments with the junctional complex. (Hull BE, Straehelin LA. Reproduced from *The Journal of Cell Biology* **81**:67–82, 1979 by copyright permission of The Rockefeller University Press.)

epithelial cells *can* be motile via a circumferential terminal web contractile ring analogous to the contractile ring of dividing cells.

Adjacent intestinal epithelial cells are attached by a characteristic *junctional complex* consisting of an apical tight junction (zonula occludens), an intermediate junction (zonula adherens or belt desmosome), and multiple desmosomes (macula adherens or spot desmosomes).[1] The tight junction (Figure 3-7) extends 0.2 to 0.5 μm in depth and, with conventional glutaraldehyde-osmium tetroxide fixation, appears to be a region in which the outer leaflets of adjacent membranes focally come into close contact or even appear to be fused. Using the freeze-fracture technique, it can be seen that these focal areas of apparent fusion of adjacent membrane outer leaflets consist of a series of interconnecting P-face strands and complementary E-face grooves (Figure 3-8). This technique (freeze fracture) cleaves the membrane along its internal hydrophobic core, that is, it splits the inner and outer leaflets apart.[9] The resulting replicas reveal structural features of the external (E) and internal or protoplasmic (P) faces of the membrane. The replicas demonstrate intramembranous proteins (IMPs). Each P-face strand is thought to be composed of a chain of closely spaced 10-nm IMPs.[9]

Tight-junction structure varies with region and with cell type. The junctions between ileal absorptive cells have greater depth and more strands than do those between jejunal absorptive cells. This correlates with function in that the ileum is less permeable to paracellular flow of fluid and solute than is the jejunum. Further, the tight junctions of villous absorptive cells are uniform in depth and strand number, whereas those of crypt cells are variable in structure. Again, the villous epithelium is tight, whereas crypt epithelium is leaky to fluid and solutes. The tight (occluding) junction also forms a barrier to the movement of integral membrane proteins, preserving membrane polarity and the distinctly different composition of the apical plasma membrane as opposed to the basolateral plasma membrane.[9]

The intermediate junction (zonula adherens) is located between the tight junction and the spot desmosomes, extends 0.3 to 0.5 μm in depth, and has a true intercellular space about 20 nm wide (Figures 3-2, 3–5 through 3-7). The intercellular space contains finely fibrillar material that may consist of the specific cell-adhesion protein, uvomorulin. There is usually an increased density of the cytoplasm subjacent to the lateral plasma membranes of the intermediate junction, due to the presence of a thick mesh of 7-nm microfilaments derived from the terminal web (Figures 3-2). Indeed, this zone is the primary site of membrane insertion for filaments of the terminal web.

Spot desmosomes, the third element of the junctional complex, are located just distal to the intermediate junction and are 0.2 to 0.3 μm in length (Figures 3-2, 3-5, 3-7). In contrast to the tight junction and the intermediate junction, which are continuous, the spot desmosome is a discontinuous buttonlike form of attachment. The intercellular space of the spot desmosome is about 24 nm wide and is occupied by a central dense layer, with lucent zones on each side. The lucent zones contain a fine fibrillar material that consists of alternating light and dark lines running obliquely from the lateral membrane to the central dense layer and with a 7-nm periodicity. It is possible that the central dense layer consists of overlap of the dark lines, the composition of which is not known other than that the lines are glycoproteins.[9] Characteristic of the spot desmosome is the presence of two straight plaques of dense material on the cytoplasmic side that are parallel to the plasma membrane (Figure 3-11). The dense

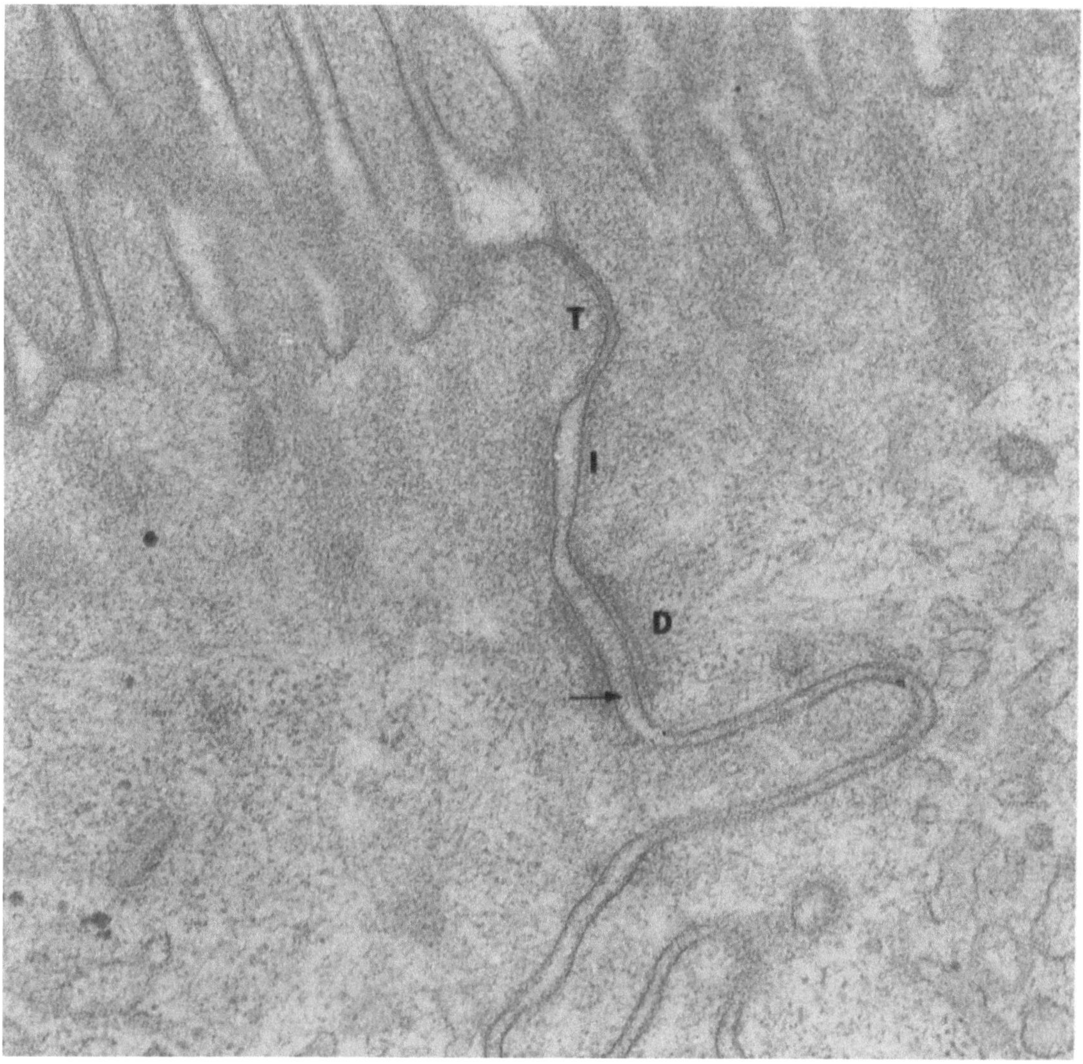

Figure 3-7. Electron micrograph of junctional complex of intestinal absorptive cells showing "apparent" areas of fusion (T) of the outer leaflets of membranes of adjoining cells in the area of the tight junction, the intermediate junction (I), and the desmosome (D). The desmosome is occupied by a dense central layer (arrow) and has two straight plaques of dense material on the cytoplasmic side of the plasma membrane. × 100,000

plaques are 15 to 20 nm thick and are separated from the lateral membrane by an 8-nm lucent zone. Numerous 10-nm tonofilaments are embedded in the plaques, loop through the terminal web, and attach to other spot desmosomes (Figures 3-7, 3-9). Smaller filaments emanate from the plaques and attach to tonofilaments and the lateral membrane. Spot desmosomes serve as "spot welds" between epithelial cells and as attachment sites for cytoskeletal filaments. Many spot desmosomes may be seen along the intercellular space of absorptive cells; they are especially numerous along the intercellular space of undifferentiated crypt cells (Figure 3-9).

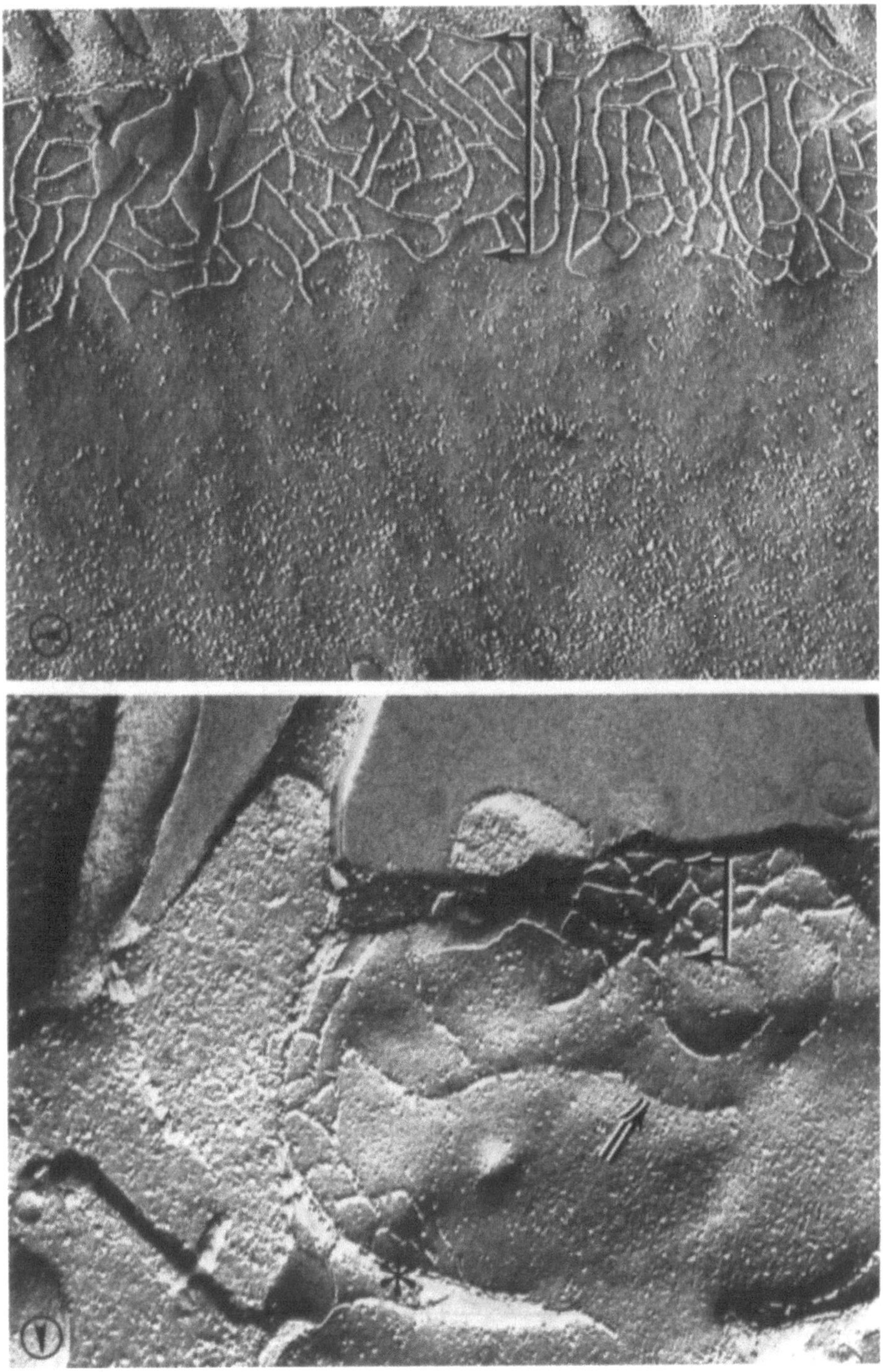

Figure 3-8. Freeze-fracture replica of tight junctions of villus (top) and crypt cells (bottom). Tight junctions of villous absorptive cells (brackets) have greater depth and are more highly organized than those of crypt cells. Apparently, isolated strands (double arrow) are common in the crypt but are rare in the villus. Cells in both crypt and villi show ladderlike specializations at three cell junctions (asterisk). × 56,000. (Reproduced from Madara JL, *Gastroenterology* **78**:963, 1980, with permission of Elsevier Science Publishing Company, Inc., and James L. Madara.)

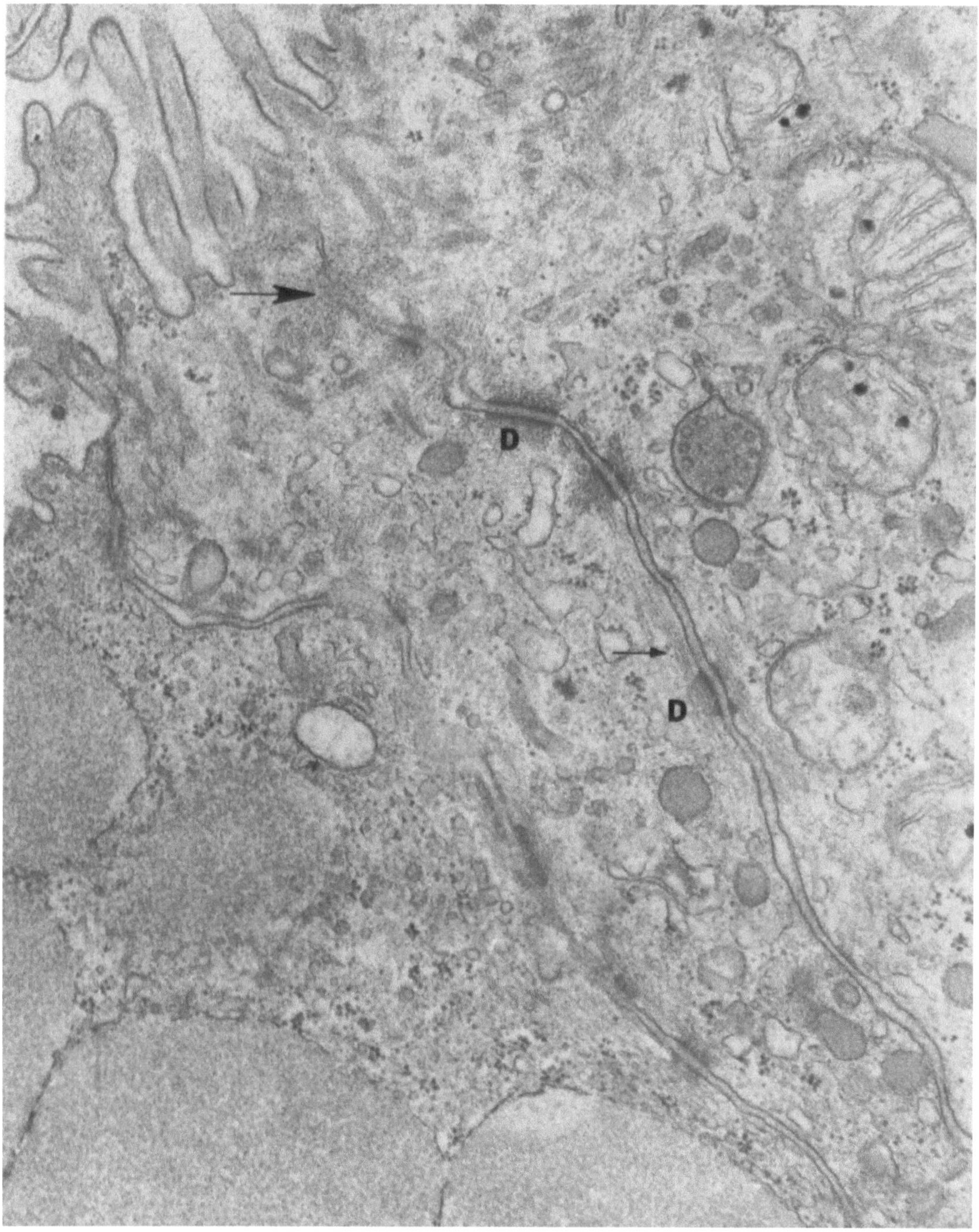

Figure 3-9. Electron micrograph of junctional complexes between undifferentiated crypt cells and between an undifferentiated crypt cell and a goblet cell. Desmosomes (D) are interconnected by tonofilaments (large arrow). Other tonofilaments appear to attach the more apical desmosomes (small arrow) with the terminal web. × 36,000. (Reproduced from Dobbins,[1] by permission of Academic Press, Inc., New York.)

Figure 3-10. Freeze-fracture replicas of microvilli of a villous absorptive cell (left) and an undifferentiated crypt cell (right). The majority of intramembranous proteins are associated with the convex membrane half that covers the microvillous core (P face). Fewer particles are found on the concave membrane half that faces the extracellular space (E face). P-face particles are more numerous on microvillous membranes of villous cells than on those of crypt cells. E-face particle density is comparable at both sites. × 97,750. (Reproduced from Madara JL, *Gastroenterology* **78**:963, 1980, with permission of Elsevier Science Publishing Company, Inc., and James L. Madera.)

The *plasma membrane* of microvilli is considerably thicker than that of most cells, including the basolateral plasma membrane of the same epithelial cell. The microvillous plasma membrane is 8.5 to 11 nm thick, with a mean thickness of 9.5 nm, whereas the basolateral plasma membrane ranges in thickness from 7.5 to 9.5 nm, with a mean of 8 nm.[1] This difference in thickness may reflect the unusual composition of the microvillous plasma membrane–high protein/lipid ratio and distinctive lipid composition.[9]

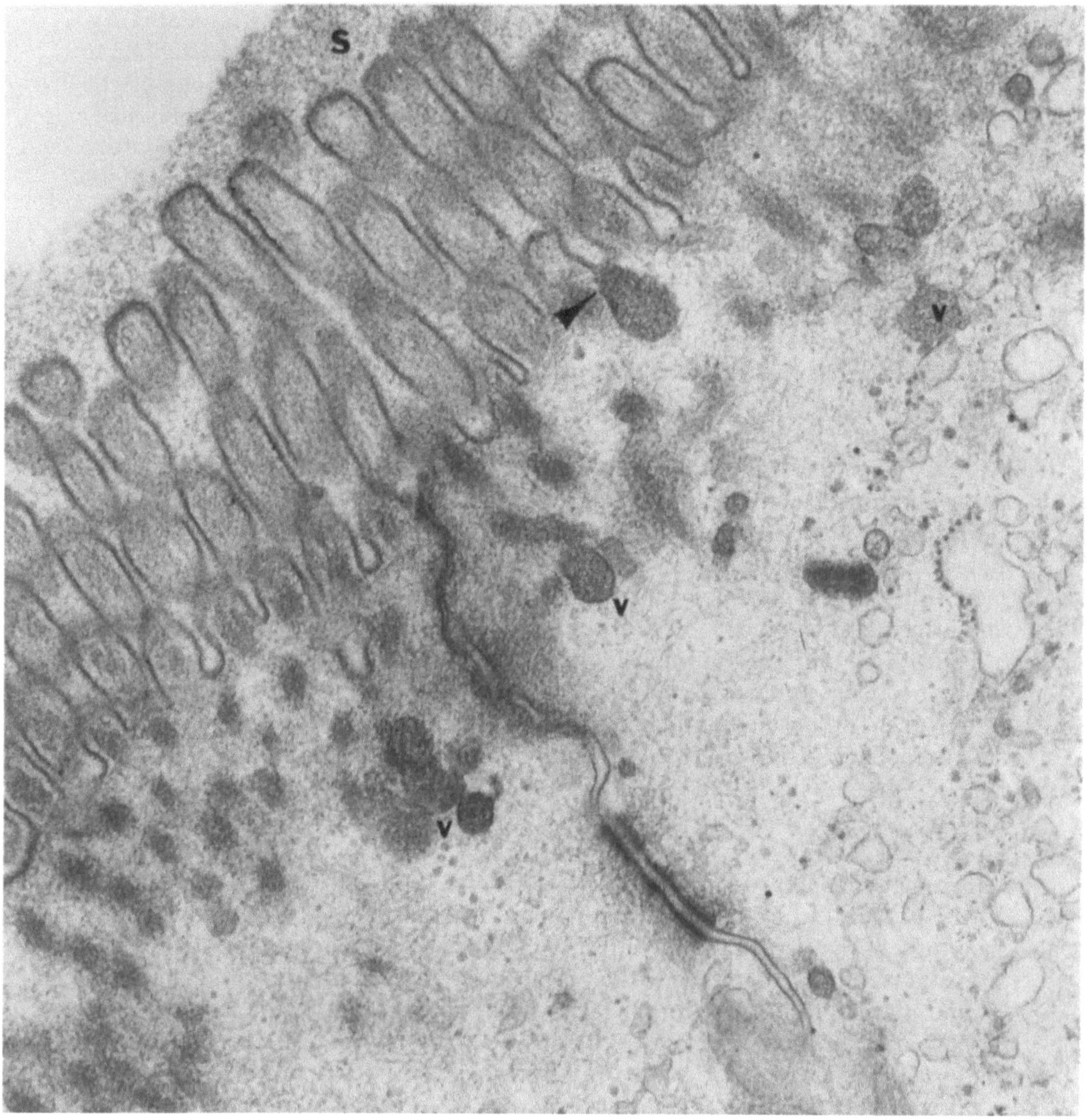

Figure 3-11. Electron micrograph of apex of two adjacent intestinal epithelial cells showing several secretory vesicles (v), the content of which bears a remarkable resemblance to the surface coat (S) over the microvilli. Note that one secretory vesicle (arrow) appears to have fused with the plasma membranes at the base of two microvilli. Note also the junctional complex, described in detail in the text. × 50,000

The surface coat (glycocalyx) is remarkably prominent on the microvillous surface. It is most prominent at microvillous tips and is sparsely present on the lateral surfaces of microvilli. Its fine structure varies with preparative procedures and with staining methods used.[1] Histochemical reactions of the surface coat indicate that its composition, in part, is that of a sulfated, weakly acidic mucopolysaccharide. It is intensely periodic acid-Schiff (PAS) positive and reacts with alcian blue or Hale's colloidal iron stain for acid mucopolysaccharides; it stains with toluidine blue at low pH.[1]

Ultrastructurally the surface coat consists of numerous fine filaments (Figures 3-2, 3-3), generally perpendicular to the plasma membrane and attached to the outer leaflet of the plasma membrane.[1,9] Freeze-fracture replicas of the microvillous membrane show the presence of numerous IMPs, considerably more than are seen on the basolateral membrane. This morphological feature correlates with the high protein (and glycoprotein) content of the microvillous plasma membrane, which has a protein/lipid ratio of 1.7:1 (Figure 3-10).

A number of enzymes, including disaccharidases and peptidases, nonenzymatic proteins (receptors), and transport proteins, are associated with the microvillous plasma membrane and its surface coat. The proteins and glycoproteins of this membrane and its surface coat are continuously synthesized in the cell, transported to the cell surface, and ultimately lost into the lumen.[1] Following translation in the endoplasmic reticulum and posttranslational processing in the Golgi complex, glycosylated proteins are transported in secretory vesicles (called surface-forming vesicles by Bonneville and Weinstock[9a]) for insertion into the microvillous membrane.[1,9] Indeed, the content of the vesicles may be indistinguishable morphologically from the surface coat[9] (Figure 3-11). High-molecular-weight microvillous membrane proteins are largely externally anchored into the membrane by a small hydrophobic tail, whereas low-molecular-weight proteins are more integral to the membrane.

The *basolateral membrane*, that portion of the absorptive cell membrane below the apical junctional complex, differs considerably from the microvillous membrane. The basolateral membrane is not as thick and contains fewer (but still abundant) IMPs, which are randomly distributed. Its biochemical composition, compared with that of the microvillous membrane, is highly enriched for sodium, potassium ATPase, and adenyl cyclase and has a protein/lipid ratio three times lower.[9]

The basal membrane rests on a *basal lamina* that consists of a sheet of fine fibrils (Figures 3-12, 3-13). It is approximately 30 nm wide and is continuous, bridging the wide basal intercellular gaps between adjacent epithelial cells. Scanning electron microscopy, after removal of the epithelium from the basal lamina, has suggested the presence of round to oval pores in the basal lamina, pores that permit passage of chylomicrons, lipoproteins, and cells such as lymphocytes.[10] These pores are probably quite transient in nature and are likely to be repaired rapidly following passage of cells and molecular aggregates such as chylomicrons. The basal lamina consists of (1) type IV collagen, which may be synthesized in the epithelial cell and is found only in the basal lamina; (2) laminin, which is synthesized in mesenchymal and epithelial cells and may serve to bind the basal lamina to epithelial cells; (3) fibronectin, which is of mesenchymal origin and acts as a binding agent; (4) glycosaminoglycans, including heparin sulfate proteoglycan and entactin; and (5) types V and VII collagens.[11] Fibronectin appears to link the basal lamina with the epithelial cell interior through a 140-kd glycoprotein receptor called integrin (a cell substrate attachment antigen), which itself is linked through talin and vinculin to cytoskeletal actin filaments.[11] The "reticular fibers" and collagen adherent to the lamina propria side of the basal lamina are probably products of lamina propria fibroblasts that are closely applied to the basal lamina in the so-called pericryptal fibroblast sheath (discussed later in the section dealing with the lamina propria).

The function of the epithelial basal lamina is only poorly understood. It may act to orient and polarize the epithelial cell (at least in vitro) and to segregate

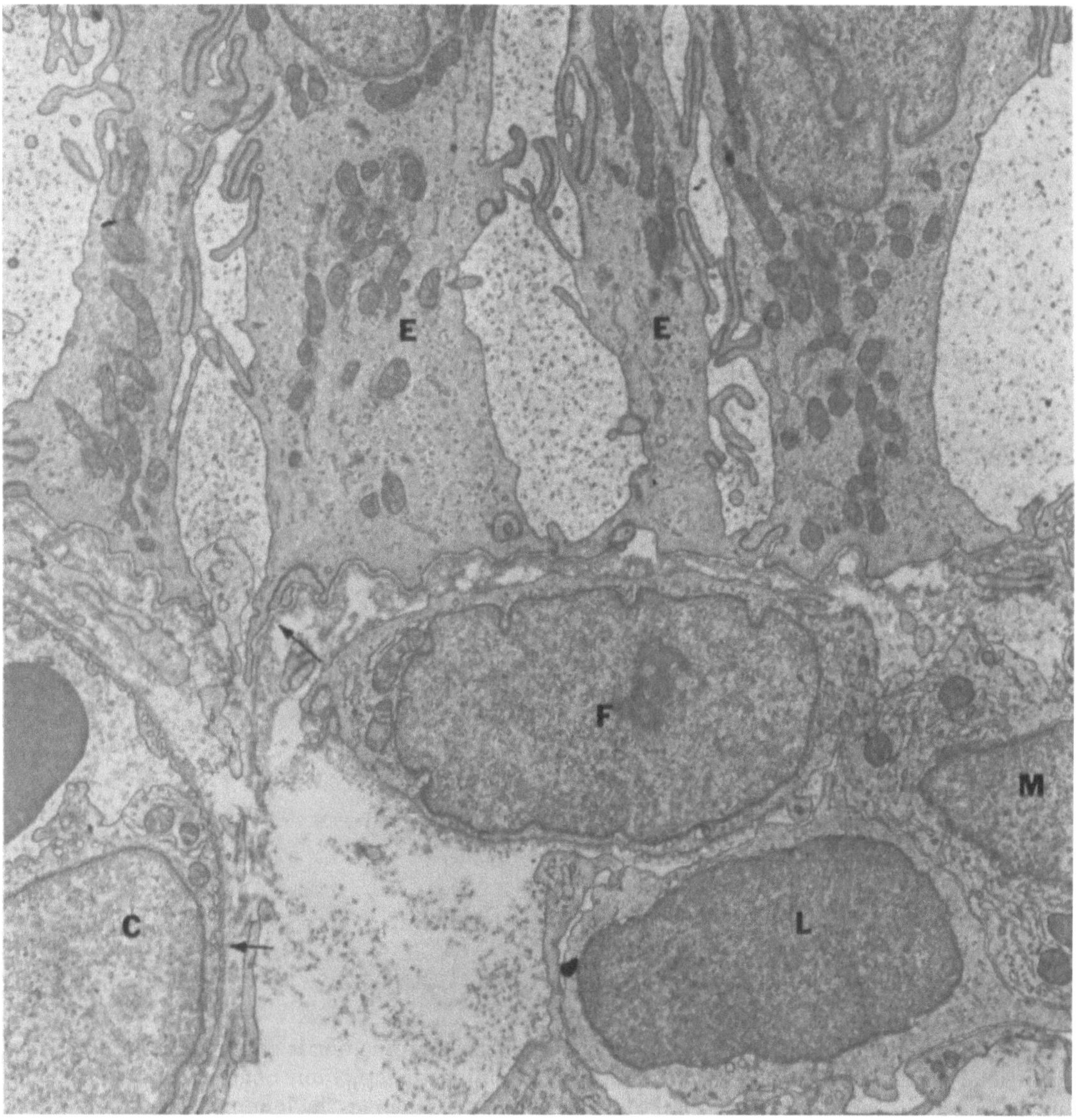

Figure 3-12. Low-magnification electron micrograph of intestinal epithelial cells showing a continuous basal lamina (arrows) below epithelial cells (E) and surrounding a blood capillary (C). Very few collagen fibers are present. Fibroblast (F), lymphocyte (L), monocyte (M). The small particles in the epithelial intercellular spaces are very-low-density lipoproteins (*J Clin Invest* **50**:2065, 1971). × 10,000

the cell into apical and basolateral components.[11] Morphologically the basal lamina consists of a lamina densa, 20 to 50 nm thick, running parallel to the basal membrane. The lamina densa is separated from the basal membrane by a 10- to 25-nm lighter area called the lamina lucidia (or rara), or the lamina rara externa. In the latter case, the lighter area just subjacent to the lamina densa facing the lamina propria is called the lamina rara interna.

The basal lamina contains a row of proteoglycan granules (chondroitin sulfate and/or heparin sulfate) in the internal lamina rara and another in the

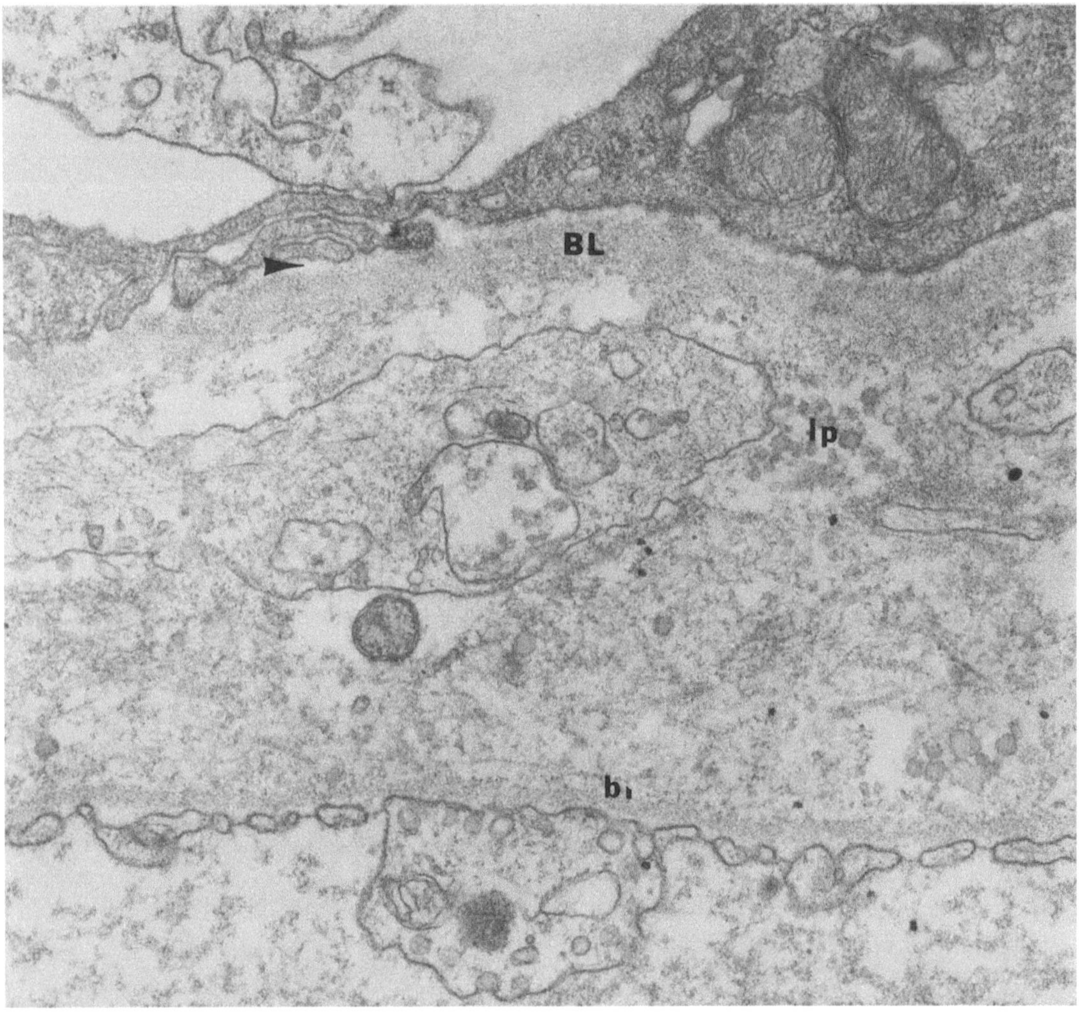

Figure 3-13. High-magnification electron micrograph showing fine fibrils of the basal lamina below (BL) jejunal epithelial cells and above (bl) fenestrated capillary endothelium. Some lipoprotein particles (lp) are free in the lamina propria. Note that the fine fibrils make up the lamina densa, and that the lamina lucida externa (arrow) separates the lamina densa from the epithelial cell basal plasma membrane. A lamina rara interna (see text) is not readily apparent. × 50,000

external lamina rara. The granules are present with a regular spacing of about 60 nm. Figure 3-14 is a putative model from Hay[11] depicting the organization of the various components of the basal lamina and their possible interactions with receptors found in the basal membrane of epithelial cells.

Goblet Cells

Goblet cells, present in both crypt and villous epithelium, are readily identified by their content of mucous granules, which fill the upper half of the cell (Figure 3-15). These cells increase progressively in number from proximal to distal

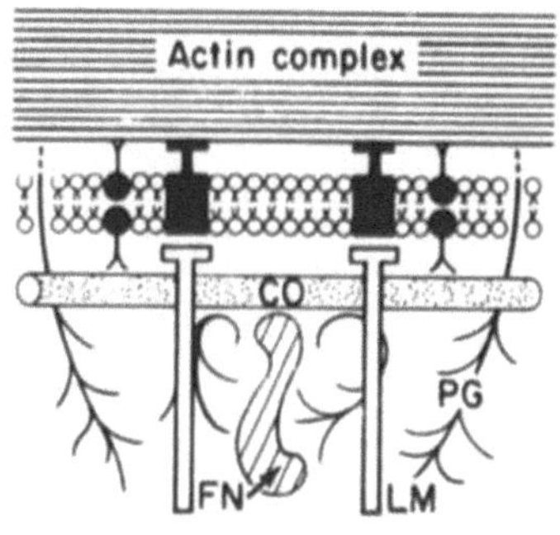

Figure 3-14. Model proposed by Hay[11] depicting the various components of the basal lamina. The model envisions membrane receptors for one or more molecules. Collagen (CO), fibronectin (FN), proteoglycan (PG), laminin (LM). This model was proposed for the interaction of matrix molecules with the corneal epithelial basal surface but is also appropriate as a model for the intestinal epithelial basal lamina. (Reproduced from Hay,[11] by copyright permission of The Rockefeller University Press.)

intestine. The microvilli of goblet cells are more irregular in shape and are less prevalent, but otherwise resemble those of absorptive cells and contain actin filaments in their core. The terminal web is discontinuous and very poorly developed but is penetrated by actin rootlet bundles of microvilli. Actin filaments are prominent in the zonula adherens of goblet cells. The secretory granules are stored as a compact apical mass limited by a dense goblet-shaped sheath or theca.[12] This sheath contains an orderly arrangement of microtubules and intermediate filaments (tonofilaments) but does not contain actin microfilaments (Figure 3-16). The microtubules are arranged vertically, like barrel staves, along the inner aspect of the supporting sheath or cytoskeleton.[12] The intermediate filaments are arranged in two continuous layers—an inner basketlike network and an outer series of circumferential bundles similar to barrel hoops.[12] Because the goblet shape of the supporting sheath is unaltered during rapid release of stored granules (induced by acetylcholine) or after treatment with colchicine (which inhibits polymerization of microtubules), it appears that the shape is maintained by intermediate filaments.

Undifferentiated Crypt Cells

The principal cell of the crypt is the undifferentiated columnar cell (Figure 3-17), a cell that divides frequently and accounts for continuous renewal of villous and crypt epithelium[9] (Figure 3-18). Most newly divided cells migrate toward the villous tip and differentiate to become absorptive cells, goblet cells, and sparsely distributed enteroendocrine cells.[13] Those cells that migrate along the villus are shed every three to five days from the cell extrusion zone at the villous tip. Other newly divided undifferentiated columnar cells migrate to the base of the crypt and differentiate into Paneth cells and enteroendocrine cells.[13] Undifferentiated crypt cells may also be the precursor of M cells and cup cells.[9]

Microvilli of undifferentiated crypt cells are shorter and more sparse than those of absorptive cells; the terminal web is less well developed; the microvillous surface coat is less prominent; and the junctional complex between undifferentiated crypt cells is less complex. The rough and smooth endoplasmic

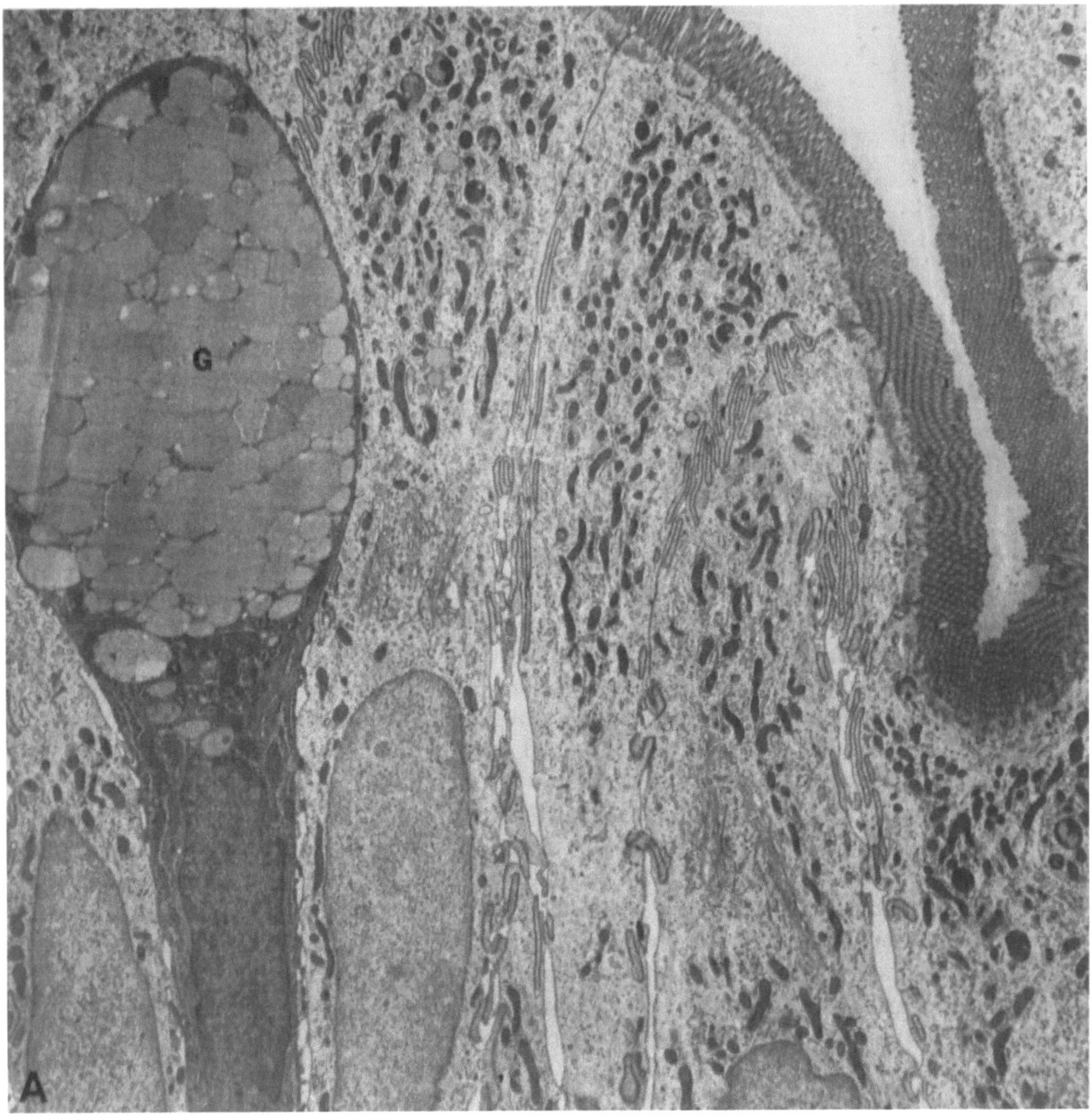

Figure 3-15. (A) Electron micrograph of intestinal epithelial cells and a goblet cell (G). The goblet cell is packed with mucous secretory granules at the apex. Note the Golgi apparatus between the nucleus and the mucous granules. × 5000. (B) High-magnification electron micrograph showing a goblet cell that has released a portion of its apical mucous granules (G). Note the increased density of cytoplasm in the compressed rim of the cell compared with that of the adjacent epithelial cells (E). Only rudimentary microvilli (arrow) are present at the apex of the goblet cell. × 41,400

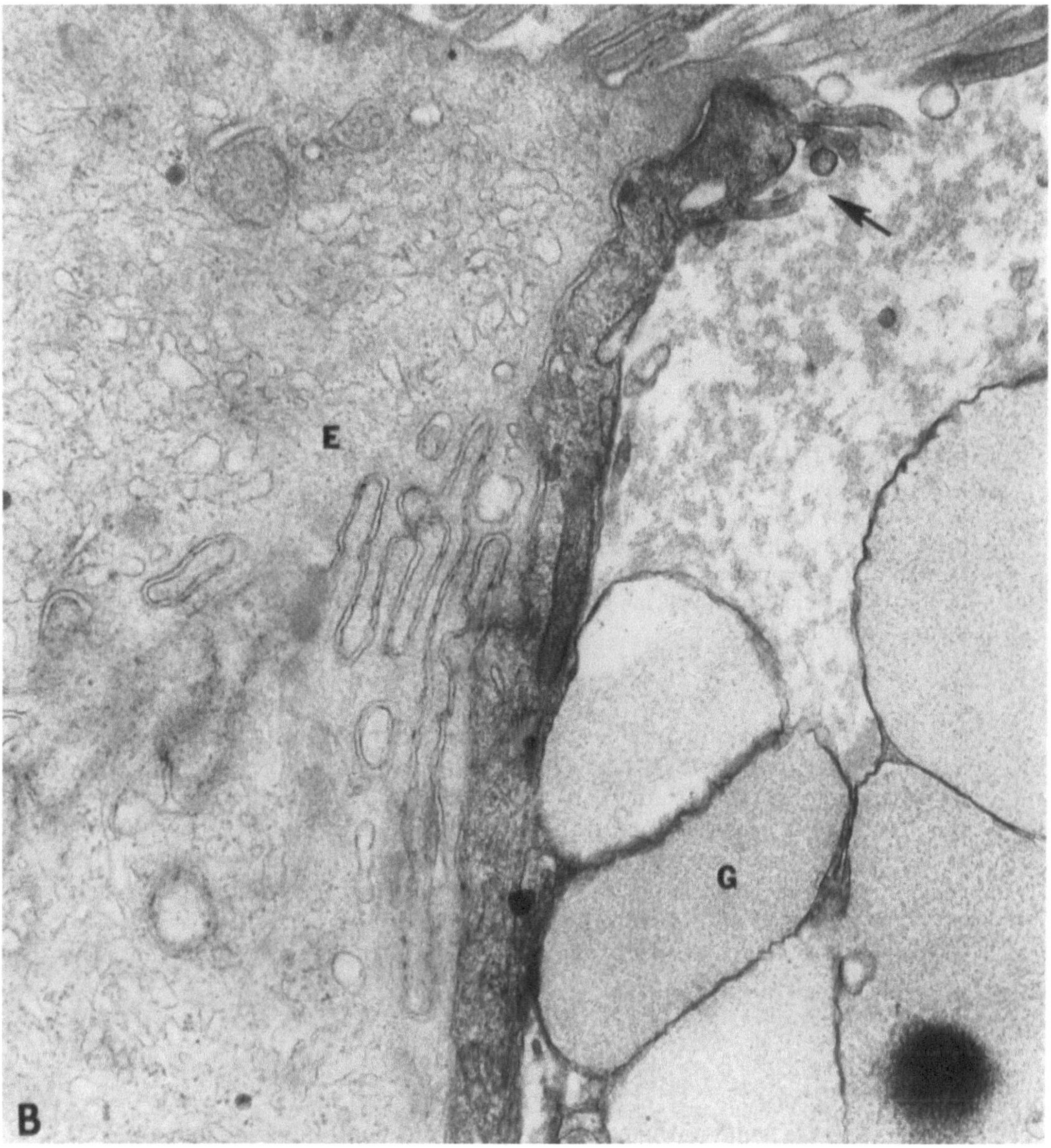

Figure 3-15B.

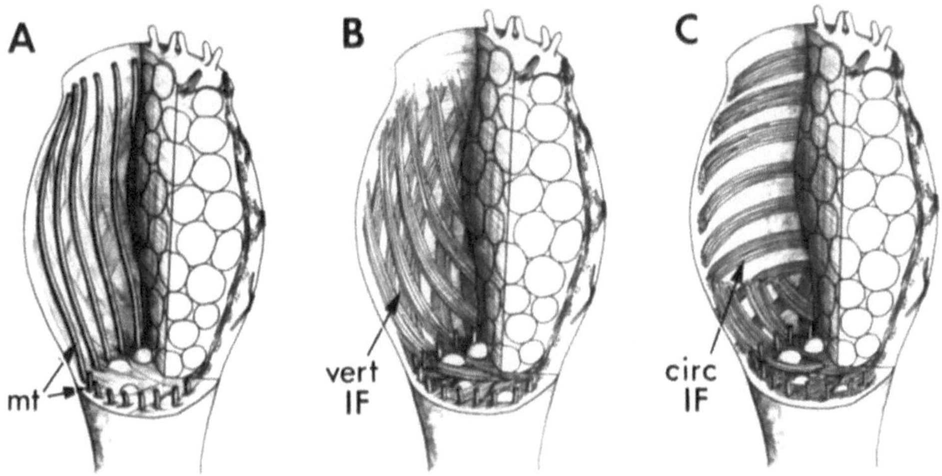

Figure 3-16. Diagrammatic view of the goblet cell theca proposed by Specian and Neutra[12] on the basis of their electron microscopy observations. (A) Vertical microtubules (mt) lie closely apposed to the central mass of stored mucous granules. (B) A basketlike network of vertically oriented intermediate filaments (vert IF) defines the cup shape of the theca. (C) Circumferential bundles of intermediate filaments (circ IF) encircle the basket. (Reproduced from Specian and Neutra,[12] by permission of Elsevier Science Publishing Company, Inc.)

reticulum is sparse, whereas ribosomes and polysomes are abundant. The Golgi apparatus is prominent, lysosomes are less common, and mitochondria are larger and more rounded than in absorptive cells. Centrioles are easily detected. Small (0.1 to 1.5 μm) secretory granules are prominent, contain glycoproteins, and enter the crypt lumen by exocytosis following cholinergic stimulation.[9] The functional role of the secretory product is not defined. Undifferentiated crypt cells are the major site of secretion of water and electrolytes into the lumen. These cells are also, like absorptive cells, involved in transport of IgA into the gut lumen.[9]

Paneth Cells

Paneth cells are found in the base of crypts of the small intestine and are occasionally present in the crypts of the appendix and right colon.[9,14] They increase in number from the duodenum to the ileum. They are pyramidal in shape and, like goblet cells, have sparse microvilli, have a poorly developed terminal web, and their apical one half is packed with secretory granules (zymogen) that closely resemble the secretory granules of pancreatic acinar cells (Figures 3-17, 3-19). The granule morphology varies with fixation at both light microscopy and electron microscopy; the granules (EM) range from homogeneously dense in appearance to those with a dense core and clear halo. The granules are enclosed within a closely applied membrane. In light microscopy preparations the granules are not preserved by acid fixatives and are best preserved in neutral buffered formalin and in mercuric-containing fixatives.[14] The granules contain acid mucopolysaccharides (in rodents), glycoproteins, and N-acetylmuramylhydrolase (lysozyme or muramidase). Lysozyme is a basic enzyme with antibacterial

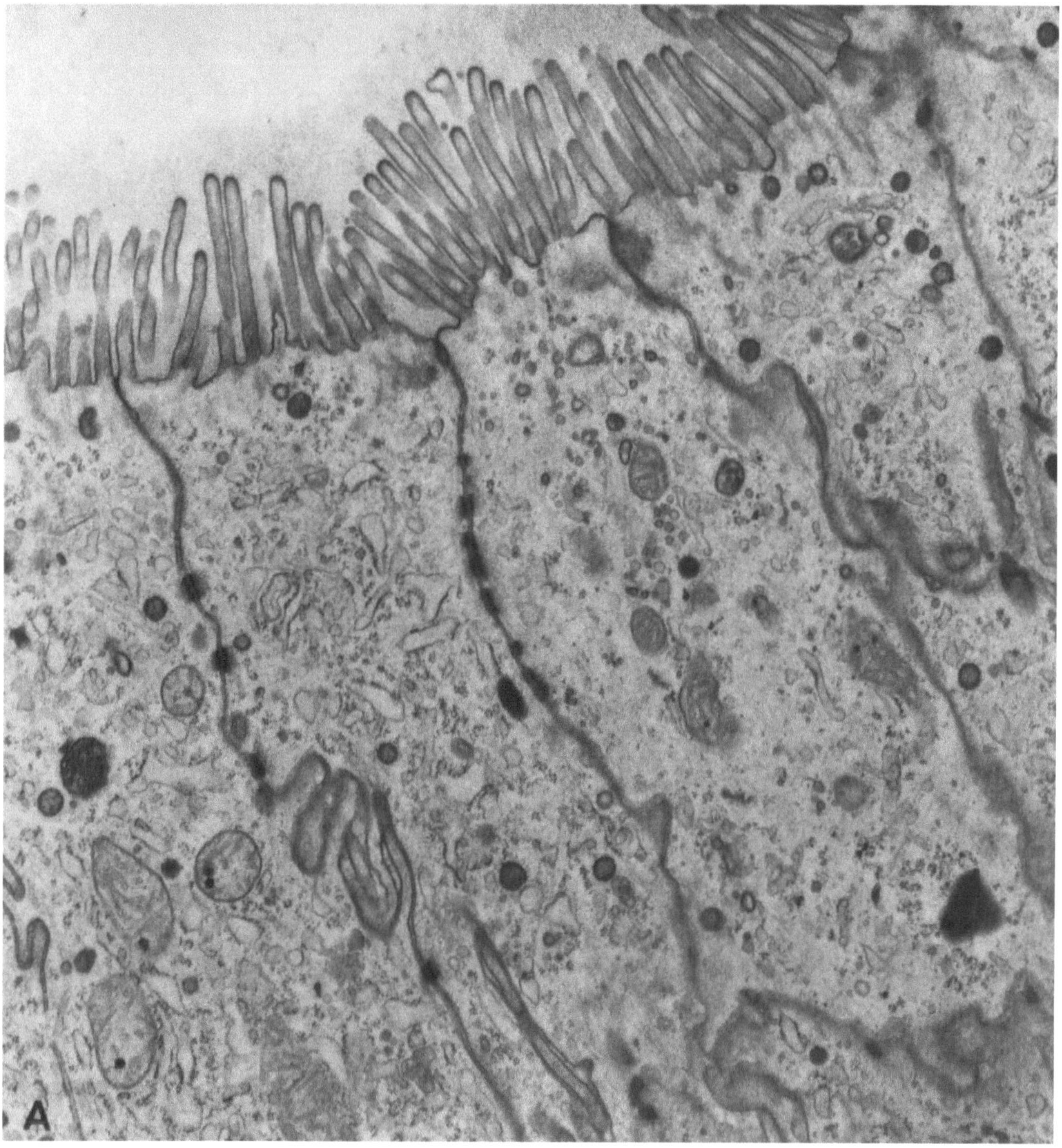

Figure 3-17. Electron micrographs of undifferentiated crypt cells. (A) Typical appearance–numerous secretory granules and irregular microvilli. See text for further details. × 20,000. (B) Base of an unusual form (U) of undifferentiated crypt cells (Trier JS. *J Cell Biol* **18**:599–602, 1963) seen in the base of crypts among Paneth cells (P). The nuclei are multilobed and there is a huge accumulation of mitochondria. The function of these cells has not been determined; they may represent stem cells. × 10,600. (C) Base of intestinal crypt, showing Paneth cells (P), undifferentiated crypt cells (U), and an enterochromaffin EC_1 cell (E). × 8,400 (Figures 3-17B and 3-17C on following pages.)

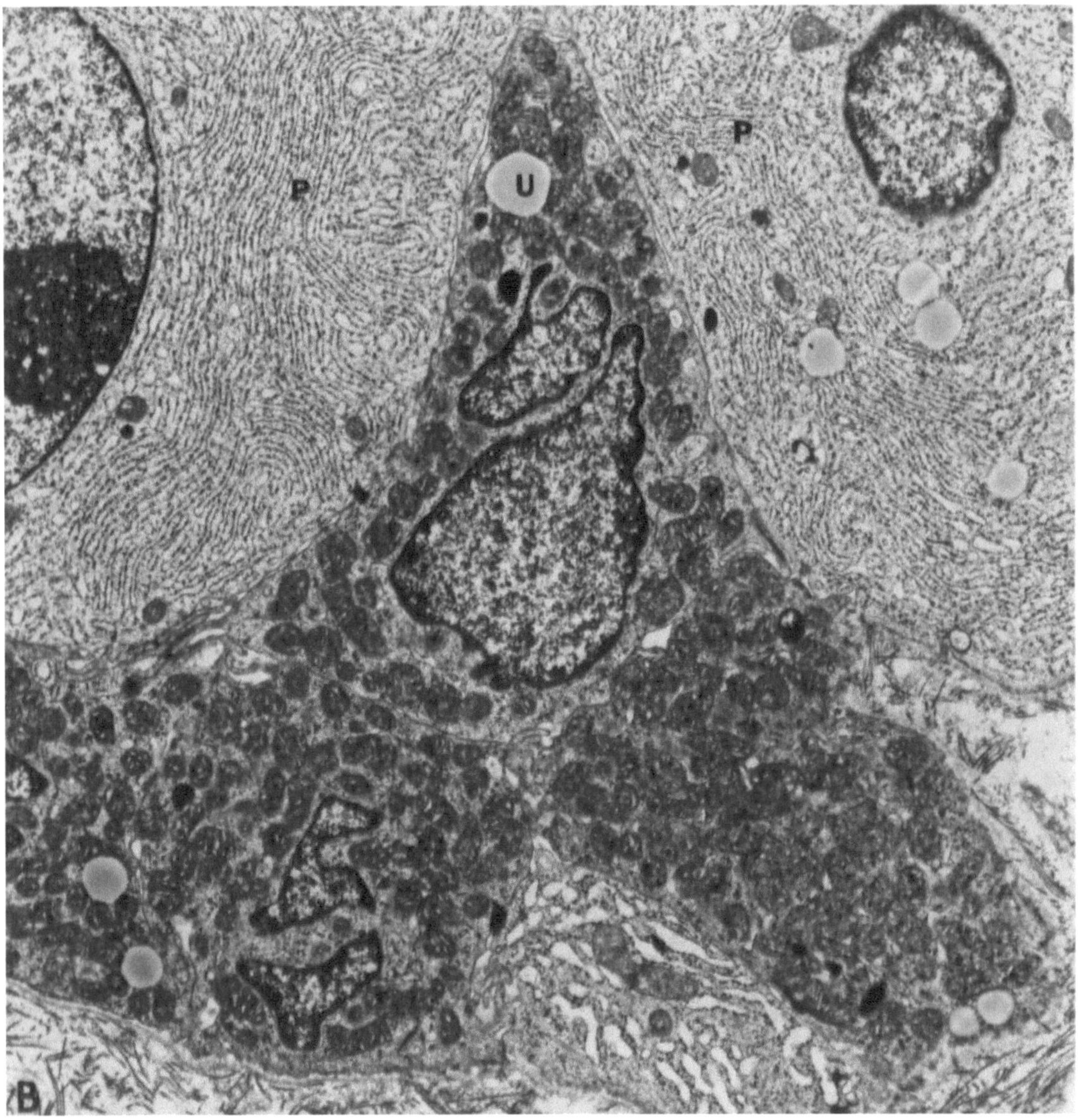

Figure 3-17B.

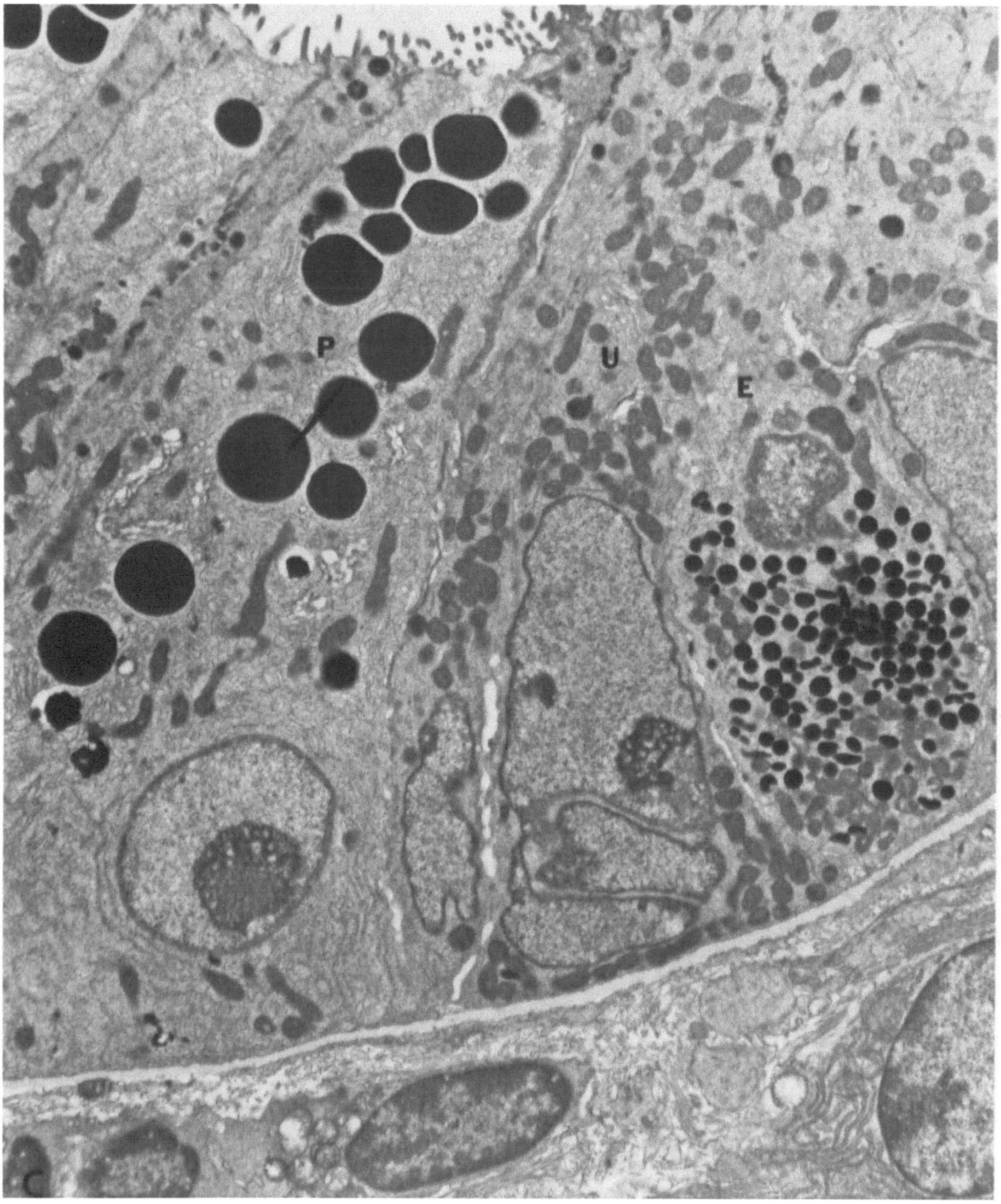

Figure 3-17C.

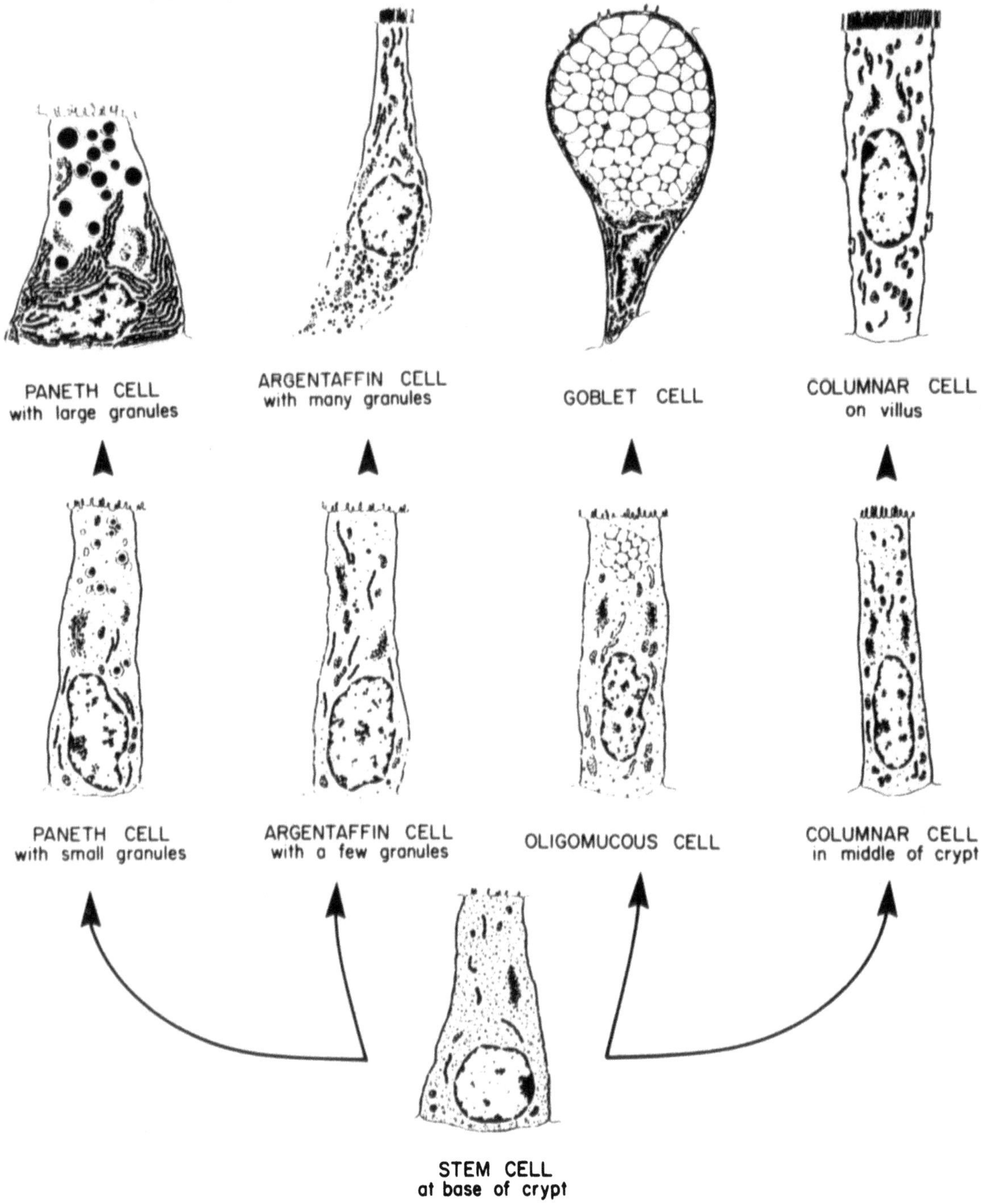

Figure 3-18. Diagram of the unitarian theory of the origin of all epithelial cells from undifferentiated crypt cells, proposed by Cheng and Leblond.[13] (Reproduced from Cheng and Leblond,[13] by permission.)

activity that acts by virtue of its ability to degrade bacterial cell walls by acting on their peptidoglycan components.[9] IgG and IgA have also been localized to human Paneth cells. Paneth cells are derived from undifferentiated crypt cells and have a large Golgi apparatus, well-developed rough endoplasmic reticulum, and mitochondria distributed throughout the cytoplasm.

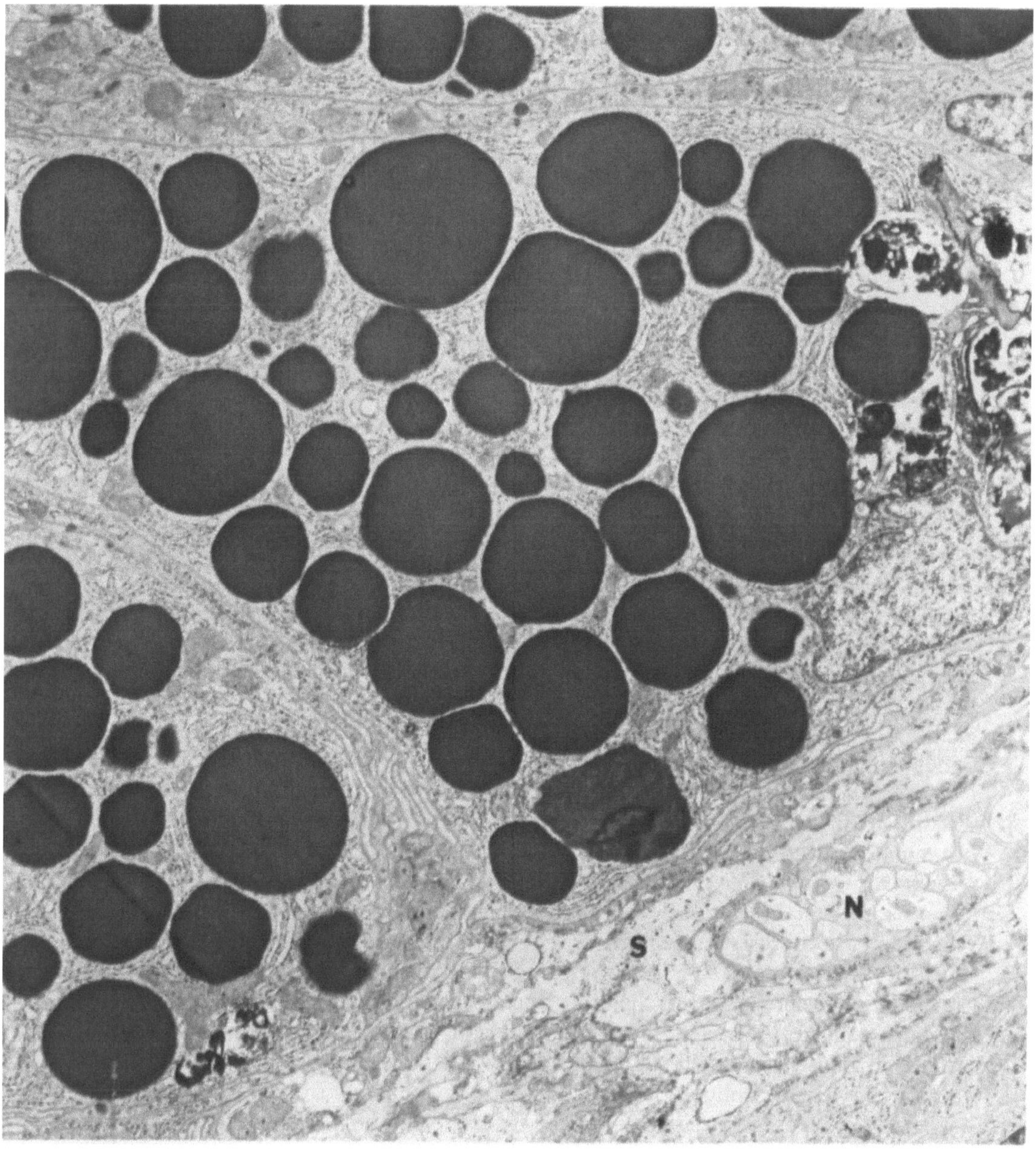

Figure 3-19. Electron micrograph of duodenal Paneth cells showing great range in appearance (size, shape, and content) of the granules. Note similarity of these cells to pancreatic acinar cells. Smooth muscle (S) and nerve fibers (N) can be seen in the lamina propria. × 10,600

Control of Paneth cell secretion is not well defined, but the cells clearly respond to cholinergic agents.[9] Their function is unknown, although their content of bacteriolytic enzyme (lysozyme) and their presumed transport (synthesis would be unique) of IgG and IgA implies an antimicrobial role.

Paneth cell changes in disease are few, and all observed changes are probably nonspecific. The most consistent change associated with Paneth cells is their presence in metaplasia (1) found adjacent to gut neoplasms of the colon, (2) in metaplasia associated with inflammatory bowel disease, and (3) intestinal metaplasia found in gastric atrophy, in the gall bladder, and in the urinary tract.[14] Paneth cells have been described as increased, decreased, and normal in number in celiac sprue. Any Paneth cell changes in celiac sprue are likely to be nonspecific and a response to the inflammatory changes in the mucosa.

Paneth cell granules are altered in appearance in acrodermatitis enteropthica prior to zinc replacement. The granules have marked heterogeneity of their dense material and contain rod-shaped and spherical inclusions.[15] Within the cytoplasm are numerous lysosomelike (acid-phosphatase-negative) inclusions morphologically up to 4 μm in length containing a fibrillar material. These changes appear to be secondary to zinc deficiency, because the Paneth cells are normal in appearance after four months of zinc therapy.[15] Neoplastic Paneth cells are exceedingly uncommon–a single duodenal adenoma consisting of Paneth cells has been described.[14]

In summary, Paneth cells are found in the base of crypts of the intestine, are not present in the normal stomach, and are few in number in the colon. Their numbers vary in inflammatory states (e.g., celiac sprue), and they appear in metaplasia that takes place when the luminal environment simulates (?) that of the intestine, that is, in colonic and gastric neoplasms and in inflammatory bowel disease. These cells secrete IgA, IgG, and a proteinaceous substance containing lysozyme into the gut lumen.

Enteroendocrine Cells

Enteroendocrine cells are widely distributed throughout the stomach, intestine, and colon and thus constitute a series of overlapping diffuse endocrine organs.[16] These cells possess few distinguishing features when examined with standard histological techniques. They have an affinity for silver stains; some are spontaneously fluorescent following formalin fixation and may have an affinity for aniline dyes and other histological stains. Chromogranin, representing a family of proteins that coexist with catecholamines in the secretory vesicles of adrenal medulla cells, appears to immunostain all gut endocrine cells. Thus, chromogranin, as well as neuron-specific enolase (NSE), can be used as a marker for mucosal endocrine cells.[17] NSE is a cytosolic protein diffusely spread in the cell and is a nongranular marker, whereas chromogranin occurs in storage granules and is an indicator of granularity. None of the staining techniques can distinguish the 15 different endocrine cells currently identified in the intestinal epithelium. These cells can be specifically identified only by use of properly controlled immunohistochemical methods.

Electron microscopy demonstrates that these cells have secretory granules with a distinctive ultrastructural appearance, and the appearance of the granules alone (with prior immunohistochemical and electron microscopy correlation) can be utilized to identify specifically many, but not all, enteroendocrine cells. Enteroendocrine cells can be classified as *open* when the apex of the cell contacts the lumen (the apex often has prominent microvilli) and as *closed* when the cell has no luminal contact.[16] In general, enteroendocrine cells have certain "typical" electron microscopy features: (1) low levels of rough endoplasmic reticulum, (2) prominent amounts of smooth endoplasmic reticulum, (3) a high content of free ribosomes, (4) dense (condensed) mitochondria that are fixation labile, (5) basally located secretory granules best fixed with glutaraldehyde, and

(6) marked variation in size and shape of their secretory granules. They may be classified into three categories:

1. Cells with well-established function
2. Cells with hypothetical or controversial function
3. Cells with unknown function.

The important electron microscopy characteristics (that may or may not be identifying) of these cells are as follows.

Cells with Well-Established Function

1. G cells (gastrin producing) are found in Brunner's glands of the duodenal bulb and are similar to gastric antral G cells (Figure 3-20). The granules are 180 to 300 nm in diameter and have a flocculent core of highly variable density. The granules are concentrated at the base of the cell, which is usually an "open" cell.
2. IG cells (intestinal gastrin-producing cells) are found in the duodenum, jejunum, and ileum. The granules are round, dense, and 190 $\pm$ 30 nm in diameter and resemble the granules seen in D_1 cells (Figure 3-21). Most intestinal gastrin seems to originate from these cells, not from the very sparse duodenal G cells.
3. D cells produce somatostatin and are characterized by the presence of round, moderately dense secretory granules measuring 300 to 400 nm in diameter (Figure 3-22). The surrounding membrane is closely applied to the granule matrix and sometimes has a "nibbled at" appearance. These cells have long cytoplasmic projections that promote their paracrine function.[18] The cell processes may be more than 40 μm long.[18] These cells are present throughout the small intestine.
4. S cells produce secretin and are present in the epithelium of the duodenum and jejunum. Their granules are round to moderately irregular in shape, 180 to 220 nm in size, and of high density; a narrow clear space often separates the granule contents from the surrounding membrane. S-cells have some similarities to D_1 cells.
5. I cells produce cholecystokinin and are found in the duodenum, jejunum, and occasionally in the ileum. They have round to slightly irregular granules 250 to 300 nm in diameter, with dense cores, and they bear a close resemblance to L cells (Figures 3-23).
6. L cells (EG cells) produce enteroglucagon and are present in the jejunum and ileum. Their granules have a round or slightly irregular core with a closely applied membrane and range in size between 250 and 300 nm. L cells are remarkably similar to I cells. The application of the Grimelius stain at electron microscopy distinguishes the two, in that I-cell granules do not stain whereas L-cell granules have a strong argyrophilic rim.
7. EC (enterochromaffin) cells are classified in three subgroups on the basis of the distinct morphology of the secretory granules. EC_1 (Figure 3-24) and EC_2 (Figure 3-24) cells are found in the intestine, while the ECn cell (Figure 3-26) is found predominantly in the gastric antrum and fundus and only occasionally in the duodenum. EC_1 cells contain serotonin (5-hydroxytryptamine) and leu-enkephalin. These cells have pleomorphic secretory granules measuring

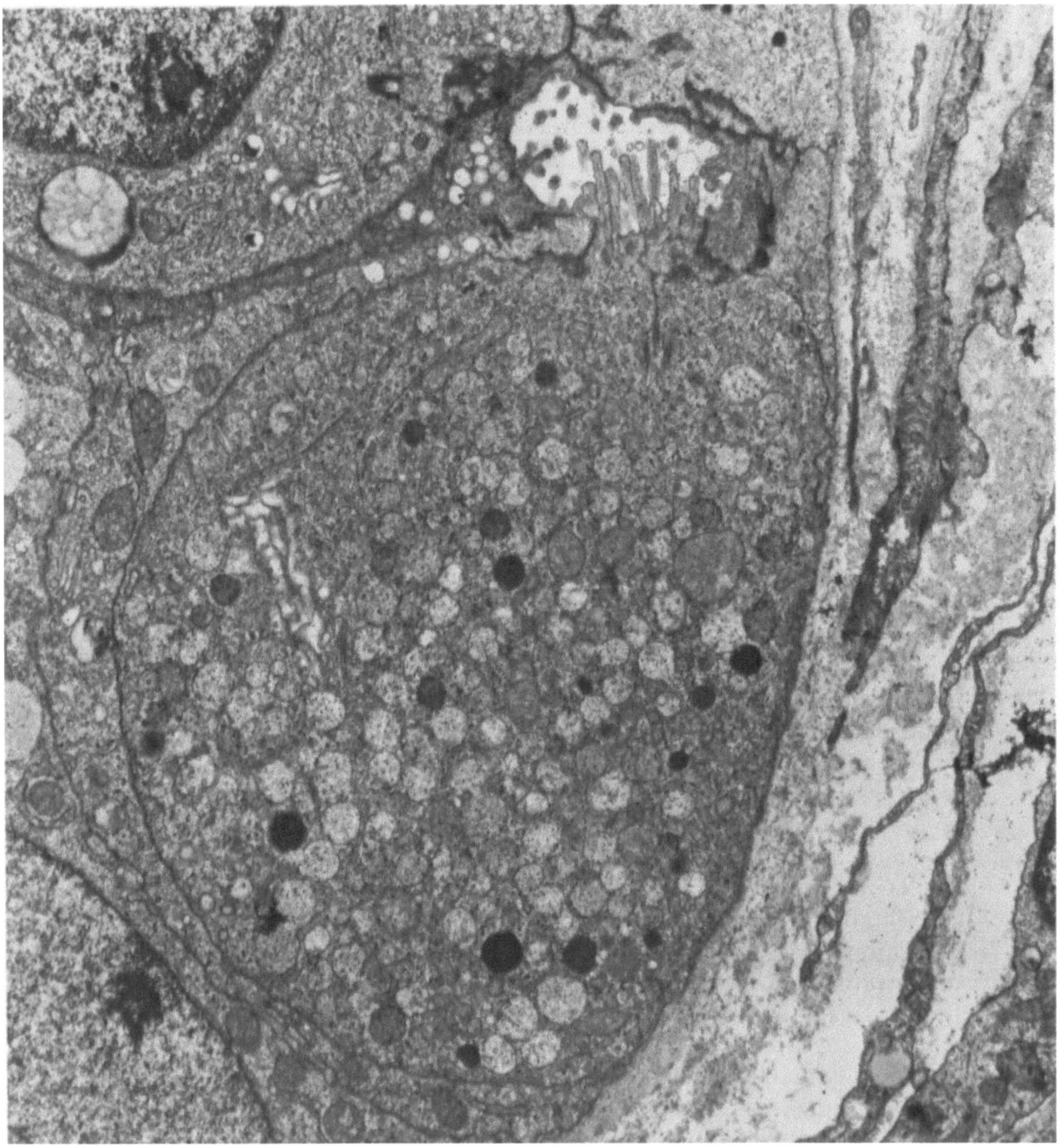

Figure 3-20. Electron micrograph of duodenal G cell immunostained with antigastrin sera and labeled with immunogold. Note intense localization of immunogold particles over the granules. This is an "open" G cell. Note flocculent core and variability in density of the granules. × 14,500

300 to 400 nm and a high-density core with a closely applied membrane. The granules may be elongated, round, oval, or reniform in shape and may have a thin peripheral halo.

8. Mo cells contain motilin, are found in the jejunum and ileum, and contain granules of the D_1 type. The granules are small, dense, and uniformly round and have a mean diameter of 160 nm.[19] Motilin cells contain numerous microfilaments. The Mo cell, along with IG and TG (tetragastrin) cells, has only

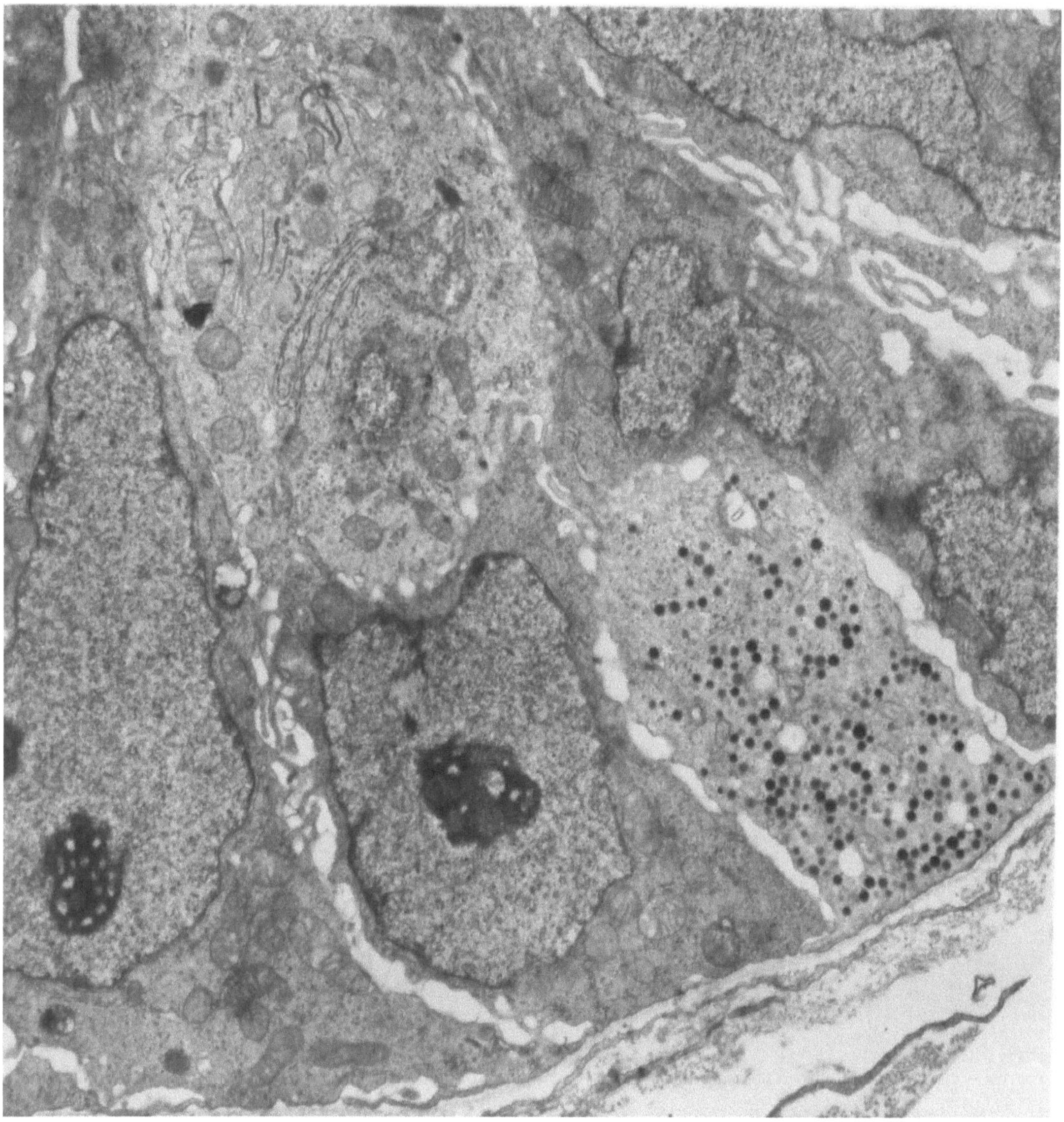

Figure 3-21. Electron micrograph of duodenal endocrine cell with small granules (it includes IG, S, Mo, D_1, and P cells). Immunolabeling is required to identify the cell type clearly. The granule size in this cell ranges from 120 to 160 nm. × 14,500

recently been accepted as an established cell type on the basis of light and electron microscopy immunohistochemical staining.[16]

Cells with Hypothetical or Controversial Function

1. EC_2 cells (Figure 3-24) are found in the duodenum and jejunum and contain serotinin, possibly motilin, and leu-enkaphalin.[19] Their granules tend to be larger than those of EC_1 cells—400 nm or more—and are more angular and

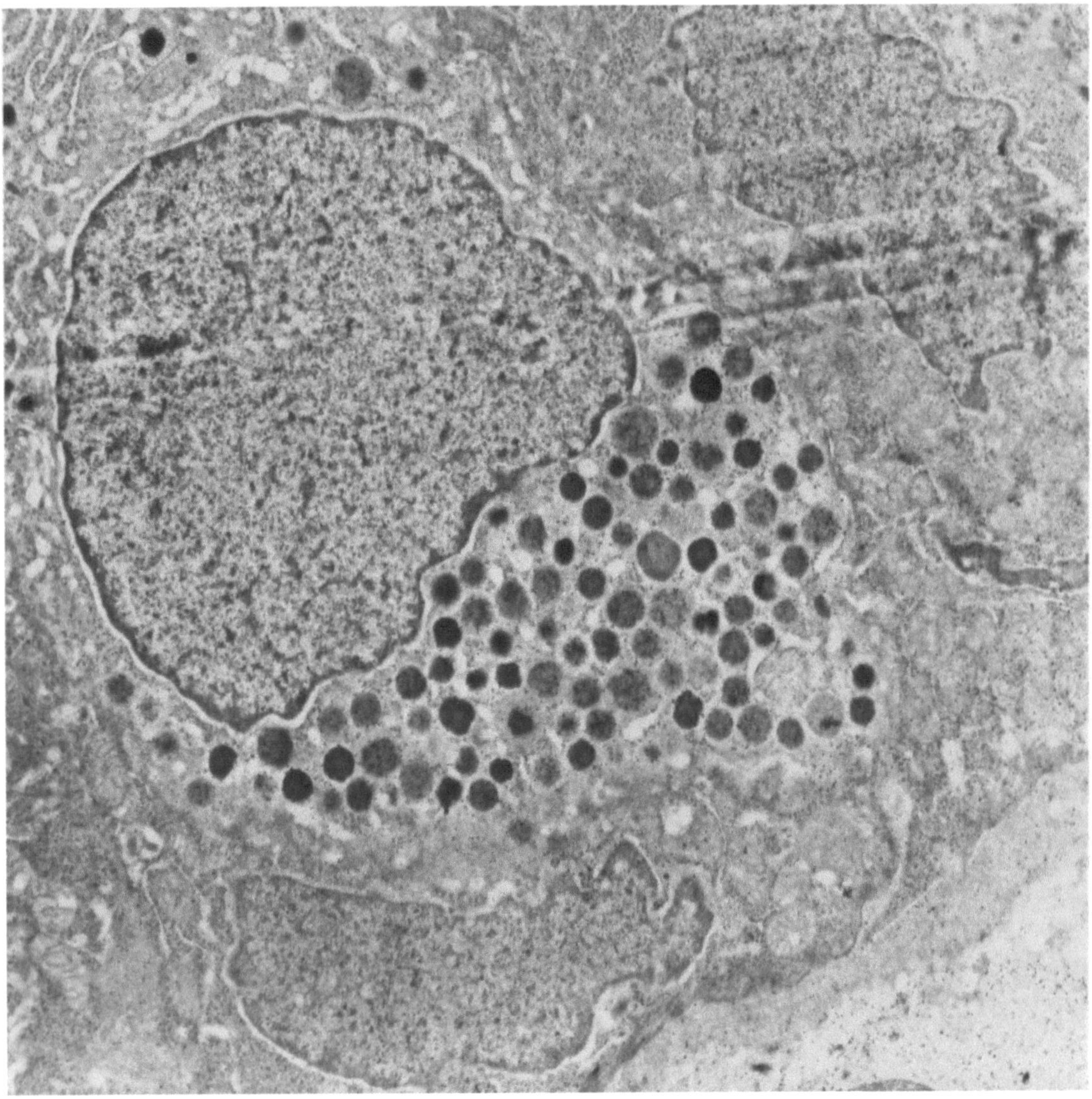

Figure 3-22. Electron micrograph of duodenal D cell immunostained with antisomatostatin sera and labeled with immunogold. There is intense localization of gold particles over the granules. The granules are moderately dense with mild variations in density. × 14,500

irregular in shape. In contrast to EC_1 granules, which have a high density, EC_2 cell granules have a medium to high density.

2. K cells are found in the duodenum and jejunum and sometimes in the ileum and contain granules that are round to somewhat irregular in shape and up to 350 nm in size. The granules often have dark, eccentrically placed cores surrounded by a less dense matrix (Figure 3-25). These cells probably produce gastric inhibitory polypeptide (GIP).

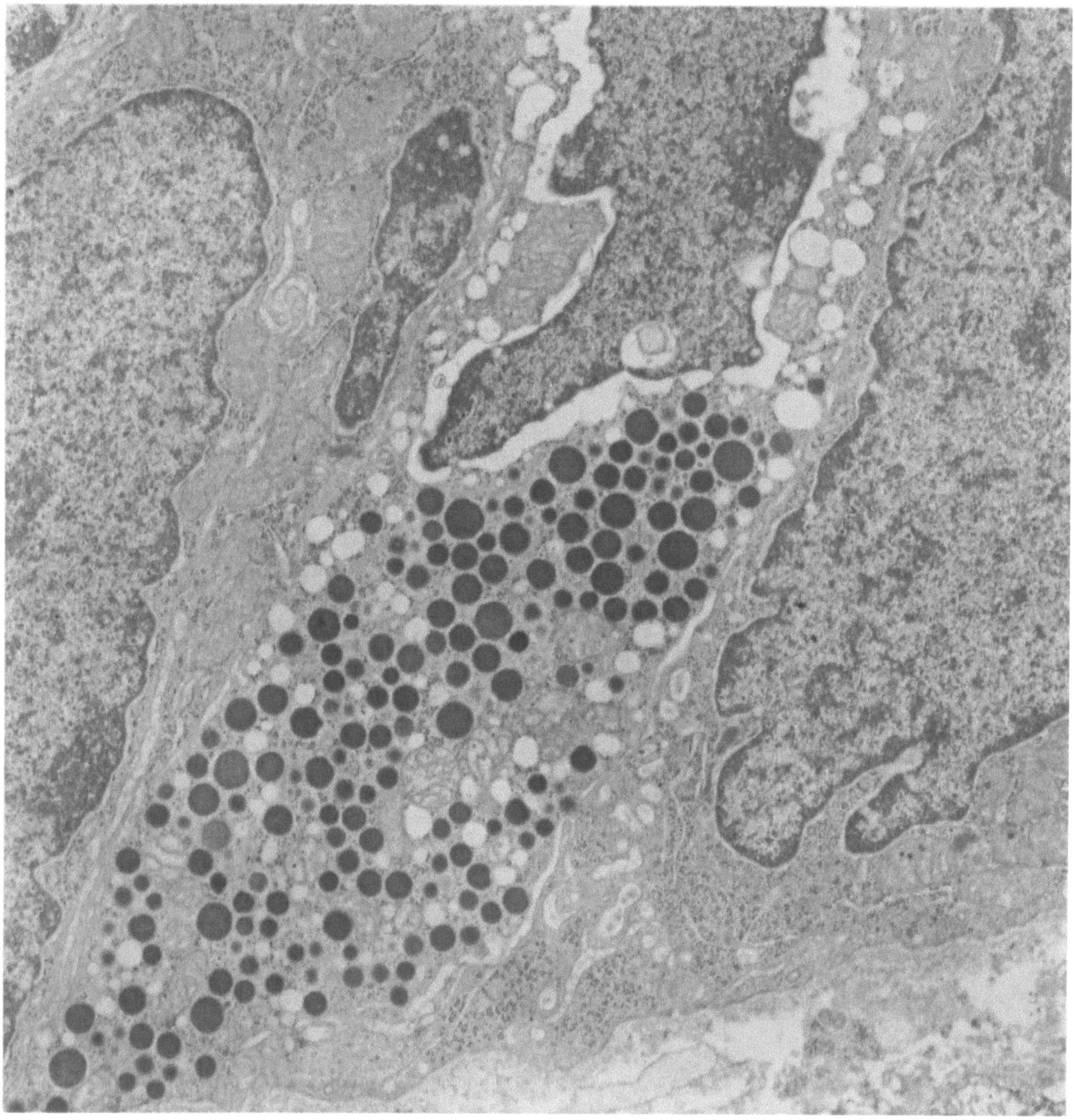

Figure 3-23. Electron micrograph of a duodenal endocrine cell that has the relatively large, uniformly dense granules seen in I, L, and N cells. Immunostaining is required to distinguish the three cell types. The granules in this cell range in size from 280 to 400 nm. × 16,200

3. D_1 cells are found throughout the intestine and are characterized by the presence of small round secretory granules measuring 140 to 190 nm with moderate to high-density cores and a narrow halo. They are quite similar to S cells (Figure 3-21) and may represent (along with ECn cells) a heterogeneous group of cells.
4. N cells produce neurotensin and are found in the ileum. They have round, moderately dense granules up to 300 nm in size. They are very similar to L

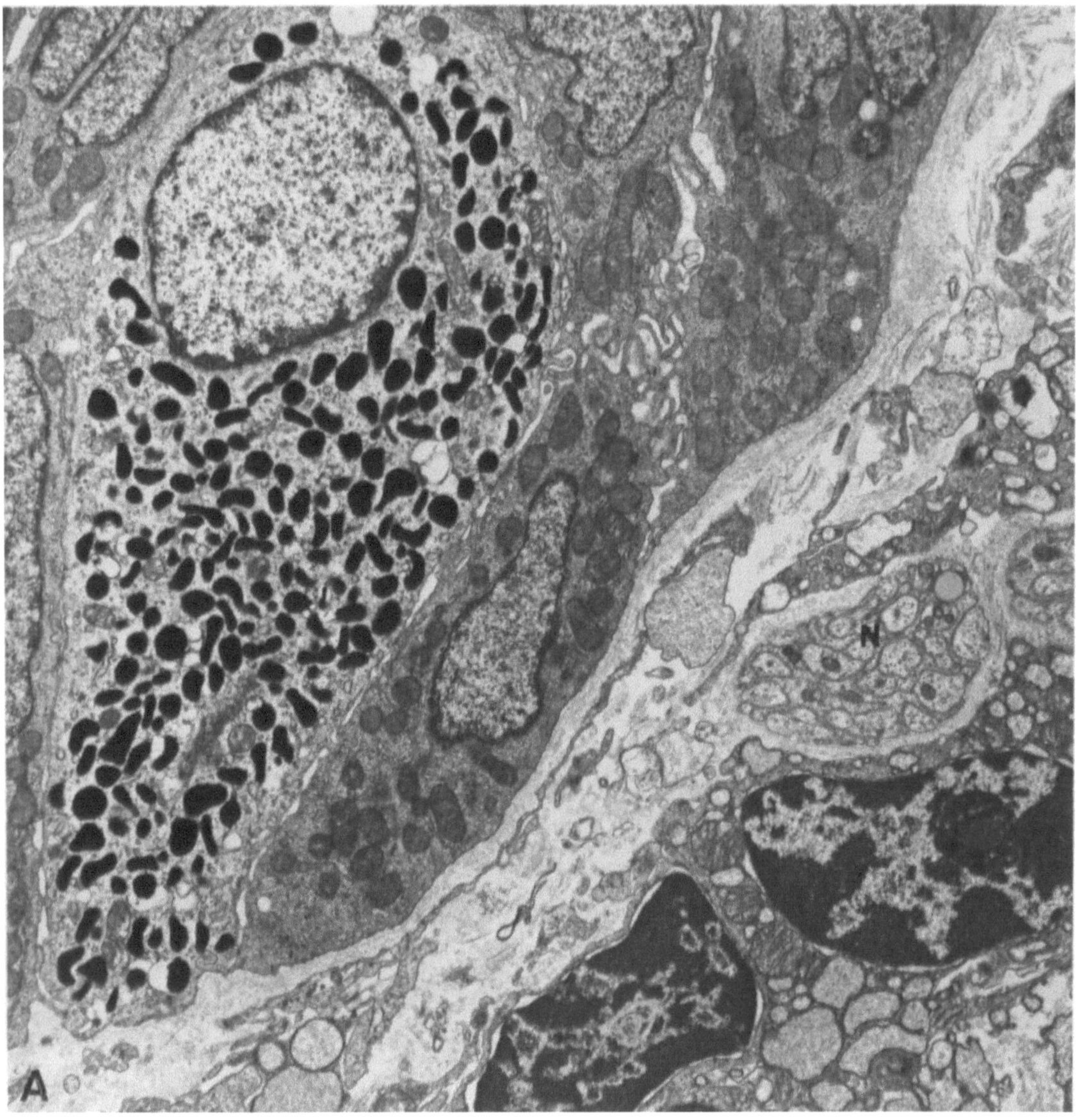

Figure 3-24. Electron micrographs. (A) Duodenal EC_1 cell that contains pleomorphic-shaped, uniformly dense secretory granules of high density. N, nerve fibers. × 14,500. (B) Duodenal EC_2 cell that contains granules that are larger, more variable in shape, and of lower density than the granules of EC_1 cells. Nerve fibers (N). × 10,000

cells (Figure 3-23) and can be distinguished from L cells only on the basis of immunohistochemical staining for neurotensin.

5. P cells, present in the duodenal mucosa, have very small round granules 90 to 150 nm in diameter. Their matrix is of medium density and is separated from the limiting membrane by a narrow clear space. They are similar to D_1 cells (Figure 3-21) and may produce a bombesinlike substance.
6. TG cells react with antibodies to the C-terminal tetrapeptide of gastrin/cholecystokinin (CCK) and have not been ultrastructurally defined.

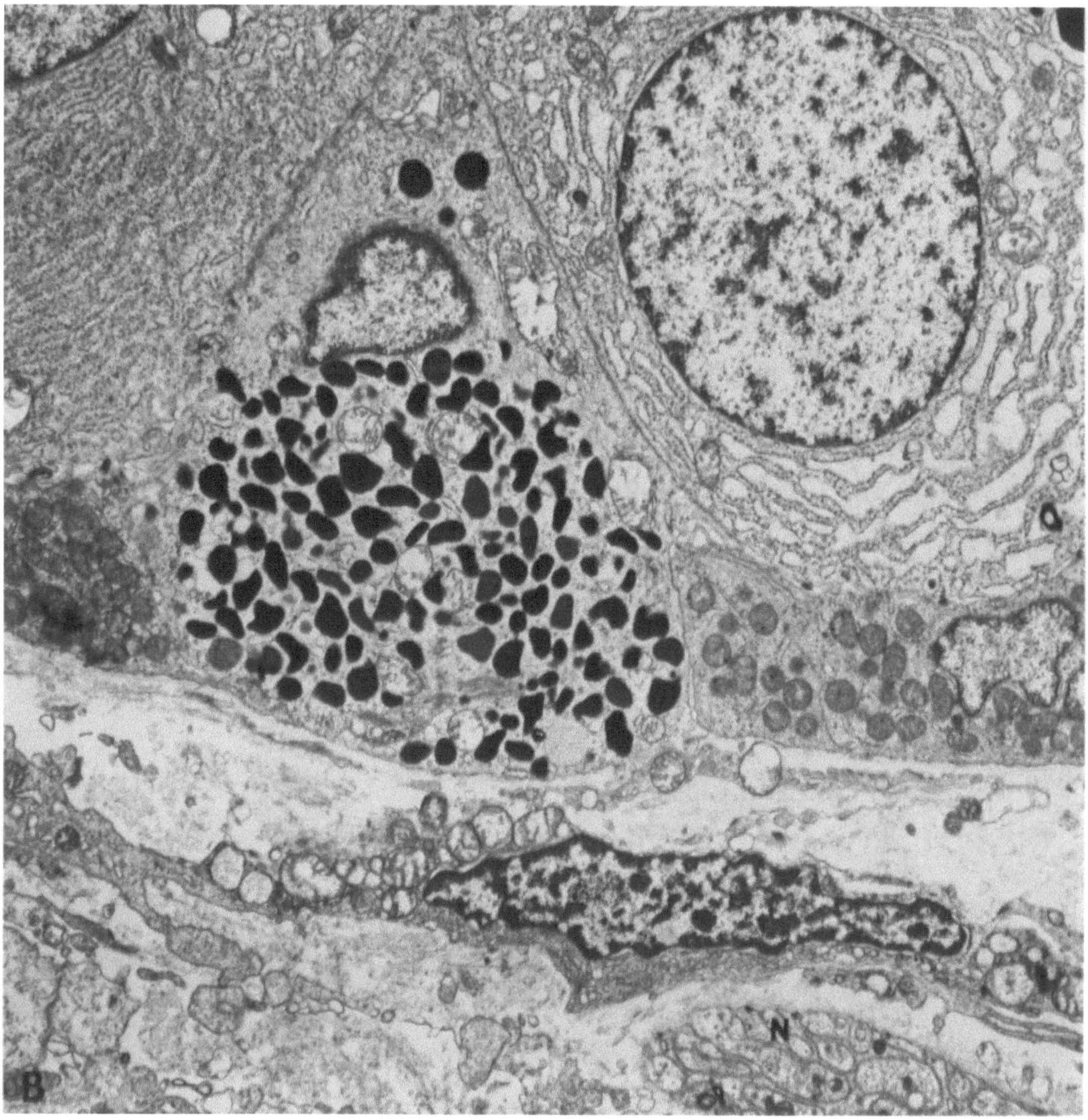

Figure 3-24B.

7. ECn cells contain serotonin and include all those cells once called gastric EC cells. They are occasionally found in the duodenum. Their granules are small, dense, and elongated, measuring 200 to 300 nm in their greatest diameter (Figure 3-26). Their functional significance has not been determined.

Cells with Unknown Function

Cells with unknown function are not found in the small intestine but include the ECL and X cell in the gastric fundic mucosa.

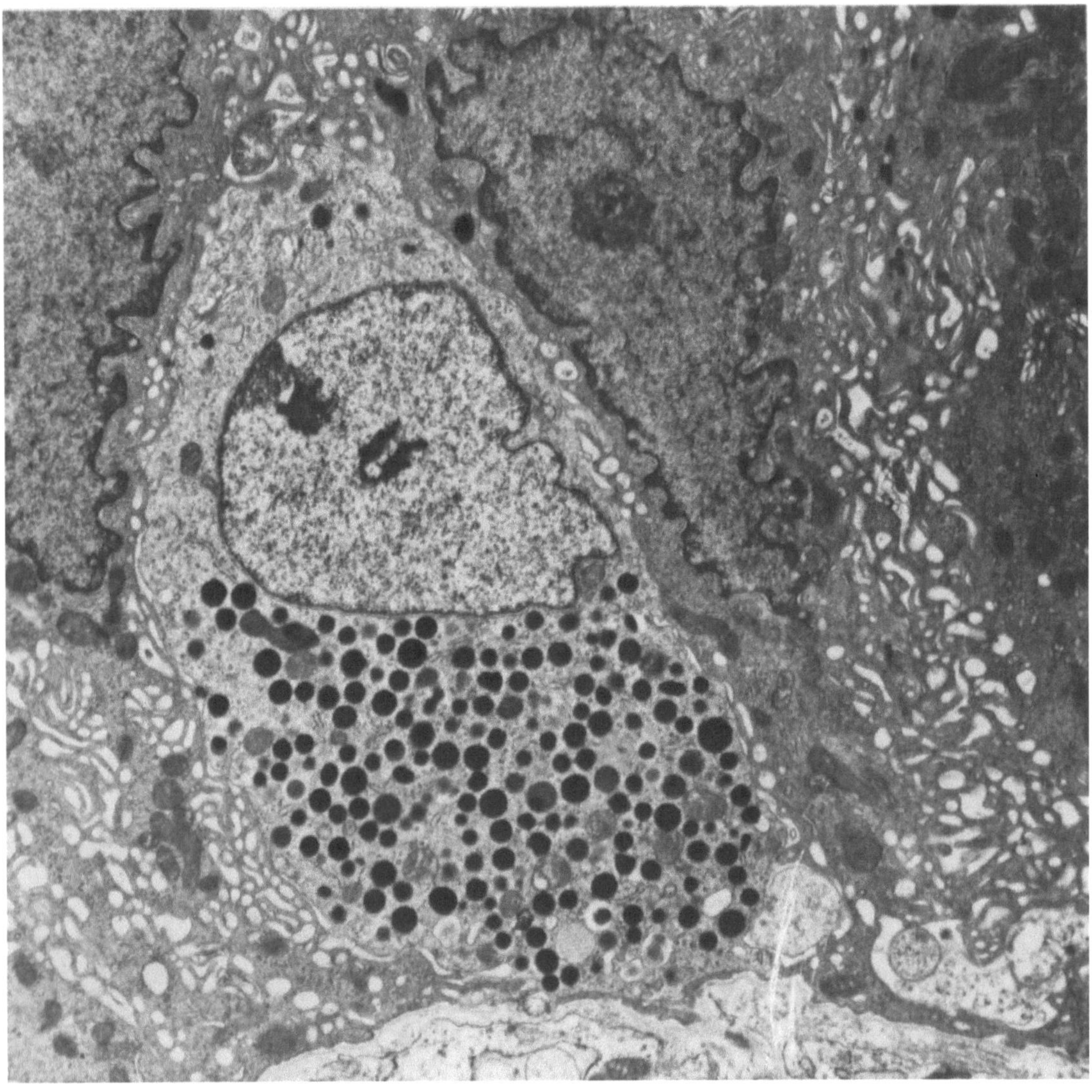

Figure 3-25. Electron micrograph of a presumed duodenal K cell that contains some granules with dark eccentrically placed cores surrounded by a less dense matrix. × 10,600

Distribution of Enteroendocrine Cells

Sjolund et al.[20] and Ferri et al.[21] have determined the regional and topographical distribution of endocrine cells in the human small intestine using immunohistochemical staining (Tables 3-1, 3-2) of sections of rapidly frozen tissues fixed in gaseous diethyl-pyrocarbonate or formaldehyde and embedded in paraffin or Araldite in vacuo. The duodenum and jejunum contained ample numbers of endocrine cells, while the ileum had fewer cells. EC cells were present in large numbers throughout the intestine, whereas L cells and D cells were present in all regions of the intestine, although few L cells were present in the duodenum. Pancreatic polypeptide cells were not present within the intestine. There were

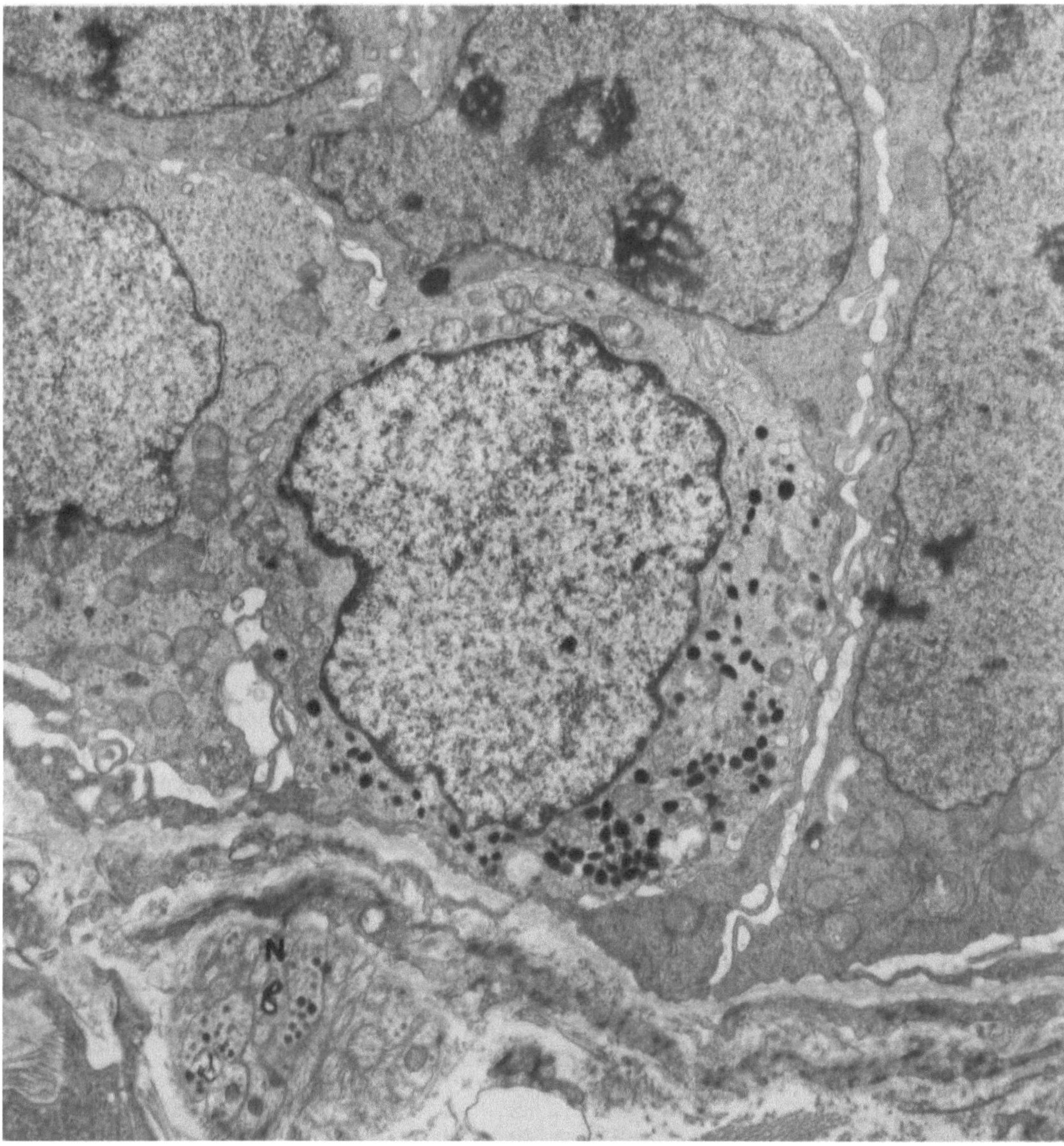

Figure 3-26. Electron micrograph of duodenal ECn cell containing small, dense, elongated granules. Nerve fibers (N) with large, dense core, secretory vesicles. × 14,500

few L, neurotensin or substance P-storing cells. G cells were few in number and virtually confined to the proximal duodenum. A notable finding was the (markedly patchy) presence of cells containing β-endorphin, pro-α-MSH (melanocyte-stimulating hormone), or β-lipotrophin, or a combination of the three. With the exception of G cells, all cells present within the duodenum were also present in the jejunum. The predominant cell types in the ileum were the EC, L, and N cells. VIP (vasoactive intestinal peptide)-containing cells were not present. S cells and N cells were found mainly within the epithelium over the apical por-

Table 3-1. Distribution and Relative Frequency of Endocrine Cells in Epithelium of Individual Villi

Antiserum	Villus	Crypt	Antiserum	Villus	Crypt
Gastrin 17	+	+	Pro-γ-melanocyte-stimulating hormone	+	+
Tetragastrin	++	+++	Motilin	+	+
Cholecystokinin	++	++	Neurotensin	++	+
Glucagon/glicentin (L)	+	+++	Secretin	++	+
Gastric inhibitory polypeptide	+	+++	Somatostatin	+	++
Glicentin	+	++	Substance P	+	+
β-Endorphin	+	+	5-Hydroxytryptamine	+	+++
β-Lipotrophin	+	+			

Modified from Sjolund et al.[20] Reprinted with permission from, "Endocrine Cells in Human Intestine: An Immunocytochemical Study," by K. Sjolund, G. Sanden, R. Hakanson, et al. *Gastroenterology* **85**:1120–1130, Copyright © 1983 by the American Gastroenterological Association.

tions of villi; Mo cells and CCK cells were in villous and crypt epithelium, whereas the other endocrine cells predominated in the crypts (Table 3-1).

Bloom et al.[21] separated epithelial cells from the lamina propria in human intestine and showed that enteroglucagon, GIP, and neurotensin were localized in endocrine cells and were found only in the epithelium whereas VIP, substance P, and bombesin were virtually restricted to nonepithelial layers and were localized exclusively in neurons. Somatostatin was present both in epithelium and lamina propria but could be immunostained only in endocrine cells.[21] Sjoland et al.[20] were able to demonstrate small numbers of substance P-staining epithelial cells (a subset of enterochromaffin cells? EC_2) in the ileum.

Table 3-2. Distribution and Number of Endocrine Cells in the Intestine

	Duodenum		Jejunum		Ileum		
Endocrine Cells	*N*	%	*N*	%	*N*	%	Total
5-Hydroxytryptamine	100	17	240	40	260	43	600
Substance P (EC_1)	2	3	4	6	60	91	66
Somatostatin (D)	40	28.5	60	43	40	28.5	140
Glucagon/glicentin (L)	5	2	30	11	230	87	265
Pro-γ-melanocyte-stimulating hormone	2	4	4	9	40	87	46
β-Lipotrophin	10	20	20	40	20	40	50
β-Endorphin	2	50	2	50	0	0	4
Gastrin/cholecystokinin	90	24	230	60	60	16	380
Gastrin (IG)	2	100	0	0	0	0	2
Cholecystokinin (I)	40	32	80	64	5	4	125
Gastric inhibitory polypeptide (K)	80	36	140	62	5	2	225
Motilin (Mo)	20	44.5	20	44.5	5	11	45
Secretin (S)	30	43	40	57	0	0	70
Neurotensin (N)	5	3	40	23	130	74	175

Number (*N*) of immunocytochemically demonstrable endocrine cells in the intestine (in millions) and distribution in the intestine (percent) as reported by Sjolund.[24] The substance P-containing cells represent a subpopulation of the 5-HT-containing EC cells. β-Endorphin cells are a subpopulation of the β-lipotrophin immunoreactive cells; the pro-γ-MSH immunoreactivity was present in cells without β-lipotrophin reactivity. Gastrin cells and CCK cells are included in the group of gastrin/CCK (TG) cells.

Bloom and Polak[22] have prepared a helpful review of the physiological functions of intestinal hormonal peptides, neurotransmitters, and putative hormones.

Identifying Enteroendocrine Cells

In the best of all worlds, each of the enteroendocrine cells would be readily identifiable on the basis of morphological criteria alone. Sadly this is not the case, although many of the cells are so identifiable. We recently reviewed the literature and all of the micrographs of endocrine cells (approximately 500) photographed in our laboratory and find that intestinal endocrine cells can be classified into six groupings.

1. G cells (Figure 3-20) are readily identified by the flocculent core of their granules. The granules tend to be small (160 nm) and dense and stain for G-34, or large (240 nm) and lucent and stain for G-17.
2. IG, D_1, Mo, S, P, and TG cells (Figure 3-21) are all similar in appearance and have small (100–220 nm) dense core granules, often with a halo about the core. They can be distinguished only by specific immunostaining.
3. D cells (Figure 3-22) have moderately and variably dense granules (300–400 nm) and a membrane surrounding the granules that often has a "nibbled at" appearance. The cells tend to have long cytoplasmic processes.
4. I, L, and N cells (Figure 3-23) have granules (250–300 nm) with very dense cores and a granule membrane that is usually indistinguishable from the dense core material.
5. EC_1 (Figure 3-24), EC_2 (Figure 3-24), and ECn (Figure 3-26) cells are usually readily identified by their characteristic granules (see descriptions earlier in this section).
6. K cells (Figure 3-25) have dense granules (200–400 nm) that often contain eccentrically placed densities distinguishable from the remainder of the core.

This classification is very helpful in identifying gut endocrine cells, although in the final analysis one must use specific immunostaining (except for EC cells) to be certain of the cell type.

Tuft Cells

Tuft cells, also called fibrovesicular cells or caveolated cells (Figure 3-27), are present within the epithelium of the stomach, colon, and intestine of humans, dogs, mice, and rats.[23-25] These cells are attached to adjacent absorptive cells by a junctional complex typical of epithelial cells, have a narrow apical region, and extend to the basal lamina.[24] They have unusually long microvilli, and the microvilli contain an unusually large number of filaments (up to 60 in a bundle), the filaments extending as much as 5 μm into the subjacent cytoplasm. The diameter of the filaments, 7.5 nm, is compatible with that of actin. Numerous caveolae, which begin as pits at the base of microvilli, extend in a parallel array down to the level of the nucleus. These caveolae may be tortuous tubules (Figure 3-28). In contrast to adjacent M cells (Figure 3-29), tuft cells do not ingest horseradish peroxidase.[23] The origin and function of tuft cells has not been determined.[24]

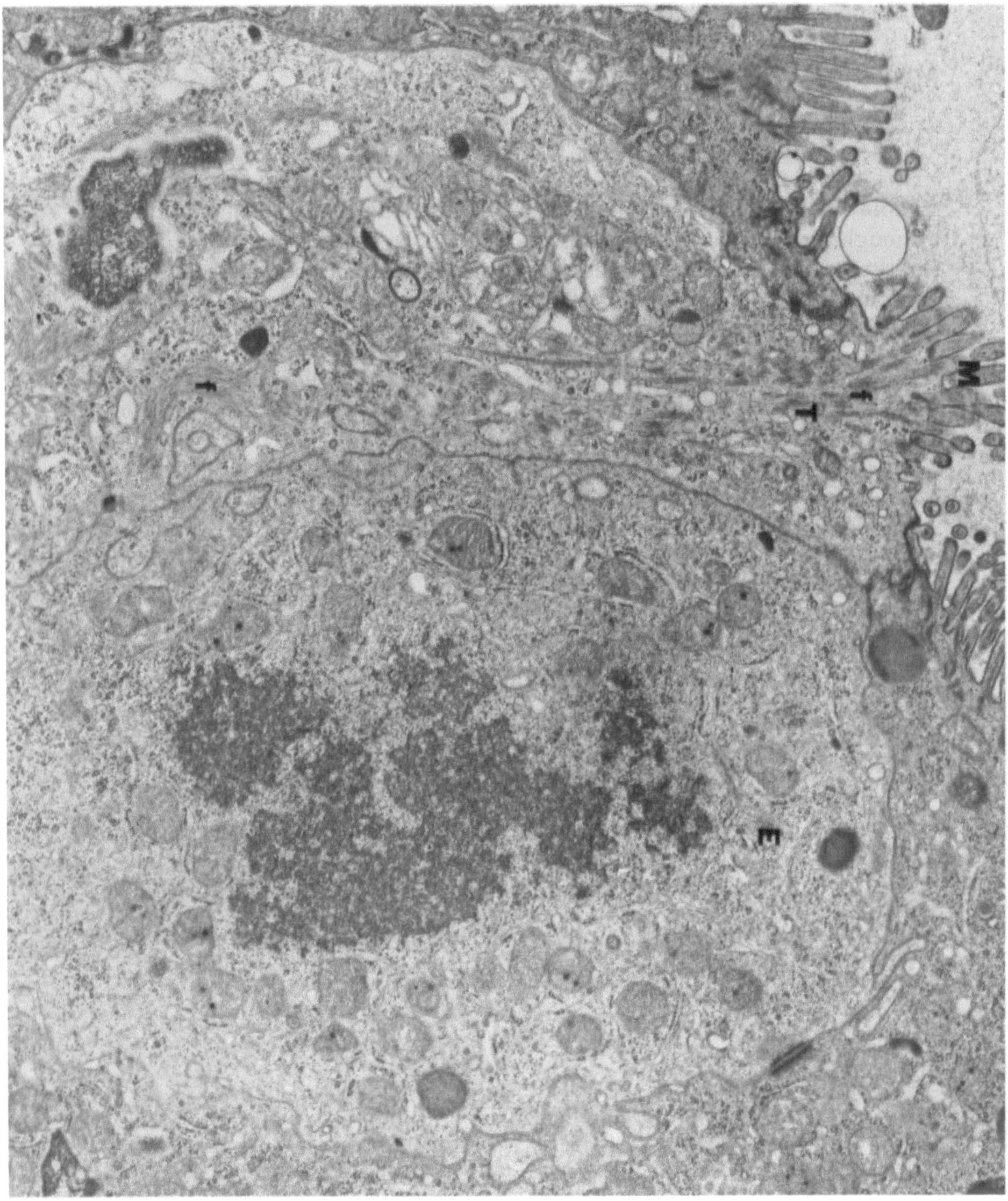

Figure 3-27. Electron micrograph showing a typical Tuft cell (T) adjacent to an epithelial cell in mitosis (E). Note the tuft of microvilli (M) at the apex of the tuft cell and the numerous, long filaments (f) extending from the microvilli deep into the tuft cell cytoplasm. Numerous vesicles (caveolae) begin at the pits of microvilli and extend to the level of the nucleus. × 15,400

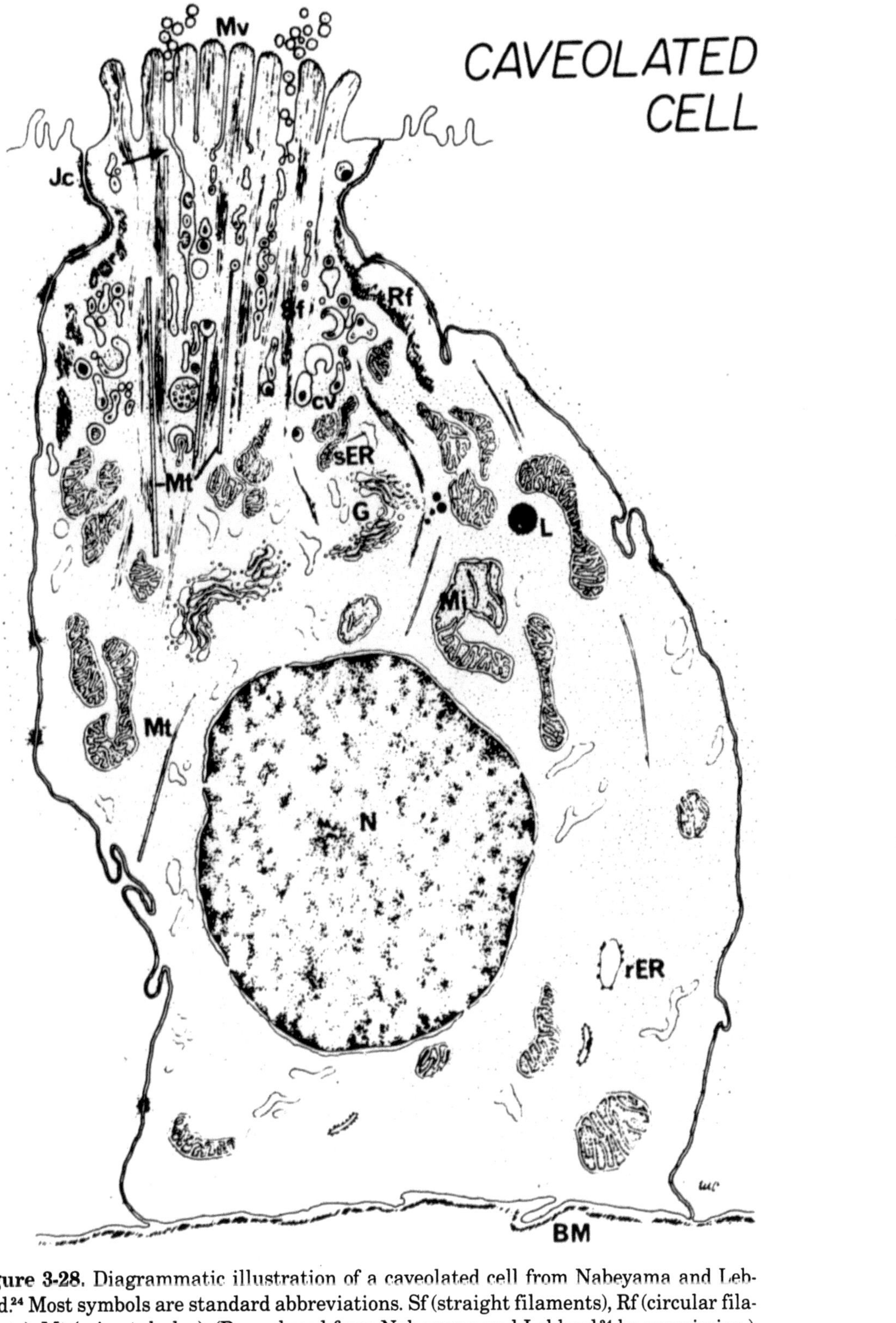

Figure 3-28. Diagrammatic illustration of a caveolated cell from Nabeyama and Leblond.[24] Most symbols are standard abbreviations. Sf (straight filaments), Rf (circular filaments), Mt (microtubules). (Reproduced from Nabeyama and Leblond,[24] by permission.)

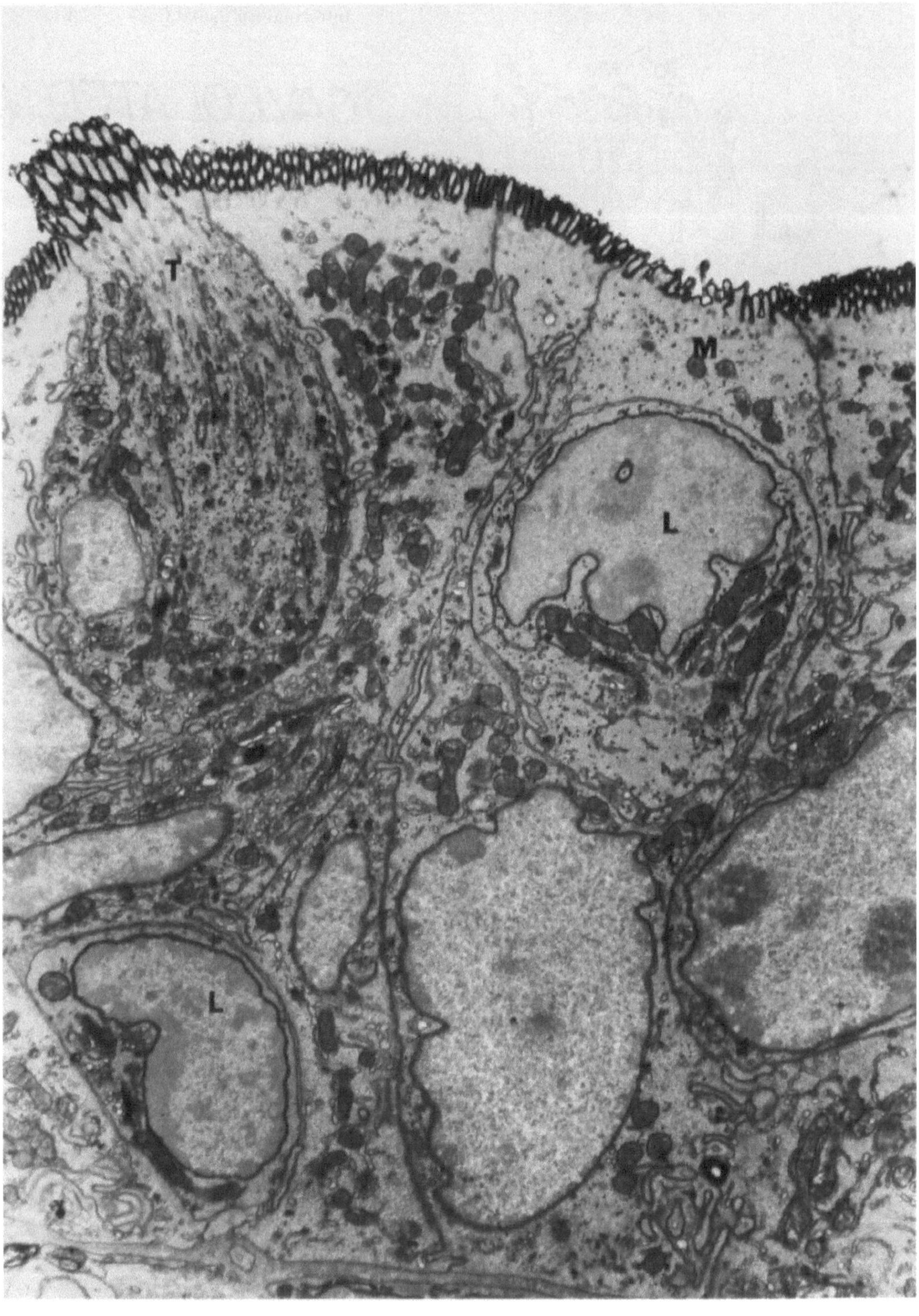

Figure 3-29. Electron micrograph of mouse intestinal (ileal) epithelium following luminal exposure to horseradish peroxidase. The tuft cell (T) does not ingest the horseradish peroxidase (black reaction product seen over the microvilli); the nearby M cell does so. Note that the microvilli over the M cell are irregular and stubby and that the cytoplasm is quite attenuated. Intraepithelial lymphocytes (L) are present. × 9,500. (Reproduced from Owen,[23] © by Williams & Wilkins, 1977.)

Cup Cells

The cup cell, first recognized by Madara in 1982 as a distinct cell found in the ileal epithelium of rabbit, monkey, and guinea pig, has not been found in human intestinal epithelium. The cells were readily recognized as pale-staining cells in 1-μm toluidine blue stained sections but could not be unequivocally identified in paraffin-embedded H & E stained ileal mucosa[26] (Figure 3-30).

Cup cells are characterized, when compared with absorptive cells, by shorter microvilli with a cuplike concavity, faint cytoplasmic staining, small mitochondria with a dense matrix and large intracristal spaces, few multivesicular bodies, highly variable numbers of lysosomes, a terminal web with a poorly demarcated inferior margin, prominent cytoplasmic intermediate filaments, and an unusual array of microvillous intramembrane particles. Cup cells have limited ability to pinocytose and do not transport horseradish peroxidase, have little brush border alkaline phosphatase, and absorb lipid poorly. There is an unusual tendency for heavy metal stains to precipitate along the plasma membrane of cup cells.[26]

Cup cells have many similarities to enteroendocrine cells, but they do not contain dense granules. Like enteroendocrine cells, cup cells have low cytoplasmic density, are rich in intermediate filaments, and have similar mitochondria. Although there are a number of explanations as to why mitochondria assume a "condensed-state" structure in both cell types, the most likely one is that these cells have a high respiratory rate and are very metabolically active.[26]

The preferential localization of cup cells to the ileum suggests that these cells have a function, not yet defined, restricted to the ileum. Both bile salt absorption and vitamin B_{12} absorption are ileal functions, although there is no reason to think that cup cells are involved selectively in these specific functions. The linear array of homogeneous IMPs found on cup cell microvillus P faces is further evidence that such cells may be responsible for a single highly selective intestinal function. Such a linearity of integral membrane proteins in other cells has often been associated with highly specific cell functions such as those of endocytosis of immunoglobulins by suckling rat intestinal epithelial cells.[26]

Madara and Carlson[27] have observed a selective association of an unidentified bacterium with the microvilli of cup cells and have also shown that the surface coat of cup cell microvilli is more prominent than that of absorptive cells, although the glycocalyx of both cells contains abundant anionic sites as determined by ruthenium red staining. Cationized ferritin is internalized by cup cells, similar to their handling of horseradish peroxidase (see foregoing). Their microvillous membranes, similar to those of other intestinal epithelial cells that are depleted of intramembrane particles (e.g., goblet cells and M cells), are enriched in morphologically detectable cholesterol (using filipin probes and freeze-fracture techniques). Noting that one virulence factor for intestinal bacterial pathogens is the ability of the pathogen to adhere to the intestinal surface, Madara and Carlson[27] suggested that cup cells (containing binding sites for selected bacteria) might play a role in initiation of some bacterial diseases of the intestine.

M Cells

M cells (membranous epithelial cells) are specialized epithelial cells (Figure 3-29) overlying lymphoid follicles of human intestine, especially the Peyer's

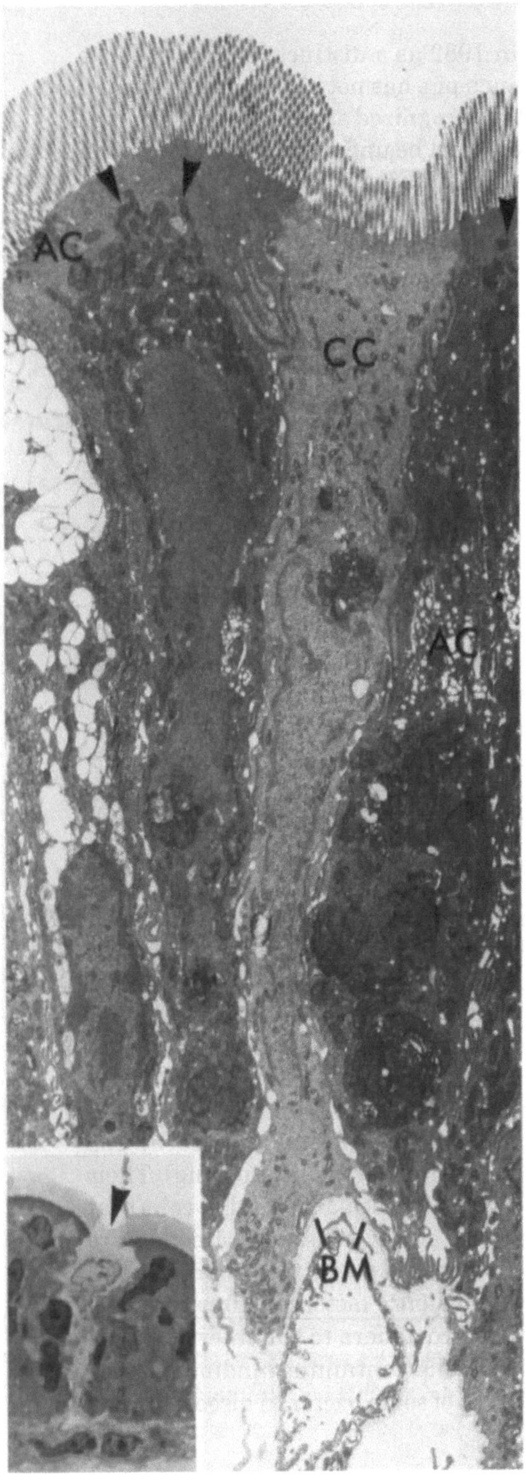

Figure 3-30. Electron micrograph of an ileal (guinea pig) cup cell (CC) between two absorptive cells (AC). The cup cell extends to the basement membrane (BM). The terminal web of the cup cell is poorly demarcated in contrast to that of the adjacent absorptive cells (outlined by arrows). × 4,000. Inset shows light microscopy appearance of cup cell (arrow) when stained with toluidine blue. × 550. (Reproduced from Madara,[26] by permission of Elsevier Science Publishing Company, Inc.)

patches of the terminal ileum.[28,29] They are called membranous (M) cells because they are very flat compared with adjacent absorptive cells (to which they are attached by the standard junctional complex of epithelial cells), forming a "membrane" between the lumen and underlying clusters of intraepithelial lymphocytes and occasional macrophages. M cells are characterized by the presence of irregular, even stubby, microvilli; attenuated cytoplasm; and prominent apical vesicles. The terminal web and the surface coat (glycocalyx) of their microvilli are less well developed than those of absorptive cells. The cytoplasm of M cells contains few lysosomes. M cells have the ability to bind and take up large quantities of horseradish peroxidase,[23] some bacteria,[30] and reovirus[31] and to transport the material intact to the intraepithelial space where it may be antigenically active following subsequent uptake by lymphocytes and/or macrophages. This function of M cells is similar in degree to that of ileal absorptive cells of neonatal animals, whereas uptake of macromolecules of adult absorptive cells can be demonstrated only to a very limited degree.[32]

M cells, also called lymphoepithelial cells, are also found in the lung over bronchial-associated lymphoid tissues.[30] The M cell has a characteristic scanning electron microscopy appearance. It has sparse microvilli that contrast sharply with the regular carpet of microvilli over absorptive cells, and irregular ridges with or without central concavities. The apical membrane of the M cell also differs markedly from that of the absorptive cell in that it has less alkaline phosphatase, more esterase, more avid binding of wheat-germ agglutinin, and less intense staining with ruthenium red.[30] Because M cells take up and transport some viral and bacterial antigens, the presence of the class II histocompatibility antigen HLA-DR (Ia in mouse) would have major functional significance, because then the M cell not only could transport antigen to the underlying lymphoid cells but could present the antigen directly to those cells.[33] We have shown, using immunoelectron microscopy, that M cells express HLA-DR on the apical surface membrane and more weakly on the lateral surface membrane[33] (Figure 3-31).

Why some antigens such as reovirus type I and some bacteria adhere specifically to M cells is not clear. Some invasive organisms penetrate the intestinal barrier directly through intestinal absorptive cells, for example, *Salmonella typhimurium*.[30] *Giardia murius* (and presumably *hominis*) does not adhere to the M cell. The presence of Peyer's patches (and thus of M cells) is not essential to a normal intestinal immune response, although the response is delayed in their absence.[34] M cells, like absorptive cells, goblet cells, Paneth cells, and enteroendocrine cells, arise from undifferentiated crypt cells.[30]

Intraepithelial Lymphocytes and Immune Cells of the Lamina Propria

The lamina propria (connective tissue core) consists of that portion of the intestinal mucosa found between the epithelium and the muscularis mucosae. It contains fibroblasts, connective tissue components such as collagen and elastic fibers, plasma cells, lymphocytes, monocytes and macrophages, a few eosinophils and rare PMN leukocytes, mast cells, unmyelinated nerve fibers, blood and lymph vessels, and intrinsic smooth-muscle cells that contract the villi periodically.

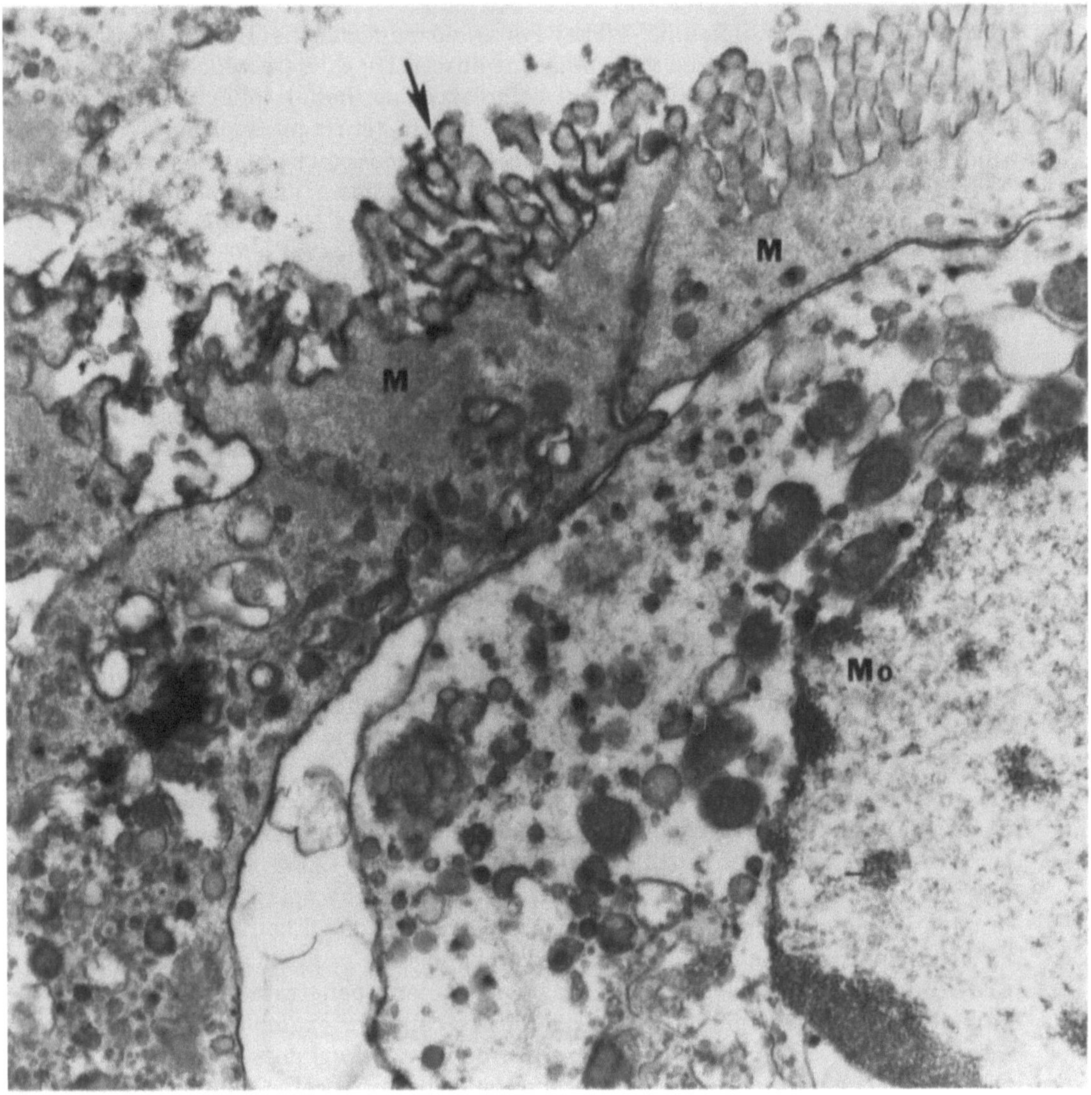

Figure 3-31. Immunoelectron microscopy localization of the monoclonal antibody I2 (HLA-DR) to one (arrow) of two adjacent M cells in human ileum. Note faint staining of microvilli and of lateral surface of the M cell with reaction product (diaminobenzidine). Mononuclear cell (Mo). × 27,800

Of all the various cell types present within the lamina propria, it is remarkable that only lymphocytes are likely to gain access to the normal epithelium.[35,36] These cells are located within the epithelium, above the basal lamina, and between epithelial cells, that is, they are interepithelial cell and intraepithelial (IEL) in location. The presence of lymphocytes within epithelial cells (emperipolesis) has not been demonstrated. IEL cells are found largely at the base of epithelial cells, below epithelial nuclei. In the normal intestine these cells are usually medium sized (5–9 μm in diameter) and contain sparse cytoplasmic organelles including mitochondria, lysosomal granules, ribosomes (both isolated and polyribosomes), endoplasmic reticulum (rough predominating over smooth by a ratio of 2:1), and a small Golgi complex.[37] The nucleus contains large amounts of heterochromatin and lesser amounts of euchromatin. Nucleoli are often present. The cytoplasm tends to be less dense than that of adjacent epithelial cells, and endocytic vesicles are often present along the plasma membrane. IEL cells tend to be rounded to elongated in shape and in active inflammatory states often have numerous elongated cytoplasmic projections that make intimate contact with adjacent epithelial cell membranes. In the noninflamed intestine, IEL cells tend to be tightly compressed between epithelial cells. IEL cells do not form junctional contacts (desmosomes and/or tight junctions) with adjacent epithelial cells (Figure 3-32).

The nature of IEL cell "granules" may be of importance in defining IEL cell function. Rodent IEL cells have very large granules, some of which contain histamine, and there is conflicting evidence that rodent IEL cells may be mast cell derived.[36] Human IEL cells are *not* mast cell derived and they contain small granules.[36] The granules are membrane bound and are clearly lysosomes.

The granules are azurophilic in Giemsa and Wright-stained preparations, they stain with alcian blue (pH 2.2), and they are metachromatic with toluidine blue (pH 4).[36] The lysosomal nature of lymphocyte granules is established by their content of acid phosphatase (AP).[36] Lymphocyte granules have also been shown to contain beta-glucuronidase and alpha-naphthyl acetate esterase (ANAE). Structures that are the site of both ANAE and AP activity represent the Gall body, a cluster of primary lysosomes surrounding a lipid droplet. The distribution of lysosomes in lymphocytes is of limited value in defining their function. The majority of human T cells that possess surface receptors for IgM contain Gall bodies that stain for ANAE in a localized or dotlike fashion, whereas the majority of human T cells that possess surface receptors for IgG have lysosomes scattered diffusely throughout the cytoplasm.[36]

Lymphocytes do not contain peroxidases, an enzyme found in the myelomonocytic series. All subclasses of T lymphocytes examined have been shown to contain, at least in part, lysosomal granules including cells with helper and suppressor phenotype, natural killer phenotype, and cytotoxic phenotype.[36] B-lymphocytes do not contain granules and have very little AP and ANAE activity. It is likely that the presence of lysosomal granules (including multivesicular bodies) is necessary for the cytolytic activity of lymphocytes, whether the activity is antibody-mediated cellular cytotoxicity (ADCC) or natural killer (NK) mediated. Indeed, the presence of granules may represent a marker for *all* human cytolytic lymphocytes.[36]

In 1974 Guy-Grand demonstrated in the rat that the vast majority of IEL cells were T cells and that, whereas both B cells and T cells home to the gut mucosa, only T cells home to the epithelium.[36] In 1976 Meuwissen et al. showed in human colon and intestine that the vast majority of IEL were T cells, defined by immunoperoxidase localization of IEL cells using an anti-T cell globulin.[36] This finding has been repeatedly confirmed with immunoperoxidase and immunofluorescent localization of a variety of monoclonal antibodies to T cells.[36] Macrophages (Figure 3-33) and mast cells (Figure 3-34) are occasionally found within the epithelium, while epithelial polymorphoneuclear leukocytes and eosinophils are quite unusual except in inflammatory states.[36]

IEL cells, virtually all of which are T cells, do not express class II histocompatibility antigens (HLA-DR, Ia-like), even in inflammatory states. Moreover, 80% to 90% of intraepithelial T cells express the phenotype of suppressor cells

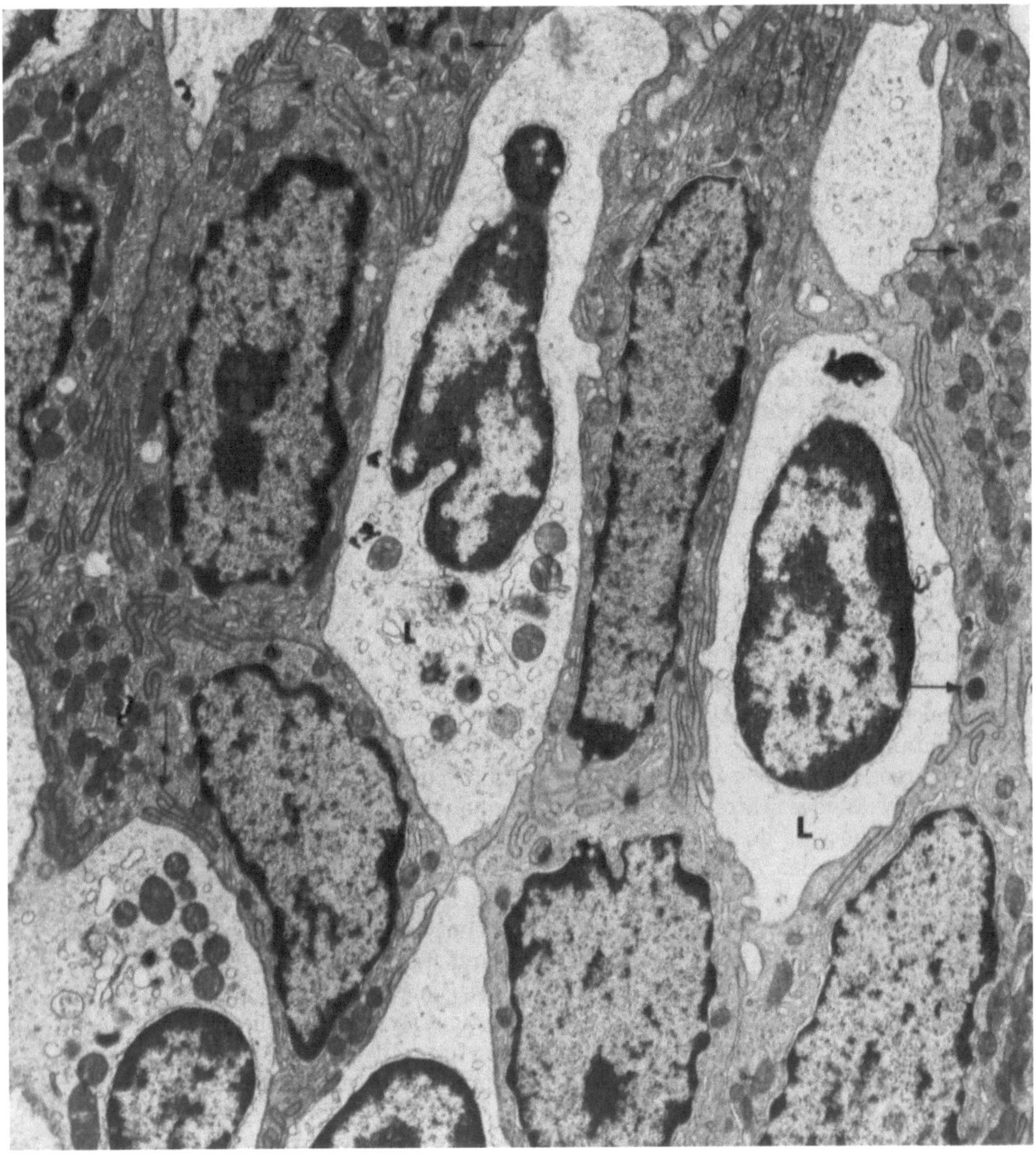

Figure 3-32. Electron micrograph showing a number of epithelial lymphocytes (L), characterized by low-density cytoplasm with sparse organelles, compressed between epithelial cells that have a dense cytoplasm and numerous organelles. Note the peroxisomelike structures (arrows) in one of the epithelial cells. See text for details of epithelial lymphocyte structure. × 10,000

(T8, Leu 2a), whereas only 10% to 20% express the helper phenotype (T4, Leu 3a).[36] In contrast, lamina propria lymphocytes (LPLs) are predominantly of the helper phenotype (60–70% T4, Leu 3a). IEL cells express class I HLA antigens but do not express T-cell activation antigen (Tac) or complement receptors (c3b) and rarely express antigens characteristic of natural killer (NK) cells (HNK-1, Leu 7, Leu 11, T10, M1).

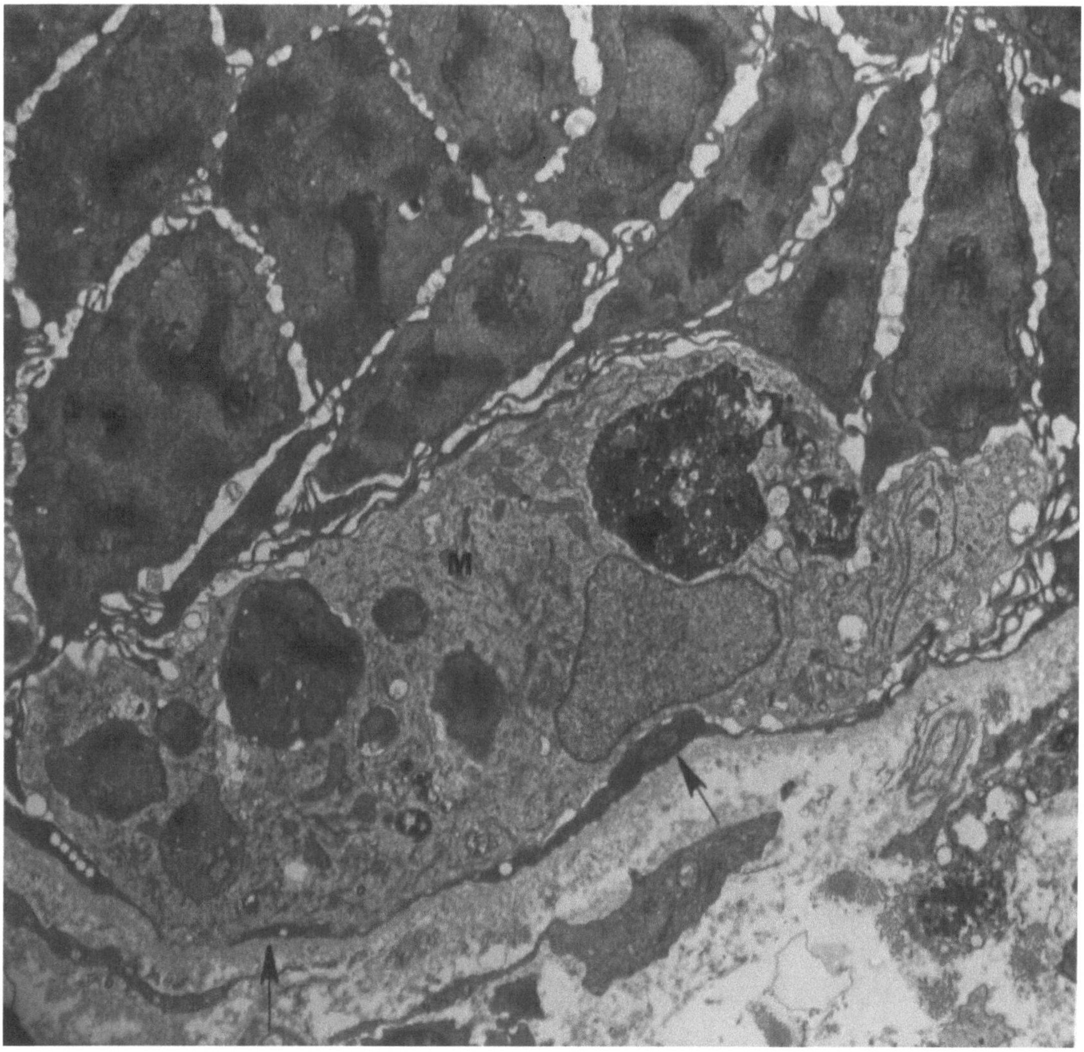

Figure 3-33. Electron micrograph showing presence of an intraepithelial macrophage (M) in the intestinal biopsy specimen of a patient with celiac sprue. Note the long epithelial cell processes between the macrophage and the basal lamina (arrows). × 8,400

The number of IEL cells per 100 epithelial cells in normal jejunum ranges from 9 to 39, with a mean of 21.1 and standard deviation of 7.5. We found a mean of 20 IEL cells per 100 epithelial cells in normal jejunum, 13 per 100 cells in normal ileum, and 5 per 100 in normal colon.[36]

The cellular immune content of the lamina propria is strikingly different from that of the epithelium, and there clearly are differences in immune function of IEL cells and LPLs. Lamina propria lymphocytes consist predominantly of the helper phenotype with a smaller component of the suppressor phenotype, the ratio of help to suppression being similar to that of peripheral blood, 2:1. Plasma cells are profusely present in the lamina propria with ratios of IgA/IgM/IgG-containing cells of 20:3:1. Macrophages, fibroblasts, occasional

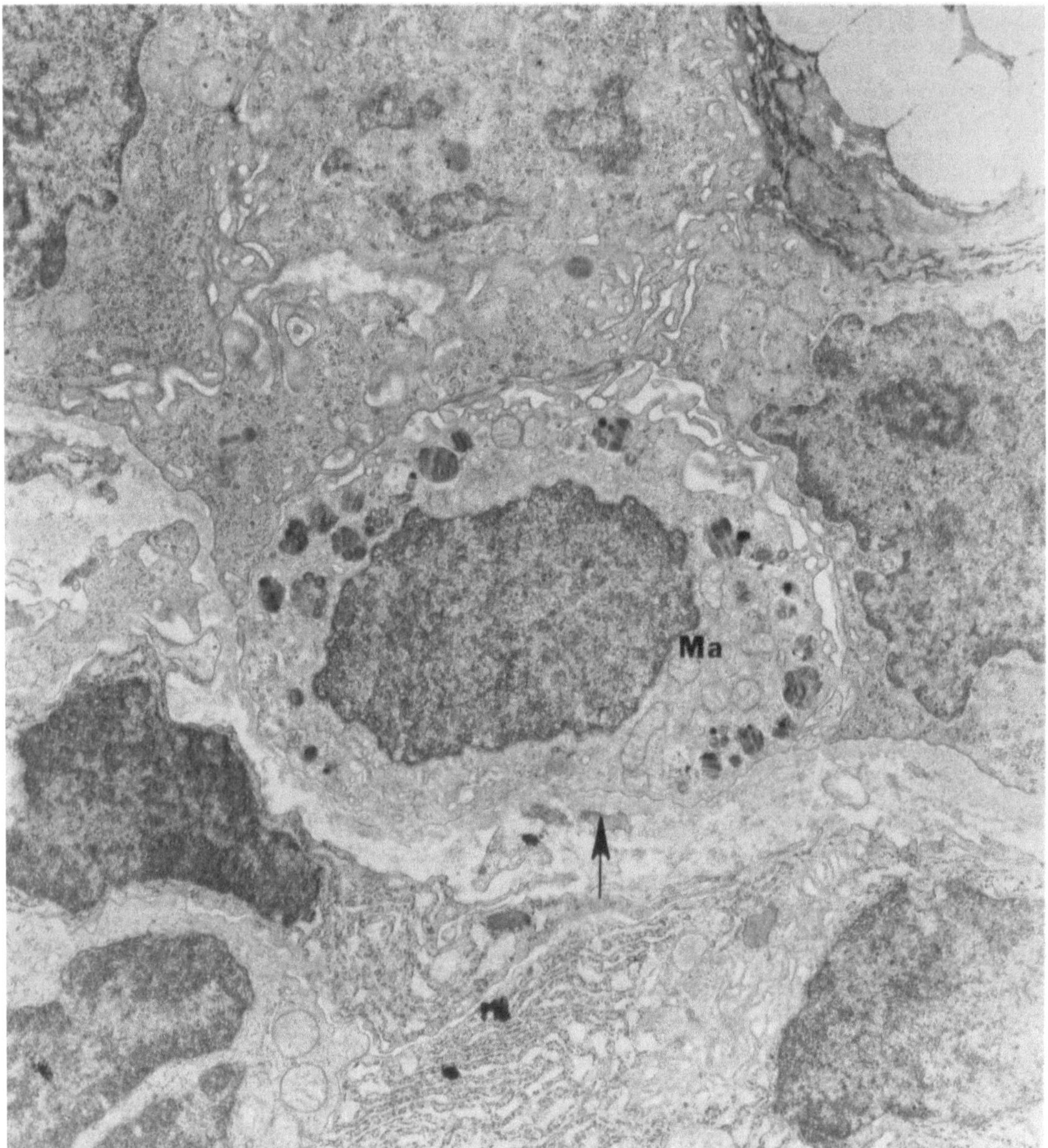

Figure 3-34. Electron micrograph showing presence of an intraepithelial mast cell (Ma) in the duodenal mucosa. Note that the mast cell is above the epithelial cell basal lamina (arrow). × 14,500

mast cells, and leukocytes are present throughout the normal lamina propria. T cells found in the lamina propria are occasionally activated on the basis of expression of class II MHC antigens and Tac antigen, and NK cells are far more easily detected in the lamina propria than in the epithelium.[36]

The specific role of IEL cells remains to be defined. The cells are largely T cells of the suppressor phenotype, and in common with other suppressor effector cells (in contrast to cytotoxic effectors), respond poorly to mitogens.[36] They do not mediate cytotoxic functions in vitro, even though one fourth of IEL cells have the

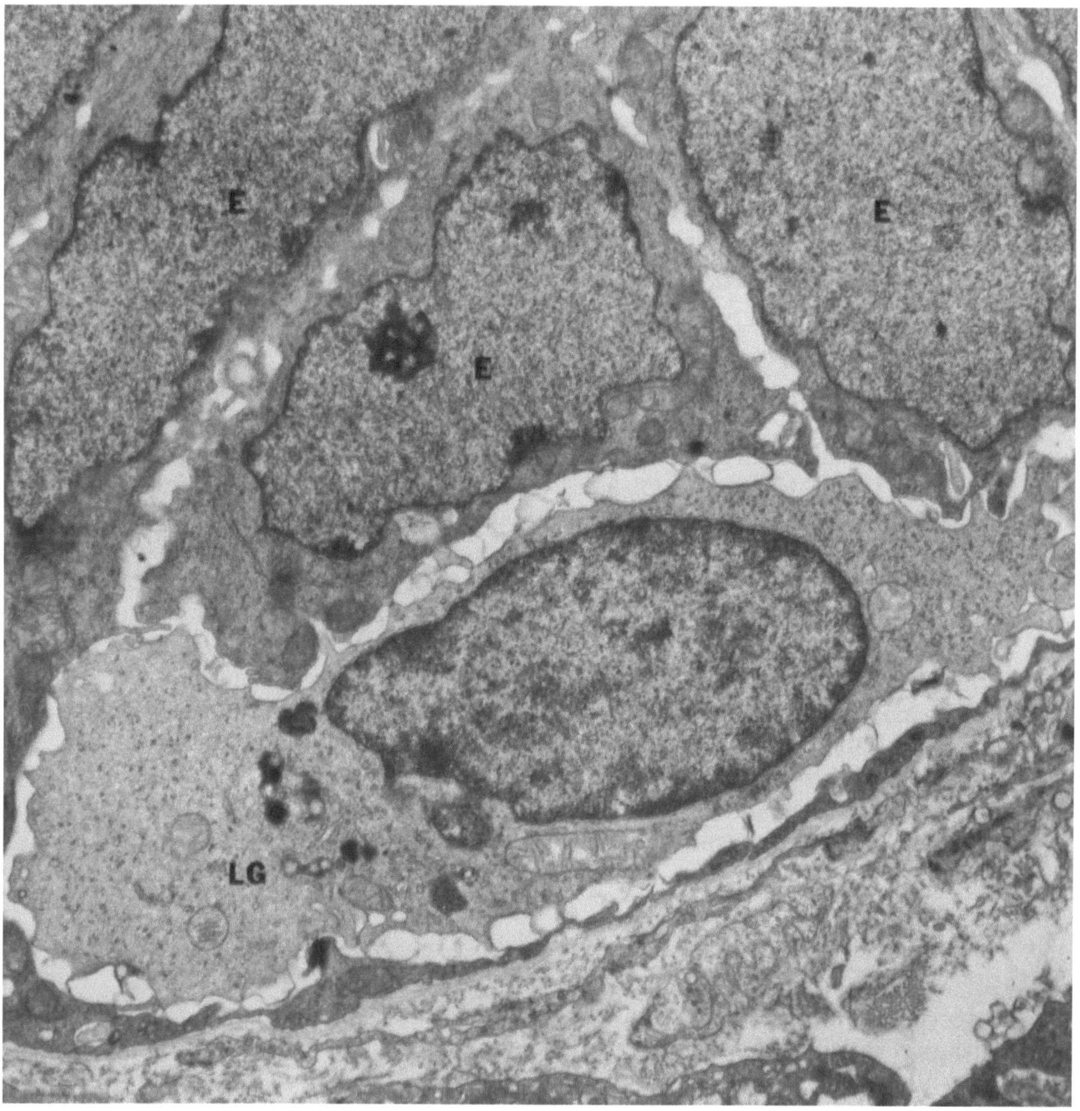

Figure 3-35. Electron micrograph of intraepithelial large granular lymphocytes (LG) in the intestine of an individual with acquired immunodeficiency syndrome (AIDS) who had intestinal giardiasis, cryptosporidiosis, and cytomegalovirus infection. Note the prominent mitochondria, large dense granules, and cytoplasmic surface projections making intimate contact with adjacent epithelial (E) cells. This cell is presumably a natural killer cell. × 19,600

morphology of large granular lymphocytes (Figure 3-35), cells that characteristically mediate natural killer function. Increased numbers of IEL cells are present in celiac sprue, cow's milk intolerance, parasitic diseases, and graft-vs.-host disease. The intestinal immune response is predominantly a helper response, particularly when LPLs are assessed. It is possible that the epithelial suppressor cells dampen this helper response, particularly when overly exuberant, although no data support this speculation at present.

Isolated rat IEL cells secrete a factor (probably gamma interferon) capable of inducing Ia antigen (HLA-DR in humans) expression by an intestinal epithelial cell line, IEC 17.[38] Importantly, lymphocyte proliferation was not essential for the secretion of the Ia-inducing factor. These data suggest that IEL cells may be involved in modulating some epithelial cell functions, and specifically Ia antigen expression. This permits the speculation that Ia antigen-positive epithelial cells have the ability to present antigen to T cells.[38] Furthermore, growth of the epithelial cell line was inhibited by supernatants from Con A (concanavalin A)-stimulated IEL cells, indicating that IEL cells may modulate epithelial cell growth.

Intraepithelial Lymphocytes in Disease States

IEL cells are markedly decreased in number in germ-free animals. When expressed in relation to number of epithelial cells, IEL cells clearly are increased in number in several disease states, including celiac sprue, tropical sprue, and the celiac sprue lesion found in dermatitis herpetiformis.[36] Failure to find a clear increase in number of IEL cells in an intestinal lesion otherwise compatible with that of untreated celiac sprue should cast serious doubt on the diagnosis. IEL cells are increased in number, but not consistently so, in other inflammatory states of the intestine such as the stasis syndromes. IEL cells have been implicated as effectors of gut epithelial cell injury in graft-vs.-host disease.[39] IEL cells are increased in number in the small intestine but not in the colon of subjects with acquired immunodeficiency syndrome (AIDS), both in the presence and absence of diarrhea and/or of obvious pathogens.[40] IEL cells are *not* increased in number in inflammatory bowel disease.[33,36]

Cells of the Lamina Propria

A special population of fibroblasts is present just under the basal lamina of epithelial cells; these fibroblasts are oriented circumferentially around crypt and villous epithelium (Figure 3-36). These cells of the *subepithelial fibroblast sheath* proliferate in the region of the crypt base and migrate along the villi parallel with epithelial cells.[41] The cells comprising this network have elongated cell bodies and long attenuated cell processes. In addition, stellate-shaped cells with large oval nuclei, scant perinuclear cytoplasm, and long cytoplasmic processes form a sparse cellular network throughout the lamina propria.[42] The cytoplasmic projections of these cells appose the cells forming the subepithelial network and the smooth-muscle cells that parallel lacteals and surround capillaries (Figure 3-37). In situ immunoperoxidase localization has shown that these cells, called myofibroblasts, contain contraction-associated proteins (isotropomyosin, isomyosin, actin, and cyclic guanosine monophosphate-dependent protein kinase) in amounts greater than those found in connective tissue fibroblasts, but less than in smooth-muscle cells.[42] It seems likely that these myofibroblasts have a contractile function, and that they may provide a supportive network for the epithelium. They may assist, along with smooth-muscle cells, in providing a pumping motion of the villus and propulsion of absorption products through the lamina propria.

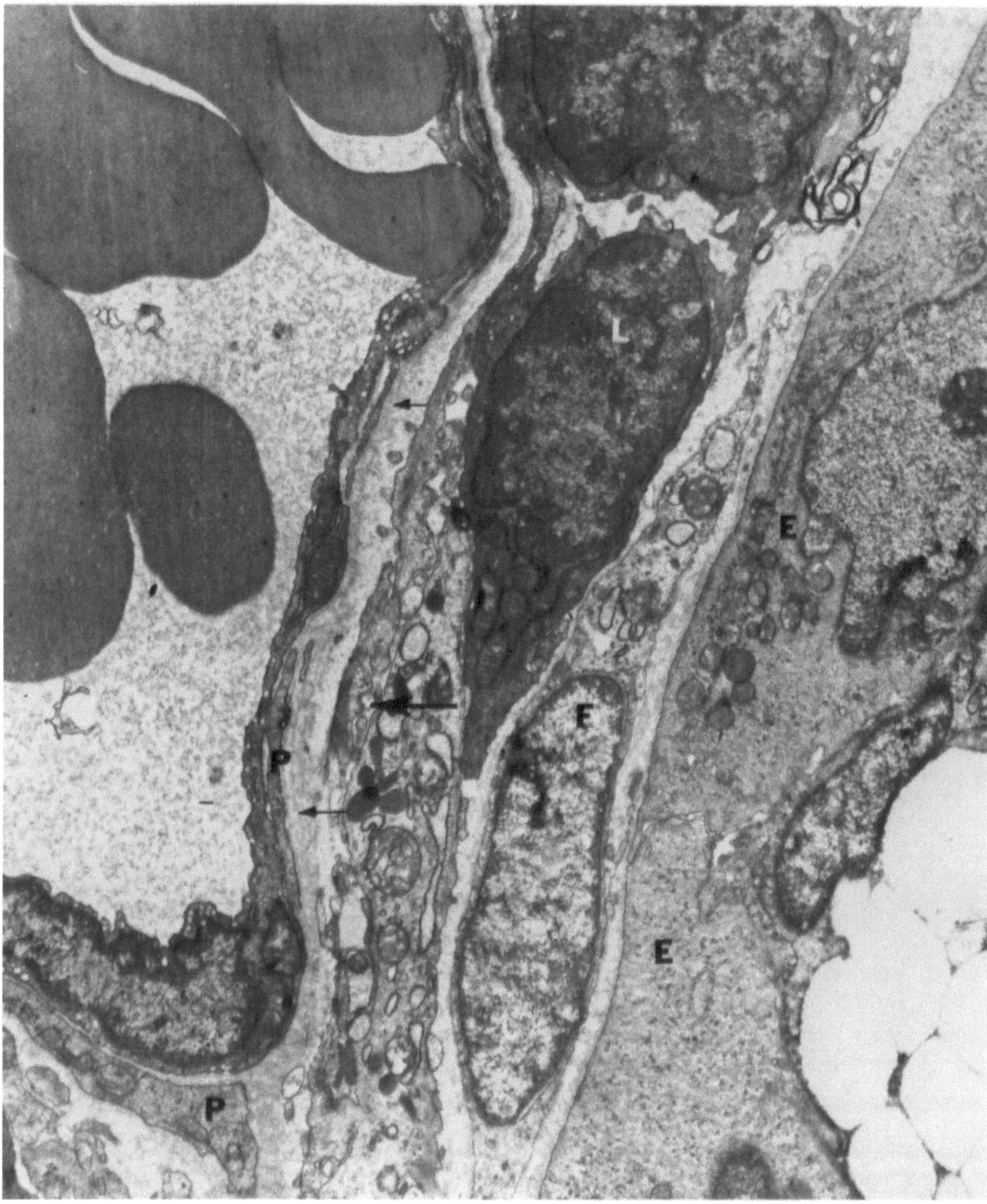

Figure 3-36. Electron micrograph of ileal epithelium showing presence of subepithelial fibroblasts (F) below epithelial cells (E). Lymphocyte (L). Note that the blood capillary (venule?) is surrounded in part by several pericapillary (P) muscle cells or pericytes encased within the capillary basal lamina (small arrows), while extrinsic to this is a myofibroblast extension (large arrow) not enveloped by a basal lamina. × 10,000

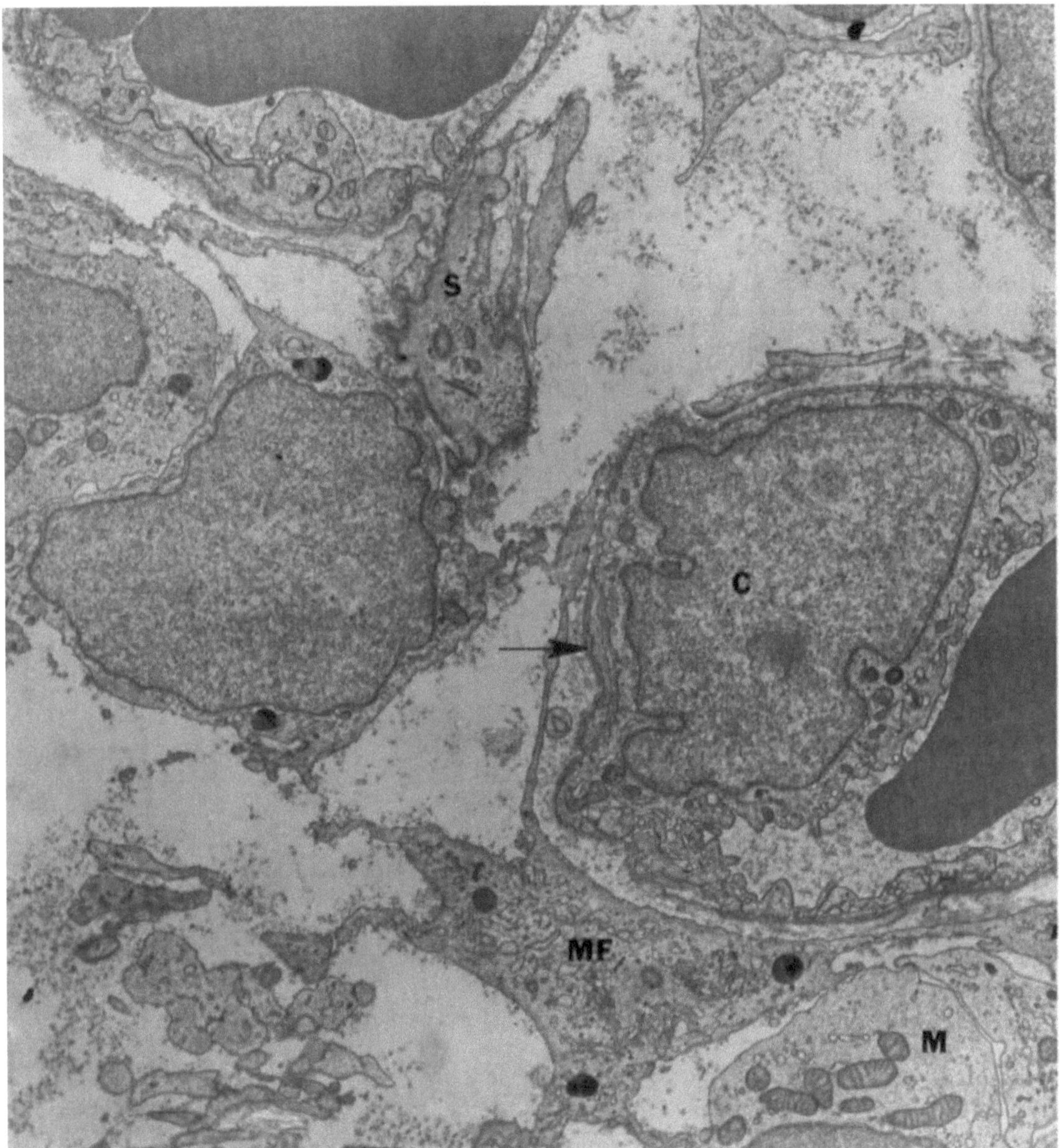

Figure 3-37. Electron micrograph showing myofibroblast (MF) in the intestinal lamina propria. Note the long cytoplasmic projections apposing an adjacent capillary (C). Note the pericyte (arrow) encased within the capillary basal lamina. Monocyte (M), intrinsic smooth muscle (S). × 10,400

The lamina propria is rich in lymphocytes (Figure 3-38) (predominantly helper T cells with a smaller component of suppressor T cells), plasma cells (Figure 3-38), phagocytic macrophages (Figures 3-39, 3-40), monocytes, and rare antigen-presenting cells with dendritic morphology (Figure 3-41), all of which are collectively referred to as the diffuse lymphoid tissue of the gut.[36] Epithelial lymphocytes (described earlier) are also a component of diffuse lymphoid tissue, but consist largely of suppressor T cells, small numbers of helper T cells, and

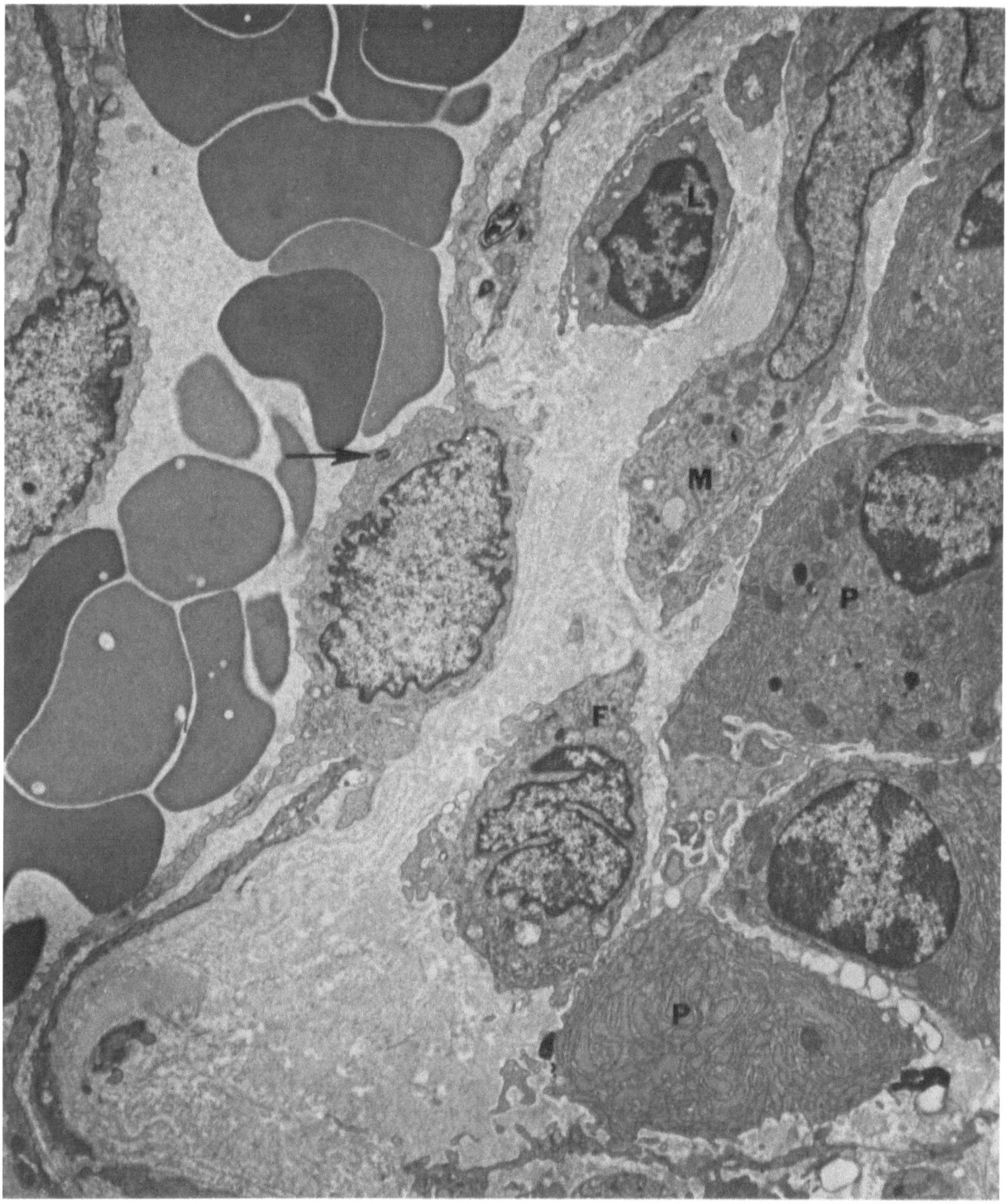

Figure 3-38. Electron micrograph of intestinal lamina propria showing a venule (note the centriole [arrow] in endothelial cell), lymphocyte (L), several plasma cells (P), a fibroblast (F), and a phagocytic macrophage (M). × 5,900

occasional natural killer cells.[36] Focal lymphoid nodules and aggregates are scattered throughout the lamina propria and contain, in addition to the lymphoid cells described above, many B cells.[43] These B cells, along with similar cells in Peyer's patches, are precursors to the plasma cells found throughout the lamina propria of the gut.[36] Eosinophils and mast cells (Figure 3-42) are common

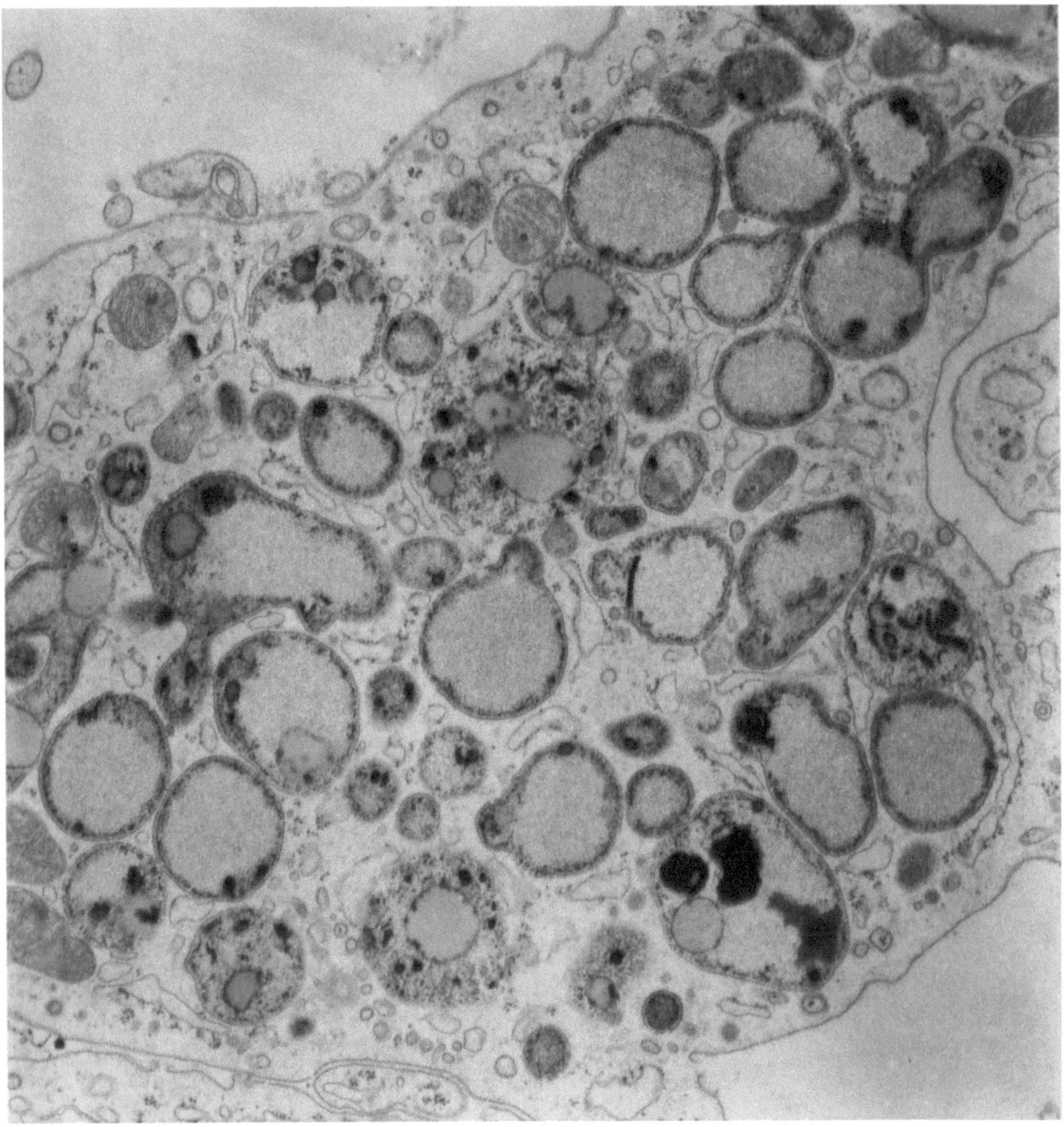

Figure 3-39. Electron micrograph of portion of a phagocytic macrophage in intestinal lamina propria of a patient with ulcerative colitis. The cell is packed with phagosomes containing a heterogeneous array of densities. × 25,000

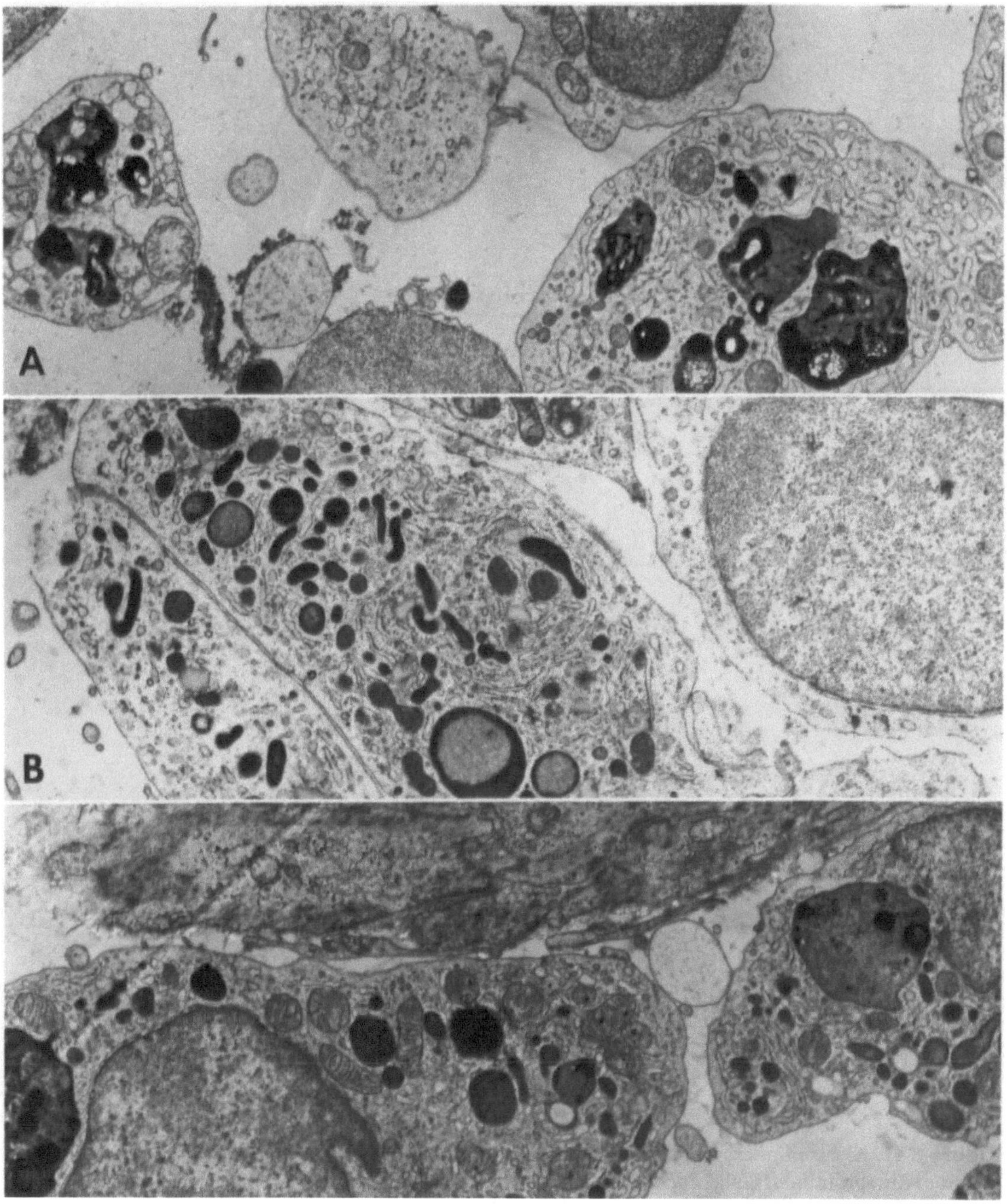

Figure 3-40. Montage of electron micrographs showing range in appearance of intestinal mucosal macrophage inclusions seen in (A) neomycin-induced malabsorption, (B) a-beta-lipoproteinemia, and (C) normal individual. × 10,500

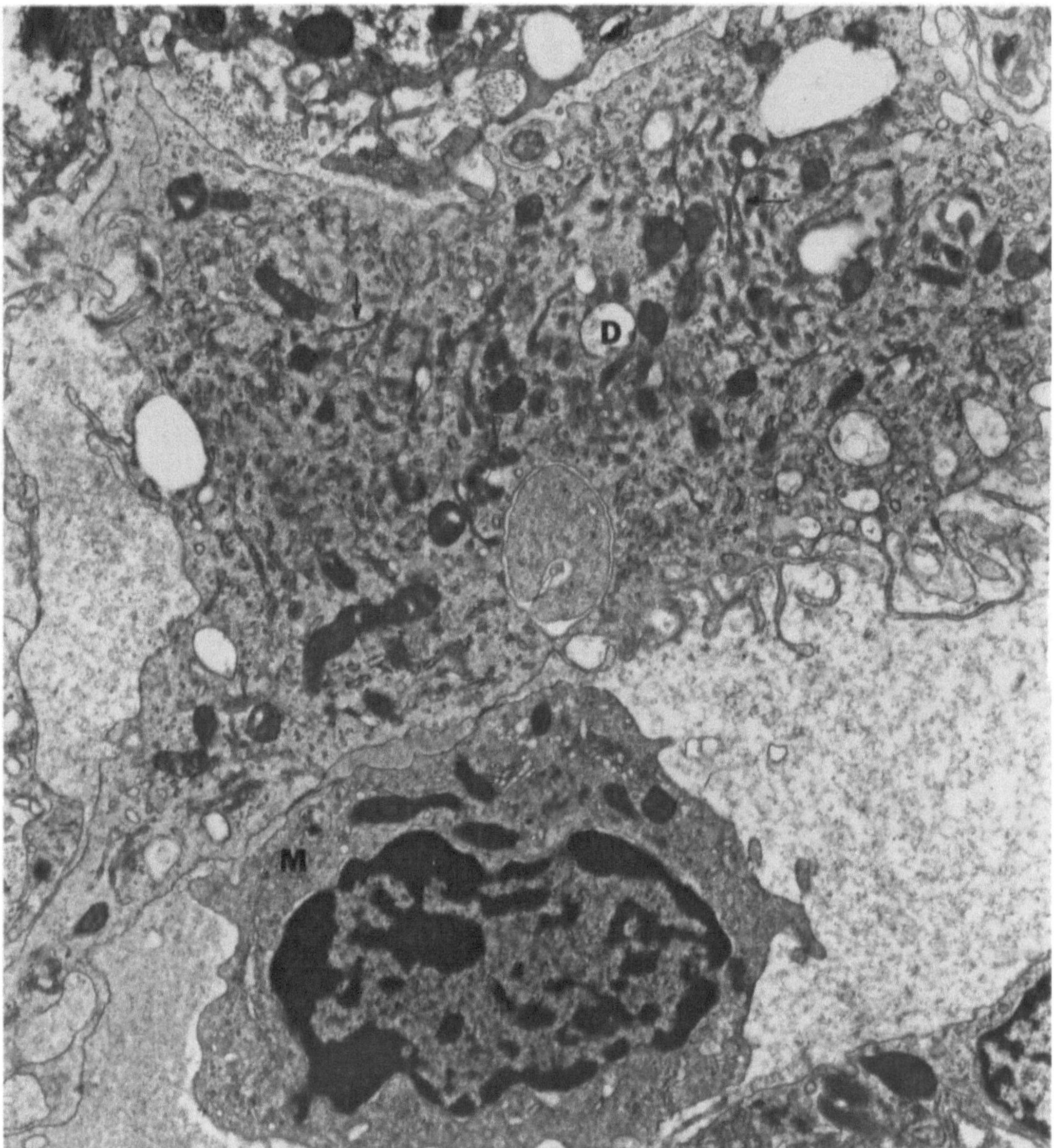

Figure 3-41. Electron micrograph showing monocyte (M) characterized by large nucleus with ample euchromatin, prominent Golgi apparatus, prominent mitochondria, numerous pseudopods, and phagocytic vacuoles adjacent to a monocyte of dendritic (D) cell type containing numerous Birbeck granules (arrows). Birbeck granules characterize dendritic cells (Langerhans) of the skin. × 14,500

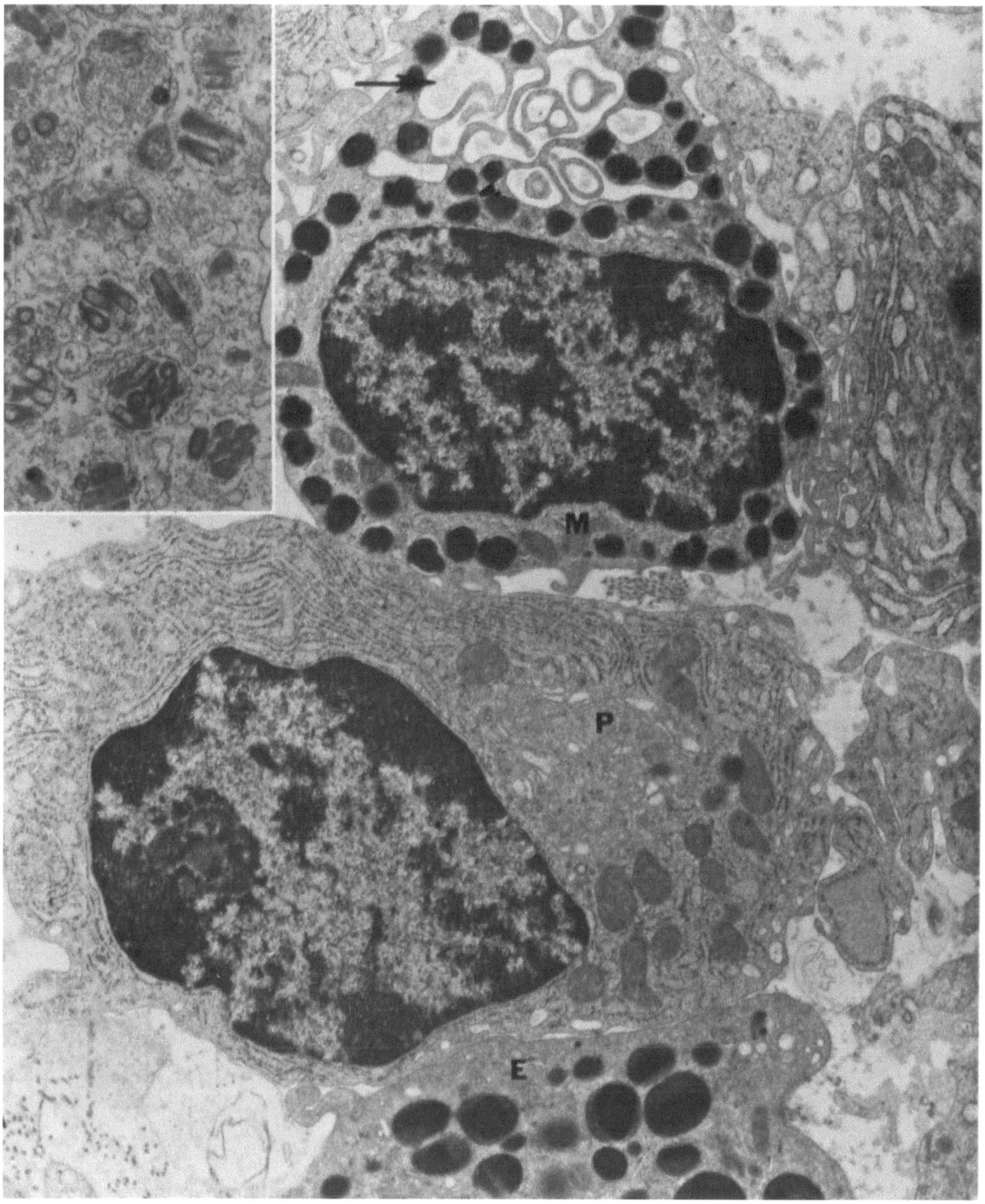

Figure 3-42. Electron micrograph showing presence of eosinophil (E) characterized by its granules, plasma cell (P), and a typical mast cell (M) with its unique granules and its "intracellular canals" (arrow), indicating recent release of many of the granules. × 15,000. Mast cell granules (in humans) often have a whorl-like and scroll-like appearance, shown better in the inset. × 31,000

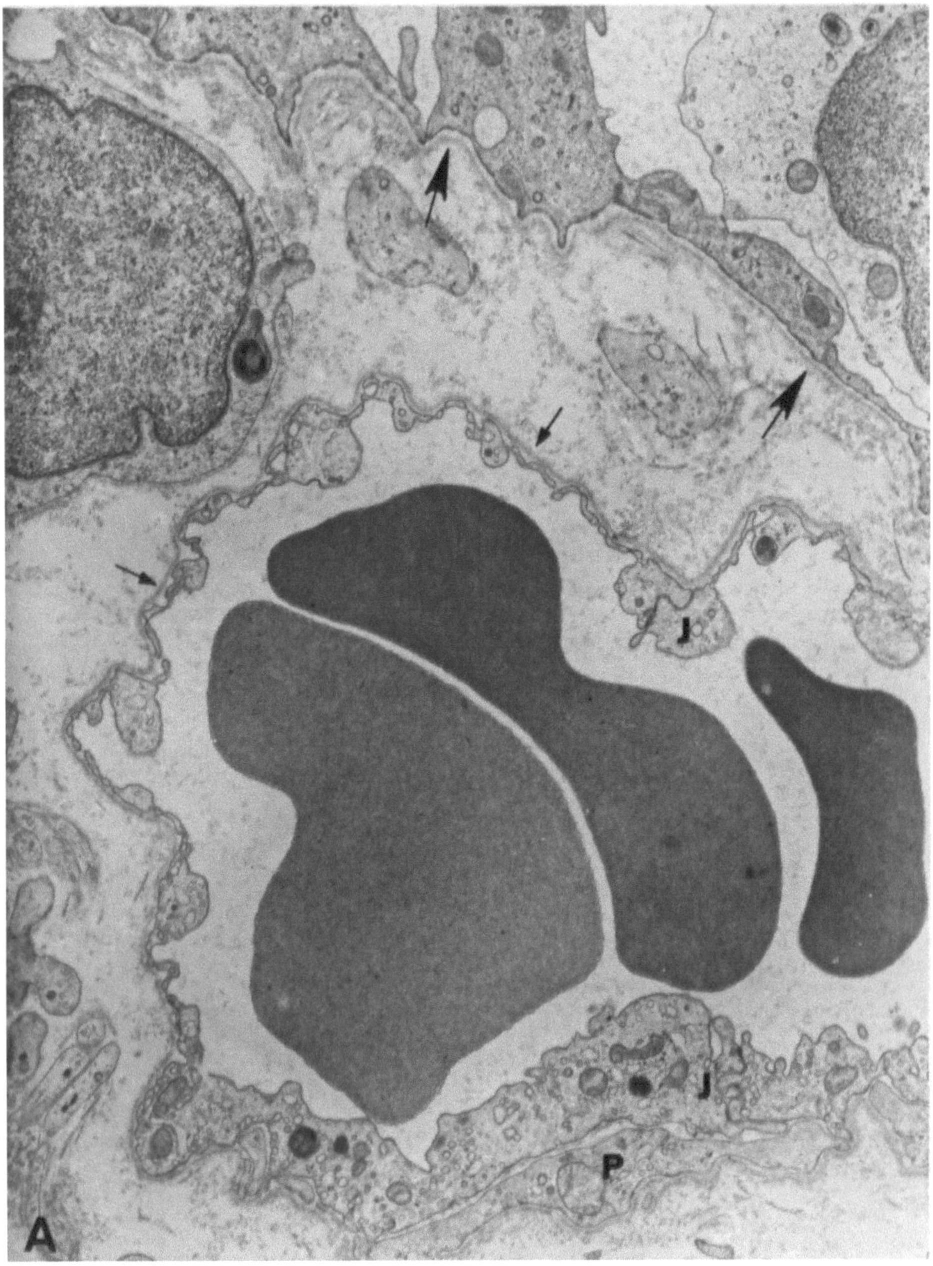

Figure 3-43. Electron micrographs. (A) Capillary in intestinal lamina propria. Note continuous basal lamina (small arrows), which is more distinct than that below epithelial cells (large arrows); numerous endothelial fenestrae, which face the epithelial cells; and the well-developed tight junctional complexes (J). Pericyte (P). × 14,000. (B) Capillary endothelium. Note the numerous Weibel-Palade bodies (arrows) and prominent caveoli (C), or transport vesicles. The basal lamina (BL) is well developed and contains numerous fine filaments and small and large collagen fibers. × 45,000

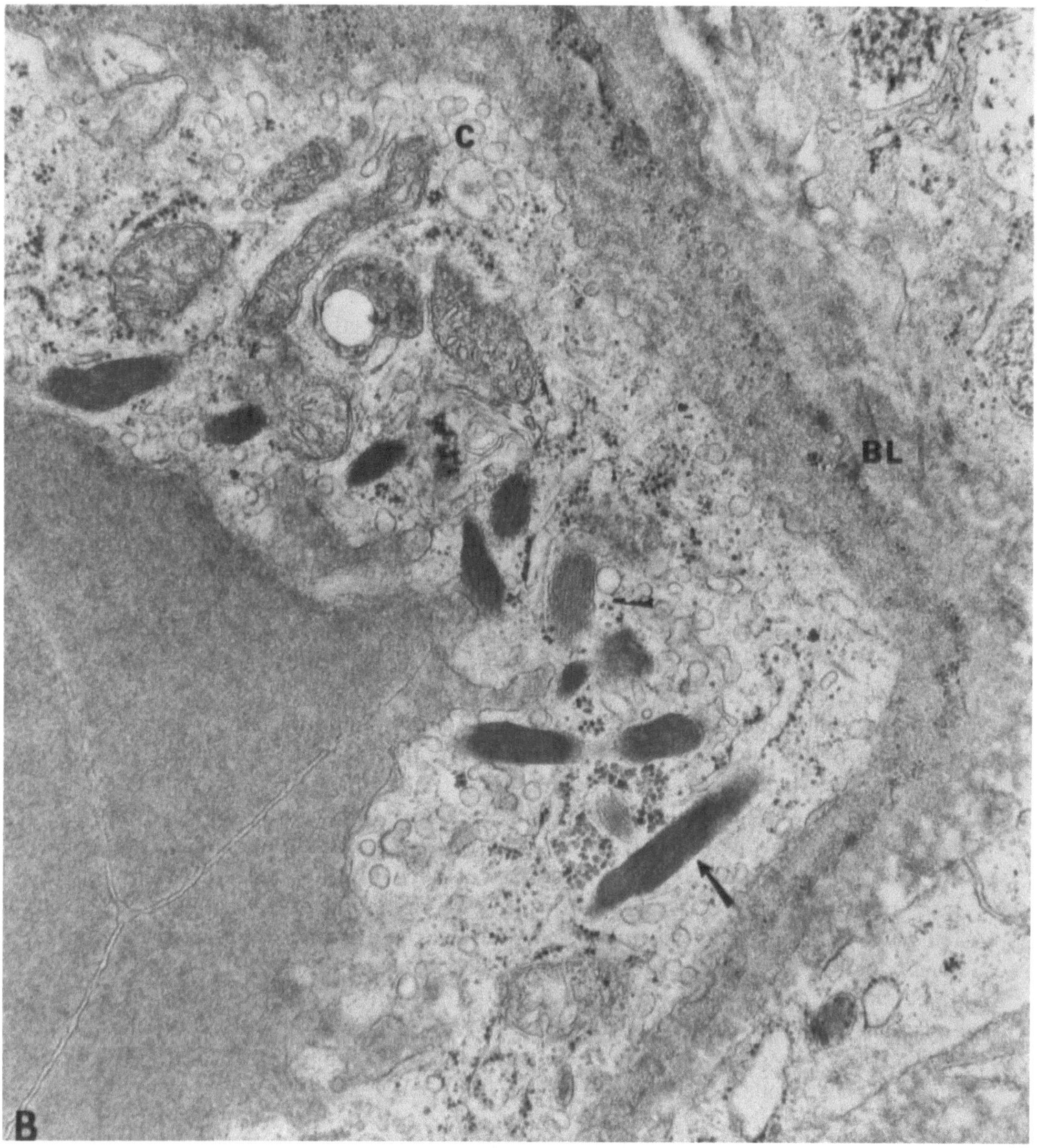

Figure 3-43B.

in the normal lamina propria, whereas the presence of more than an occasional PMN leukocyte is abnormal.

Blood vessels derived from a central arteriole give rise to a diffuse capillary network found just below the basal lamina of epithelial cells. Intestinal mucosal *capillaries* (Figure 3-43) are characterized by the presence of a well-developed basal lamina that is continuous and a fenestrated endothelium, and often by the presence of surrounding smooth-muscle-like pericytes. The pericytes are enclosed within a basal lamina that is continuous with that about the capillary endothelium. The endothelial fenestrae face absorptive cells, are covered by a thin diaphragm, and are 0.05 to 0.1 μm in diameter (Figure 3-43). These fenestrae are much more permeable than the intervening pore-free areas of endothelial cytoplasm, which contains many transport vesicles.[44] The junctional complex between capillary endothelial cells consists of tight junctions that are relatively impermeable, at least to macromolecules.[44] A rod-shaped organelle, called a Weibel-Palade body (Figure 3-43), is found in intestinal capillaries and arterioles and stains immunohistochemically for von Willebrand's factor, a clotting factor apparently manufactured in endothelial cells.

Lymphatic capillaries, called lacteals (lymphatics that transport chyle), begin as blind canals in the center of each villus. Mucosal lymphatic endothelial cells are not fenestrated (Figure 3-44) but are generously supplied with transport vesicles (caveoli) that are somewhat larger than those of blood capillaries, being a mean of 0.16 μm in lymphatics and 0.075 to 0.1 μm in blood capillaries.[45] The lymphatic endothelial basal lamina is poorly developed and discontinuous. Occasional collagen fibers are found external to mucosal lymphatic endothelia, whereas submucosal lymphatics are encased with collagen fibers many of which attach to the endothelia as "anchoring filaments" (Figure 3-44). The mucosal lymphatics do not contain pericytes but are frequently juxtaposed to smooth-muscle fibers that are normally intrinsic to the lamina propria. Mucosal lymphatic endothelial cell junctions tend also to be tight (zonulae occludentes), but are occasionally open. Macromolecules enter lacteals both through these open junctions and by transport in vesicles.[45]

Unmyelinated nerve fibers are present throughout the lamina propria (Figure 3-45). The nerves consist of autonomic motor (sympathetic and parasympathetic) and sensory fibers.[46,47] The ganglia of the sympathetic nerves are external to the gut wall (celiac and mesenteric plexi), while the parasympathetic ganglia are intramural, being located in the submucosal plexus of Meissner (Figure 3-46) and in the myenteric plexus of Auerbach. The nerve fibers innervate smooth muscle intrinsic to the villus, myofibroblasts (?), and pericytes.[42] The parasympathetic nerve endings contain neurotransmitters (neurocrine gut hormones) such as substance P, and bombesinlike hormone, as well as acetylcholine.[47] Other peptide hormones including somatostatin, gastrin, neurotensin, and cholecystokinin have been identified in nerve fibers of the gut.[46] Enkephalin released from enteric neurons may act to inhibit release of acetylcholine from cholinergic neurons. Parasympathetic stimulation generally increases muscular activity, circulation, and secretion; these activities are inhibited by sympathetic stimulation. Sympathetic neurotransmitters include norepinephrine and vasoactive intestinal peptide (VIP).

Metaplasia

Epithelium resembling that normally seen in the stomach is frequently observed in the otherwise normal-appearing proximal duodenal mucosa.[48] This metaplastic (change of one type of mature cells into another) gastric mucosa may be found to a varying extent and may, in part, be correlated with the degree of acid secretion from the stomach. James[48] systematically examined the mucosa just beyond the pylorus in patients with duodenal ulcer, pyloric ulcer,

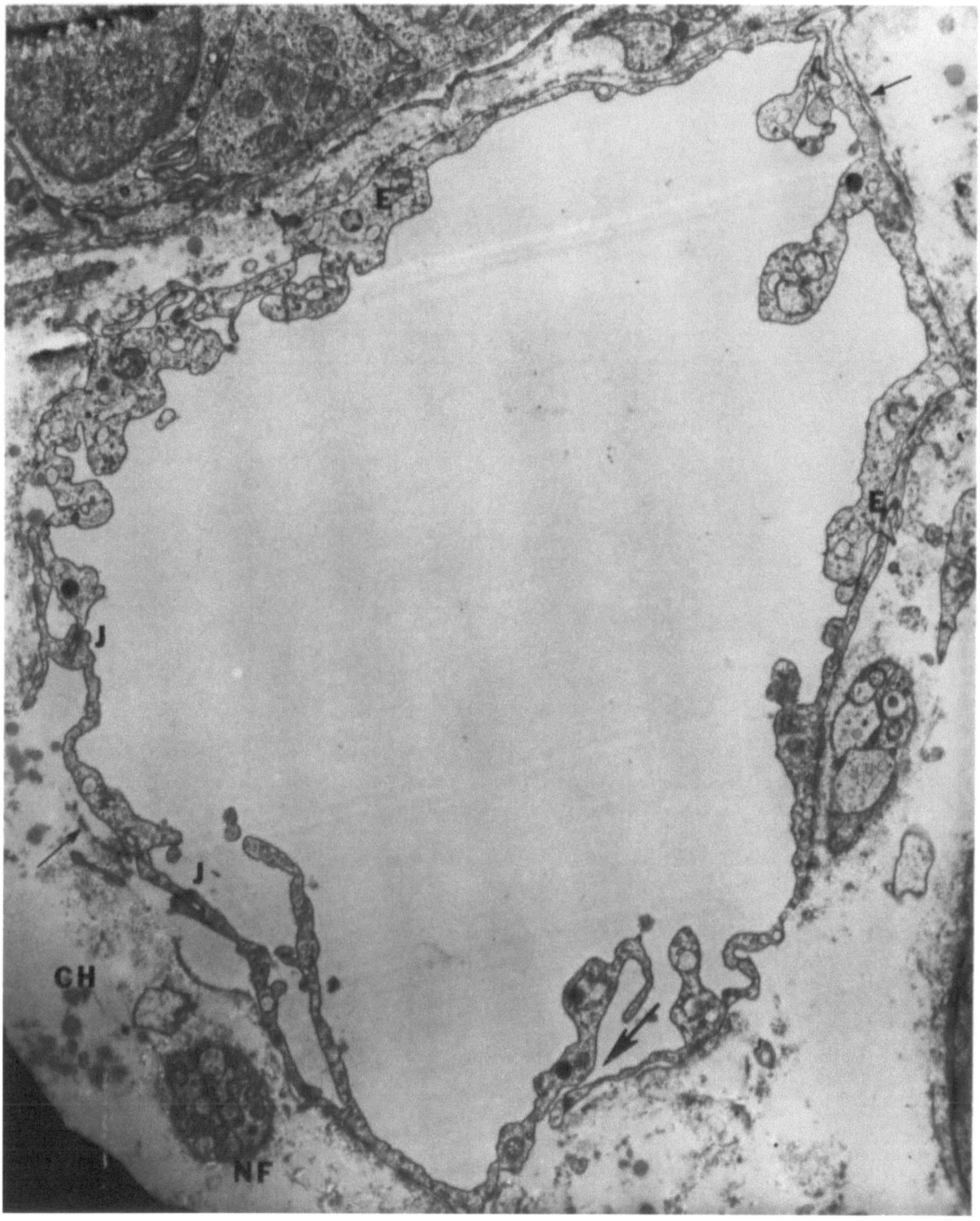

A

Figure 3-44. Electron micrographs. (A) Cross section of lacteal endothelium (E) in the lamina propria. Note the discontinuous basal lamina (small arrows), numerous tight junctional complexes (J) and a single open junction (large arrow). Nerve fibers (NF), and occasional chylomicrons (CH) are also present. × 10,600. (B) Portion of lamina propria lacteal endothelium (E). Note thin, discontinuous basal lamina (small arrow) and overlapping endothelial junctional complexes (J). There are numerous microfilaments throughout the endothelial cytoplasm; presumably these filaments and the overlapping junctional complexes permit expansion of the lumen during maximal fluid transport. Note myofibroblast cytoplasmic extension (MF). × 10,000. (C) Portion of submucosal lymphatic endothelium (E). Note that the collagen fibers (C) appear to "insert" into the abluminal plasma membrane of the endothelium as well as into the plasma membrane of the adjacent myofibroblast (MF). × 15,750 (3.44B and C on following pages)

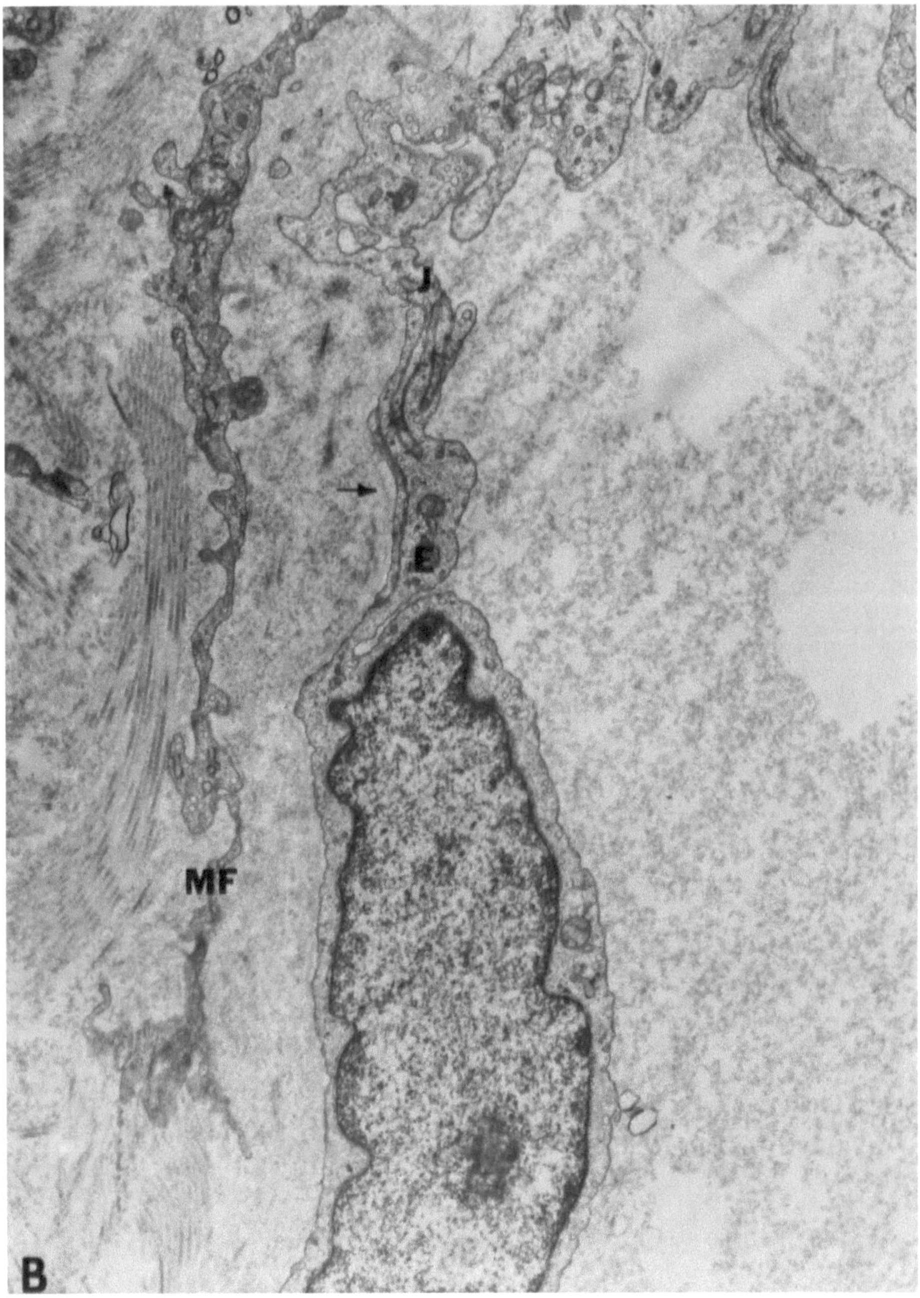

Figure 3-44B.

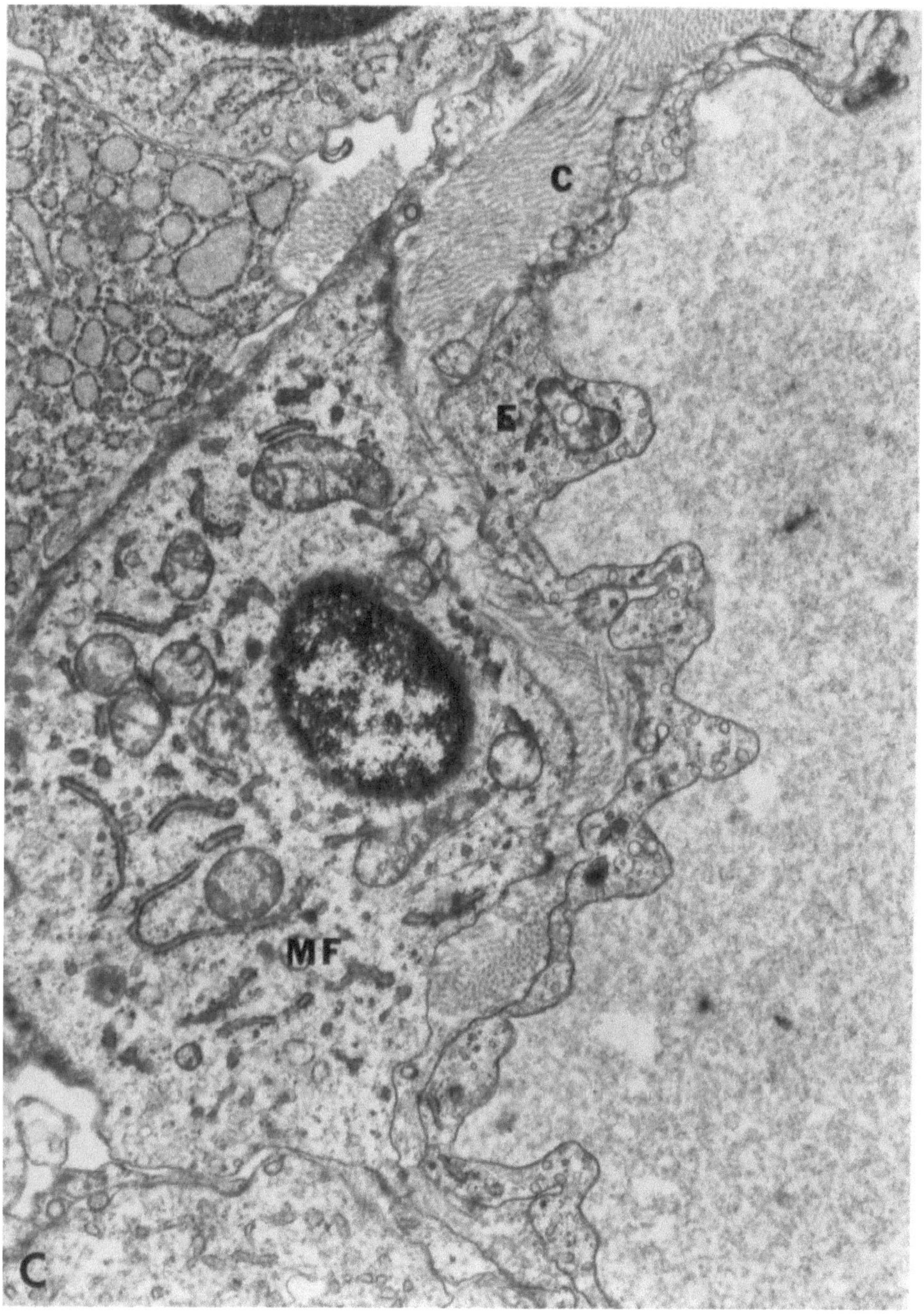

Figure 3-44C.

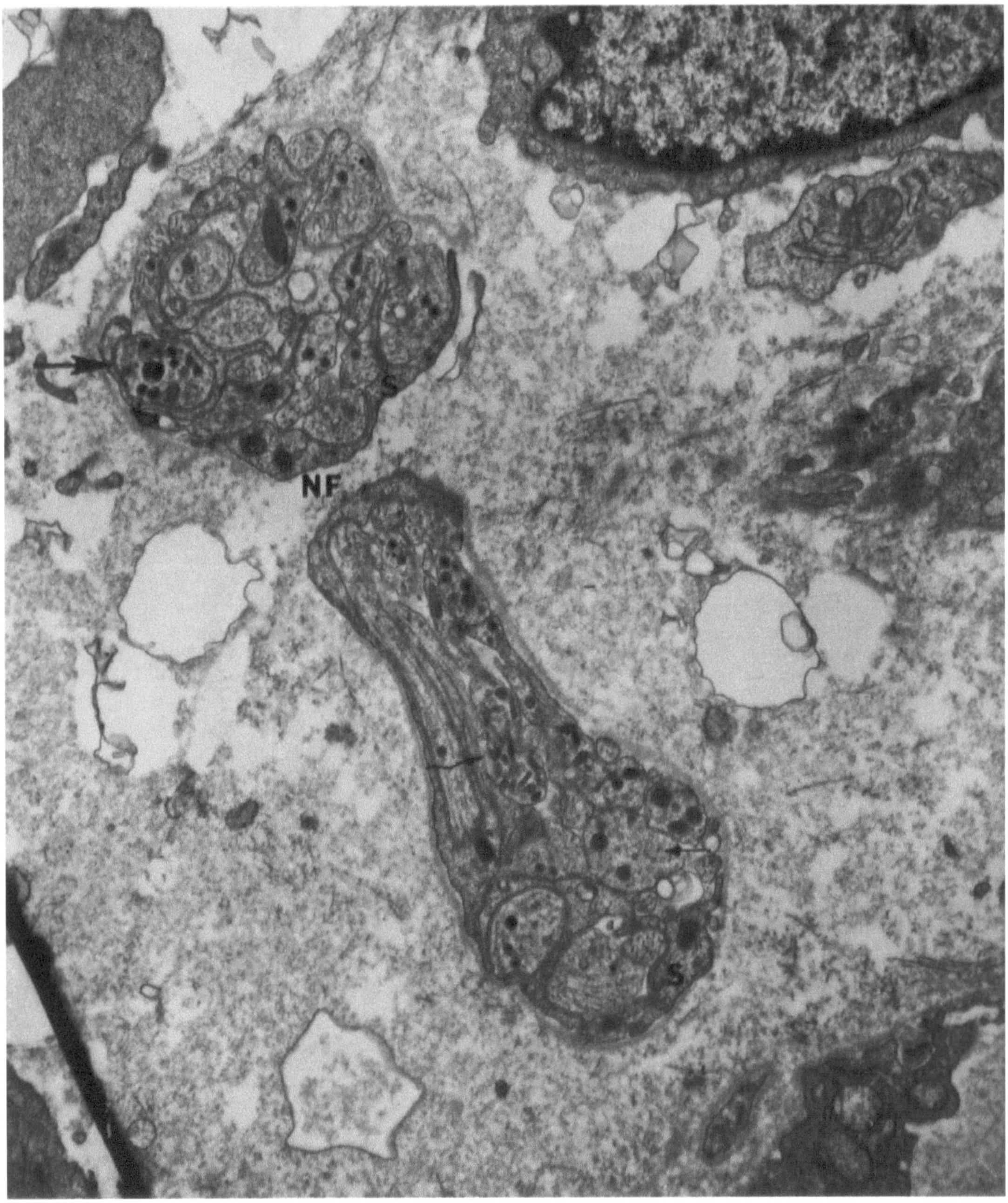

Figure 3-45. Electron micrograph showing two bundles of unmyelinated nerve fibers (NF) containing numerous nerve endings or varicosities. The nerve fibers contain prominent neurotubules (microtubules) and are encased within extensions of Schwann cell cytoplasm (S). The varicosities contain small granular vesicles (large arrow), agranular vesicles (small arrow), and elongated granular vesicles (curved arrow). The shape and content of the vesicles cannot be used to define function consistently (e.g., adrenergic, cholinergic, etc.). Note the prominent basal lamina surrounding the nerve "bundles." × 19,000

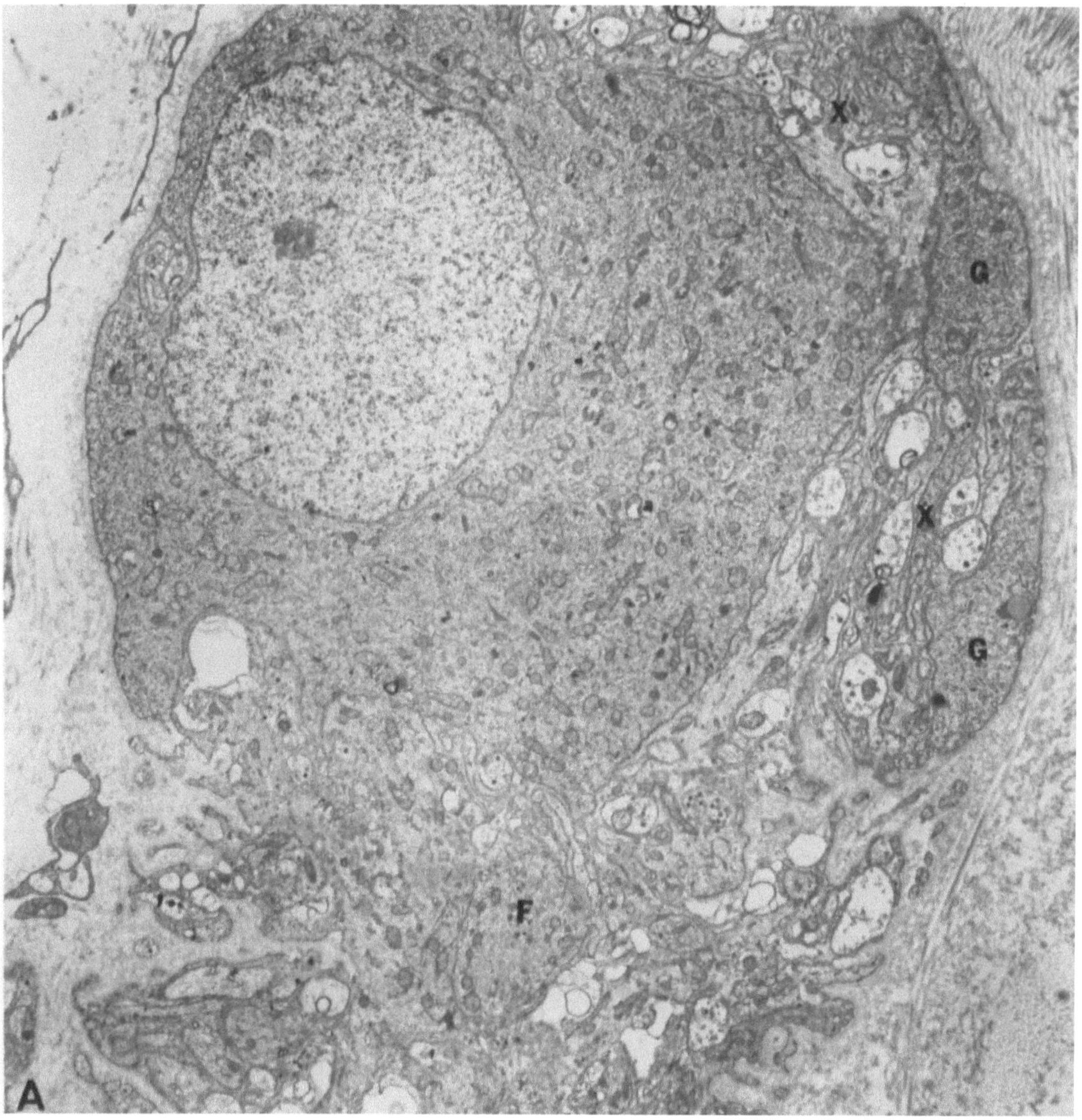

Figure 3-46. Electron micrographs. (A) Meissner plexus ganglion cell, the largest cell in the intestinal submucosa, usually with a prominent nucleus and nucleolus. There is a profuse neuropil (X) containing axons, dendrites, and processes of Schwann (glial) cells. The ganglion cell and its neuropil is enclosed within a continuous basal lamina. Note the prominent intermediate filaments (gliofilaments) (F) present in glial cell processes. These filaments, like astrocytes of the central nervous system, are rich in glial fibrillary acid protein. Note small isolated portions (G) of cytoplasm of adjacent ganglion cell(s). × 7,500. (B) Portion of cytoplasm (perikaryon) of Meissner plexus ganglion cell. Note the large surface area of the perikaryon containing mitochondria, ribosomes, lysosomes, and large accumulations of rough endoplasmic reticulum or Nissl (Ni) bodies. Parts of the cytoplasm reach the surface and are separated from the lamina propria only by its basal lamina (arrows). Note the neuropil (X) formed by dendrites, axons, and Schwann cell processes. Numerous nerve endings containing small dense core and agranular vesicles are present. A portion of a mast cell (M) and of a smooth muscle cell (Sm) are present. × 11,500

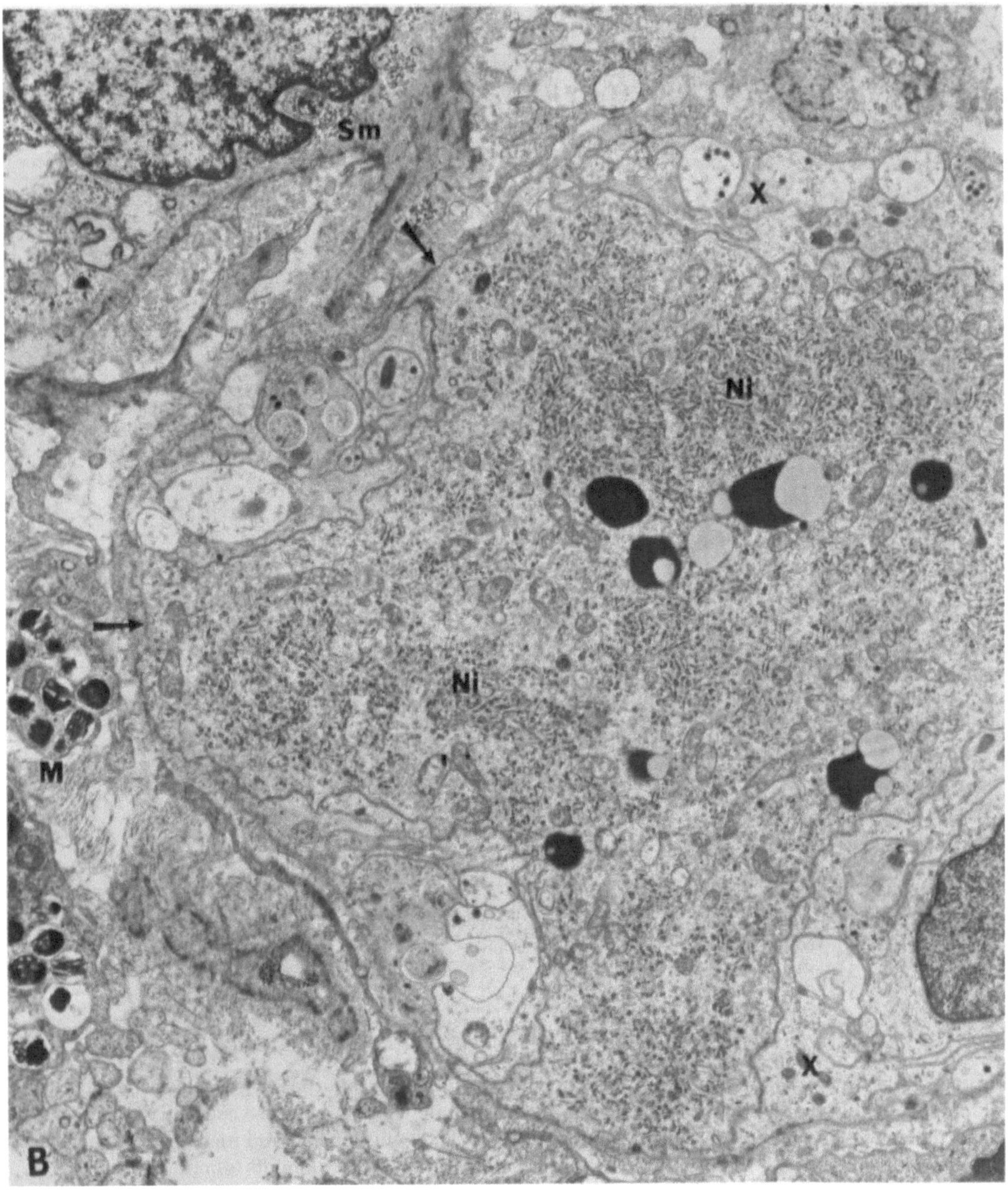

Figure 3-46B.

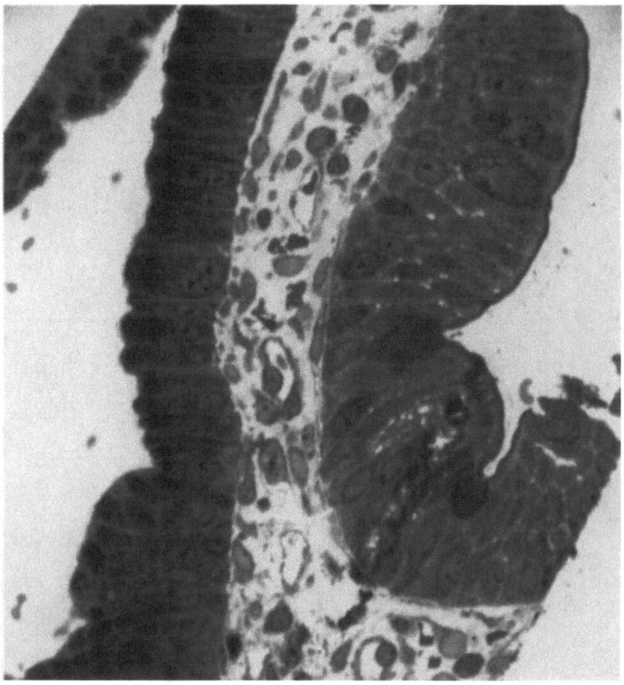

Figure 3-47. Light microscopy photograph of thick section showing portion of an intestinal villus lined by gastric surface mucous cells on the left and by normal intestinal epithelial and goblet cells on the right. × 500

gastric ulcer, and gastric carcinoma. He found that the majority of villi were covered with normal intestinal epithelium. When PAS-stained sections were examined, cells with the morphology of surface mucous cells of the stomach (lacking a brush border and containing apical mucous droplets of variable size) were readily recognized. Islands of gastric cells were common in patients with duodenal ulcer (16% of villi), less so in patients with pyloric ulcer (9% of villi), and rare in patients with gastric ulcer or carcinoma (1% of villi). When present, metaplastic surface mucous cells were more often found on "stunted" villi.

Such metaplasia is a frequent finding in the duodenal mucosa of individuals with the gastric acid hypersecretory syndrome of Zollinger-Ellison and may even be found in the proximal jejunum.[49] One patient, studied serially over a 17-month period, developed progressive morphological lesions in the duodenum. The lesions consisted initially of inflammatory cells in the lamina propria, then of pyloric gland metaplasia (see below), and finally gastric surface mucous cell metaplasia, which was also present in the proximal jejunum[49] (Figures 3-47, 3-48). In correlation with the structural changes, there was marked (functional) impairment in mucosal esterification of radiolabeled palmitic acid and in the ability of mucosal tissue to concentrate radiolabeled basic and neutral amino acids.[49]

At electron microscopy, the metaplastic surface mucous cells are similar, if not identical, to those found in the normal stomach. They contain numerous apical mucous droplets similar to those of gastric surface mucous cells and quite distinct from those of intestinal goblet cells (Figure 3-48). They have sparse microvilli, prominent ribosomes and rough endoplasmic reticulum, sizable Golgi complexes, elaborate interdigitations of lateral cell membranes, and prominent intermediate filaments. The secretory droplets are dense to stippled and often contain halfmoon-shaped densities in their periphery.

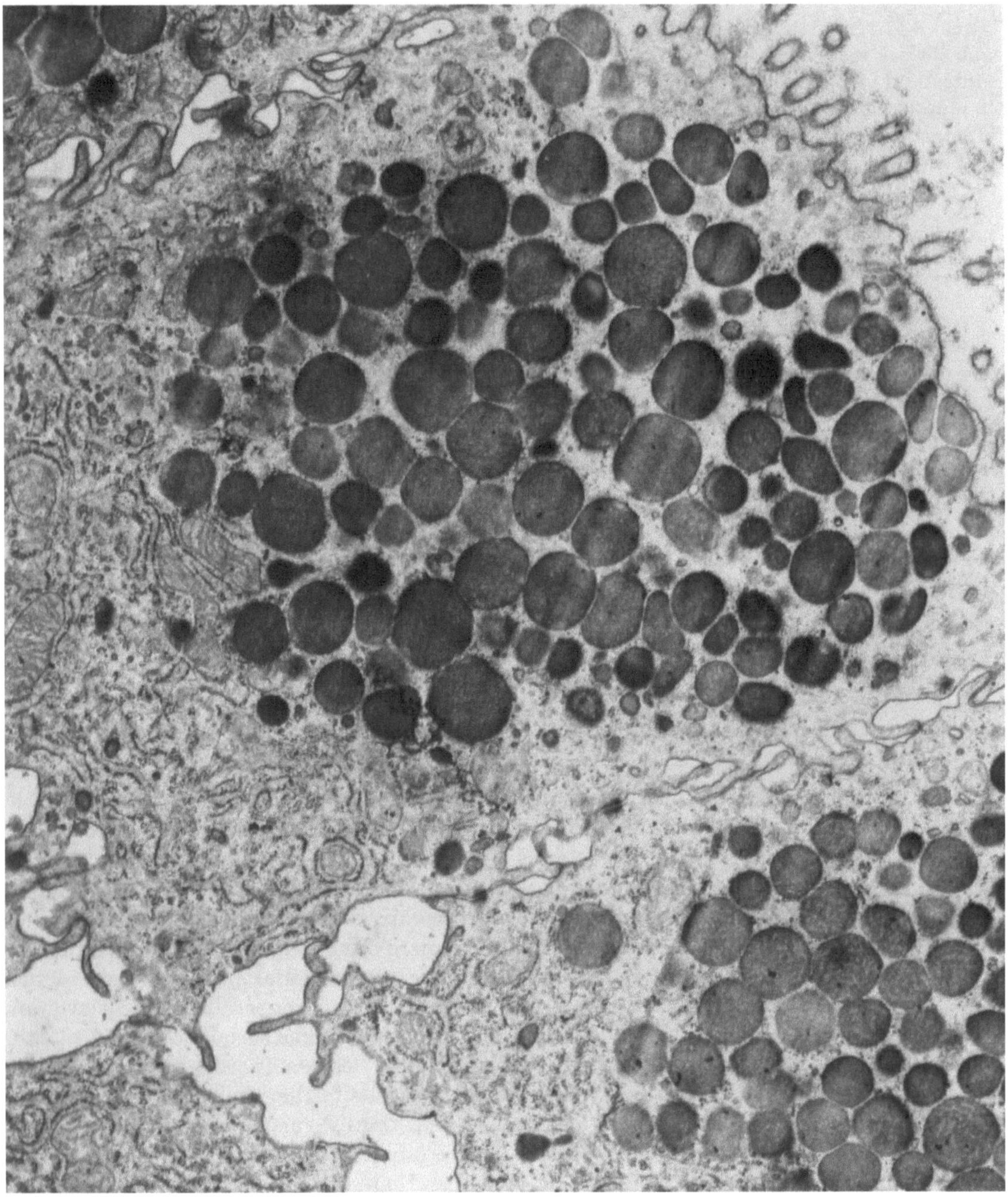

Figure 3-48. Electron micrograph of gastric surface mucous cells lining villus of intestinal biopsy specimen of an individual with Zollinger-Ellison syndrome and diarrhea. The cells are packed with mucous secretory droplets characteristic of gastric surface mucous cells, contain a prominent rough endoplasmic reticulum and prominent Golgi complex (not shown here), and have sparse microvilli. × 22,000

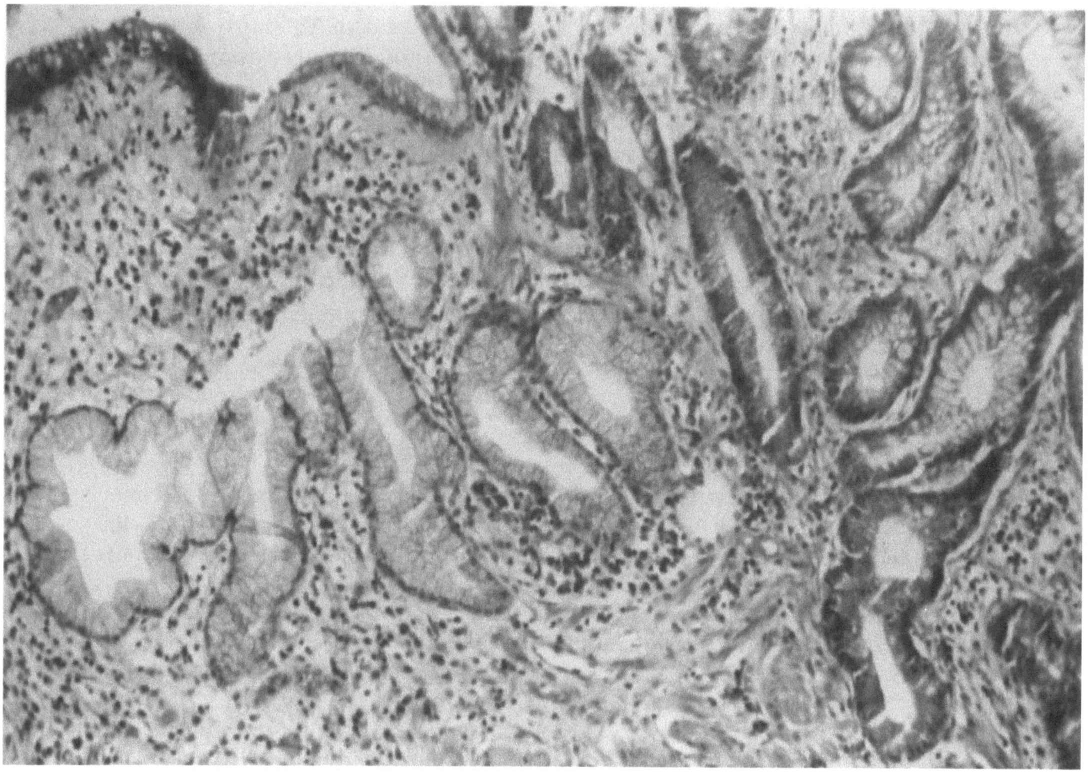

Figure 3-49. Light microscopy photograph of intestinal mucosa in a patient with Crohn's disease and with pyloric gland metaplasia in the involved mucosa. See text for details. × 250

Gastric metaplasia of the intestinal epithelium has been produced experimentally in the pig following prolonged exposure of the mucosa to an acid environment.[50] Gastric mucosa is more resistant to acid injury than is the intestinal mucosa, and it seems likely that gastric metaplasia in the intestine is a protective adaptation.

Pyloric glandular metaplasia is found in the area of crypts and appears to develop in response to chronic inflammation. It is particularly likely to develop in the presence of Crohn's disease, but is seen in many other chronic inflammatory states including Zollinger-Ellison syndrome. The crypt glands change from simple tubes to branching glands (Figure 3-49) and the epithelial cells are indistinguishable in appearance (H & E staining) from pyloric glandular epithelium of the stomach or from Brunner gland epithelium of the duodenum. The metaplastic pyloric glandular epithelial cell mucus is neutral, compared with acid mucins of intestinal goblet cells. When parietal and chief (zymogenic) cells are present, the heterotopic area should be termed gastric heterotopia or gastric fundic gland metaplasia.[51] Heterotopic fundic gland mucosa in the duodenum is occasionally detected in biopsy specimens of otherwise normal proximal duodenal mucosa and has no diagnostic significance.[52,53] Hoedemaeker[52] found islets of fully developed gastric mucosa, containing parietal cells and sometimes chief (pepsinogen) cells, in the duodenal cuff in 52 of 158 gastric resection specimens. This gastric metaplasia was seen more frequently in individuals with duodenal ulcer. Endoscopists often are unable to avoid the temptation to biopsy small nodules in the duodenal bulb or proximal duodenum and often are rewarded by the histological finding of gastric fundic metaplasia,[53] a finding of no significance. Spiller, Shousha, and Barrison[53] observed macroscopic gastric heterotopia in 8 of 900 consecutive endoscopies. The lesions consisted histologically of full-thickness fundic mucosa with well-formed glands containing parietal and chief cells. Gastric metaplasia of the proximal duodenum is much more common in men than in women (ratio of 2.6:1), is associated with histological duodenitis, and the extent of metaplasia is directly related to the severity of inflammation.[54] All of these features add further support to the hypothesis that the metaplasia is related, in part, to high gastric acidity.

References

1. Dobbins WO III: Human intestinal epithelium as a biological membrane. In Trump F (Ed), *Pathobiology of Cell Membranes.* New York, Academic Press, 1975 pp 429–469.
2. Lorenzsonn V, Korsmo H, Olsen WA: Localization of sucrase-isomaltase in the rat enterocyte. *Gastroenterology* **92**:98–105, 1987.
3. Mooseker MS: Brush border motility: Microvillar contraction in triton-treated brush borders isolated from intestinal epithelium. *J Cell Biol* **71**:417–433, 1976.
4. Tilney LG, Mooseker MS: Actin filament-membrane attachment: Are membrane particles involved? *J Cell Biol* **71**:402–416, 1976.
5. Hirokawa N, Tilney LG, Fujiwara K, et al: Organization of actin, myosin, and intermediate filaments in the brush border of intestinal epithelial cells. *J Cell Biol* **94**:425–443, 1982.
6. Burgess DR: Reactivation of intestinal epithelial cell brush border motility: ATP-dependent contraction via a terminal web contractile ring. *J Cell Biol* **95**:853–863, 1982.
7. Howe CL, Mooseker MS: Characterization of the 110-kdalton actin-, calmodulin-, and membrane-binding protein from microvilli of intestinal epithelial cells. *J Cell Biol* **97**:974–985, 1983.
8. Hull BE, Straehelin LA: The terminal web: A reevaluation of its structure and function. *J Cell Biol* **81**:67–82, 1979.
9. Madara JL, Trier JS: Functional morphology of the mucosa of the small intestine. In LR Johnson (Ed.), *Physiology of the Gastrointestinal Tract* (2nd ed. New York, Raven Press, 1987, pp 4401–4441.

9a. Weinstock M, Bonneville MA: Compartments rich in acidic carbohydrate protein complexes with electrolyte- and water-transporting cells. *Lab Invest* **24**:355–367, 1971.

10. McClugage SG, Low FN: Microdissection by ultrasonication: Porosity of the intestinal epithelial basal lamina. *Am J Anat* **171**:207–216, 1984.
11. Hay E: Extracellular matrix. *J Cell Biol* **91**:205s–223s, 1981.
12. Specian RD, Neutra MR: Cytoskeleton of intestinal goblet cells in rabbit and monkey. *Gastroenterology* **87**:1313–1325, 1984.
13. Cheng H, Leblond CP: Origin, differentiation and renewal of the four main epithelial cell types in the mouse small intestine. V. Unitarian theory. *Am J Anat* **141**:537–562, 1974.
14. Sandow MJ, Whitehead R: Progress report: The Paneth cell. *Gut* **20**:420–431, 1979.
15. Bohane TD, Cutz E, Hamilton JR, et al: Acrodermatitis enteropathica, zinc, and the Paneth cell: A case report with family studies. *Gastroentrology* **73**:587–592, 1977.
16. Lechago J: The endocrine cells of the digestive and respiratory systems and their pathology. In JMB Bloodworth (Ed), *Endocrine Pathology.* Baltimore, Williams & Wilkins, pp 513–555, 1982.

17. Facer P, Bishop AE, Lloyd RV, et al: Chromogranin: A newly recognized marker for endocrine cells of the human gastrointestinal tract. *Gastroenterology* **89**:1366–1377, 1985.
18. Larsson L-I, Goltermann N, de Magistris L, et al: Somatostatin cell processes as pathways for paracrine secretion. *Science* **205**:1393–1394, 1979.
19. Christofides ND, Bryant MG, Ghatei MA, et al: Molecular forms of motilin in the mammalian and human gut and human plasma. *Gastroenterology* **80**:292–300, 1981.
20. Sjolund K, Sanden G, Hakanson R, et al: Endocrine cells in human intestine: An immunocytochemical study. *Gastroenterology* **85**:1120–1130, 1983.
21. Ferri G-L, Adrian TE, Ghatei MA, et al: Tissue localization and relative distribution of regulatory peptides in separated layers from the human bowel. *Gastroenterology* **84**:777–786, 1983.
22. Bloom SR, Polak JM: Regulatory peptides and hormone secreting tumours. In CC Booth, G Neale (Eds), *Disorders of the Small Intestine.* Boston, Blackwell Scientific Publishing, 1985, pp 376–397.
23. Owen RL: Sequential uptake of horseradish peroxidase by lymphoid follicle epithelium of Peyer's patches in the normal unobstructed mouse intestine: An ultrastructural study. *Gastroenterology* **72**:440–451, 1977.
24. Nabeyama A, Leblond CP: Caveolated cells characterized by deep surface invaginations and abundant filaments in mouse gastrointestinal epithelia. *Am J Anat* **140**:147–166, 1974.
25. Blom H, Helander HF: Quantitative ultrastructural studies on parietal cell regeneration in experimental ulcers in rat gastric mucosa. *Gastroenterology* **80**:334–343, 1981.
26. Madara JL: Cup cells: Structure and distribution of a unique class of epithelial cells in guinea pig, rabbit, and monkey intestine. *Gastroenterology* **83**:981–994, 1982.
27. Madara JL, Carlson SL: Cup cells: Further structural characterization of the brush border and the suggestion that they may serve as an attachment site for an unidentified bacillus in guinea pig ileum. *Gastroenterology* **89**:1374–1386, 1985.
28. Bockman DE, Cooper MD: Pinocytosis by epithelium associated with lymphoid follicles in the bursa of Fabricius, appendix, and Peyer's patches: An electron microscopic study. *Am J Anat* **136**:445–477, 1973.
29. Owen RL, Jones AL: Epithelial cell specialization within human Peyer's patches: An ultrastructural study of intestinal lymphoid follicles. *Gastroenterology* **66**:189–203, 1974.
30. Wolf JL, Bye WA: The membranous epithelial cell and the mucosal immune system. *Ann Rev Med* **35**:95–112, 1984.
31. Wolf JL, Kauffman RS, Finberg R, et al: Determinants of reovirus interaction with the intestinal M cells and absorptive cells of murine intestine. *Gastroenterology* **85**:291–300, 1983.
32. Cornell R, Walker WA, Isselbacher KJ: Small intestinal absorption of horseradish peroxidase: A cytochemical study. *Lab Invest* **25**:42–48, 1971.
33. Hirata I, Austin LL, Blackwell WH, et al: Immunoelectron microscopic localization of HLA-DR antigen in control small intestine and colon and in inflammatory bowel disease. *Dig Dis Sci* **31**:1317–1330, 1986.
34. Keren DF, Hold PS, Collins HH, et al: The role of Peyer's patches in the local immune response of rabbit ileum to live bacteria. *J Immunol* **120**:1892–1896, 1978.
35. Marsh MN: Studies of intestinal lymphoid tissue. I. Electron microscopic evidence of "blast transformation" in epithelial lymphocytes of mouse small intestinal mucosa. *Gut* **16**:665–682, 1975.
36. Dobbins WO III: Progress report: Human intestinal intraepithelial lymphocytes. *Gut* **27**:972–985, 1985.
37. Austin LL, Dobbins WO III: Intraepithelial leucocytes of the intestinal mucosa in normal man and in Whipple's disease: A light and electron microscopic study. *Dig Dis Sci* **27**:311–320, 1982.

38. Cerf-Bensussan N, Quaroni A, Kurnick JT, et al: Intraepithelial lymphocytes modulate Ia expression by intestinal epithelial cells. *J Immunol* **132**:2244–2252, 1984.
39. Galluci BB, Sale GE, McDonald GB, et al: The fine structure of human rectal epithelium in acute graft-versus-host disease. *Am J Surg Pathol* **6**:293–305, 1982.
40. Weber JR, Dobbins WO III: The intestinal and rectal epithelial lymphocyte in AIDS: An electron microscopic study. *Am J Surg Pathol* **10**:627–239, 1986.
41. Parker FG, Barnes EN, Kaye GI: The pericryptal fibroblast sheath. IV. Replication, migration, and differentiation of the subepithelial fibroblasts of the crypt and villus of the rabbit jejunum. *Gastroenterology* **67**:607–621, 1974.
42. Joyce NC, Haire MF, Palade GE: Morphologic and biochemical evidence for a contractile cell network within the rat intestinal mucosa. *Gastroenterology* **92**:68–81, 1987.
43. Hirata I, Berrebi G, Austin LL, et al: Immunohistological characterization of intraepithelial and lamina propria lymphocytes in control ileum and colon and in inflammatory bowel disease. *Dig Dis Sci* **31**:593–603, 1986.
44. Clementi F, Palade GE: Intestinal capillaries: Permeability to peroxidase and ferritin. *J Cell Biol* **41**:33–58, 1969.
45. Dobbins WO III, Rollins EL: Intestinal mucosal lymphatic permeability: An electron microscopic study of endothelial vesicles and cell junctions. *J Ultrastruct Res* **33**:29–59, 1970.
46. Gabella G: Structure of muscles and nerves in the gastrointestinal tract. In LR Johnson (Ed), *Physiology of the Gastrointestinal Tract* (2nd ed). New York, Raven Press, 1987, pp 335–381.
47. Dahl JL, Bloom DD, Epstein ML, et al: Effect of chemical ablation of myenteric neurons on neurotransmitter levels in the rat jejunum. *Gastroenterology* **92**:338–344, 1987.
48. James AH: Gastric epithelium in the duodenum. *Gut* **5**:258–294, 1964.
49. Mansbach CM, Wilkins RM, Dobbins WO III, et al: Intestinal mucosal function and structure in the steatorrhea of Zollinger-Ellison syndrome. *Arch Intern Med* **121**:487–494, 1968.
50. Florey HW, Jennings DA, O'Conner RC: The reactions of the intestine of the pig to gastric juice. *J Pathol Bacteriol* **49**:105–123, 1939.
51. Yokoyama I, Kozuka S, Ito K, et al: Gastric gland metaplasia in the small and large intestine. *Gut* **18**:214–218, 1977.
52. Hoedemaeker PJ: Heterotopic gastric mucosa in the duodenum. *Digestion* **3**:165–173, 1970.
53. Spiller RC, Shousha S, Barrison IG: Heterotopic gastric tissue in the duodenum: A report of eight cases. *Dig Dis Sci* **27**:880–883, 1982.
54. Shousha S, Barrison IG, El-Sayeed W, et al: A study of incidence and relationship of intestinal metaplasia of gastric antrum and gastric metaplasia of duodenum in patients with nonulcer dyspepsia. *Dig Dis Sci* **29**:311–316, 1984.

4 Immunoperoxidase Techniques: Light and Electron Microscopy Applications

For the past four years we have utilized immunoperoxidase methods for the localization of a variety of T and B cell antigens in the intestinal and colonic mucosa utilizing both light and electron microscopy techniques (Figures 4-1 through 4-3).[1-3] We have followed the methods described by Warnke and Levy[4]; by Hsu, Raine, and Fanger[5]; by Cerf-Bensussan, Schneeberger, and Bahn[6]; and by Kurosaka and Ziff.[7] Briefly, Warnke and Levy localized antigens on frozen sections utilizing avidin-horseradish peroxidase (HRP) binding to a biotinylated secondary antibody; Hsu et al. utilized an avidin-biotin HRP complex to localize the biotinylated secondary antibody at light microscopy; and Cerf-Bensussan et al. and Kurosaka and Ziff utilized tissues briefly fixed in periodate-lysine-paraformaldehyde (PLP) and then embedded in agar[6] or in OCT (Tissue Tek) compound* for electron microscopy localization of T and B cells. We have found that all of these methods are useful, that combinations of some of the methods give improved results in the intestinal mucosa, and that there are a number of problems unique to the gut that must be guarded against when using avidin-biotin techniques. This chapter describes these methods, their modifications, and precautions as they relate to the gut mucosa.

Material and Methods: Light Microscopy

Paraffin-embedded tissues are routinely fixed in Bouin's solution prepared using paraformaldehyde and are sectioned at 4- to 6-μm thickness. Frozen sections at this thickness are obtained from fresh surgical or biopsy specimens. A 1-mm cross section of each specimen is placed in embedding medium (OCT) in an airtight plastic capsule (Beem #00) and is snap frozen in isopentane cooled in liquid nitrogen. These capsules are stored at −70 °C until sectioned. Specimens that are to be freeze dried are placed in capsules containing small amounts of saline; are snap frozen; and are then stored at −70 °C until freeze dried.[8] For

* *Tissue Tek OCT Compound* from Miles Scientific, Naperville, IL 60566, USA.

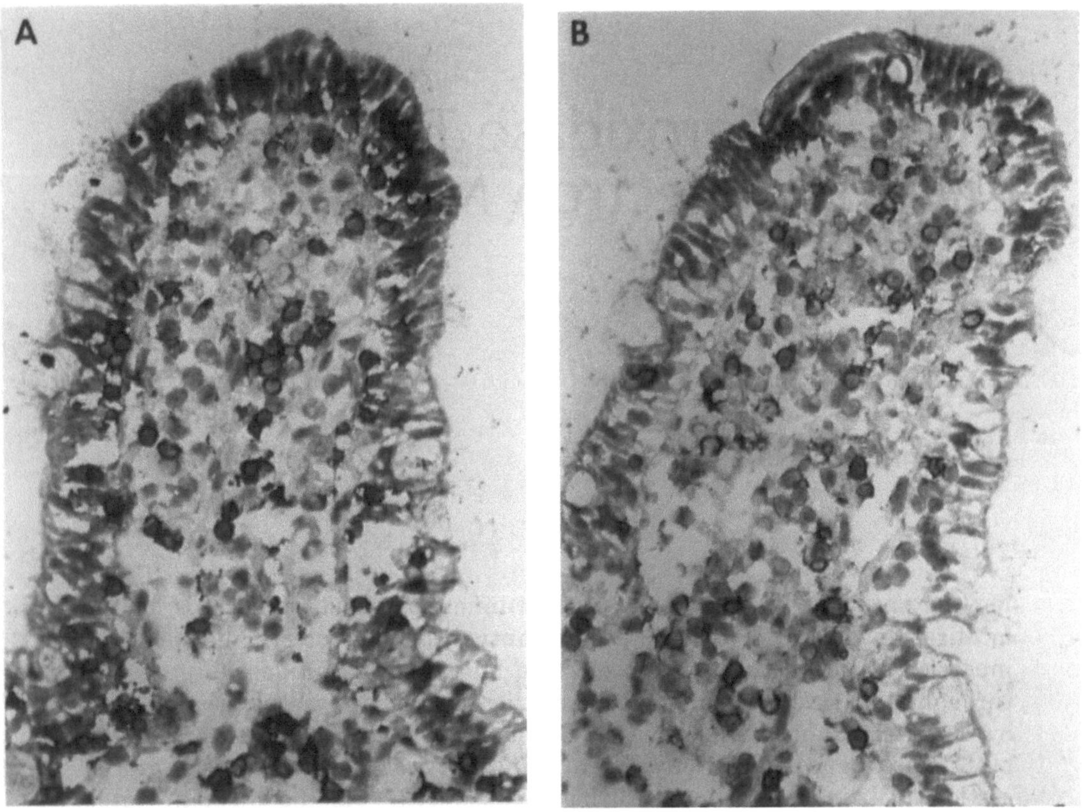

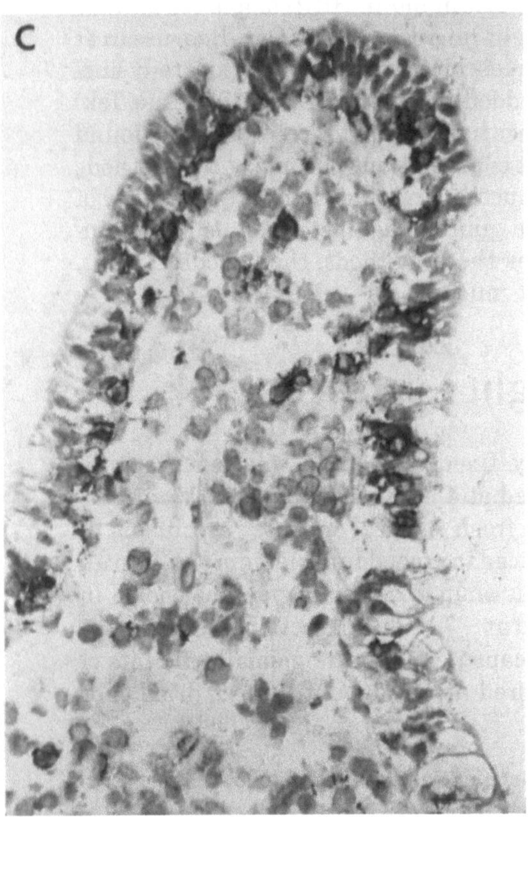

Figure 4-1. Light microscopy photographs of immunoperoxidase-stained normal intestinal mucosa in which serial sections have been stained with Coulter monoclonal antibodies to T cells (T11), helper T cells (T4), and suppressor T cells (T8). All × 250. (A) After staining with T11 there is surface staining of the majority of lymphocytes in the lamina propria and the epithelium. (B) T4 staining shows that the majority of stained lymphocytes are in the lamina propria; few are present in the epithelium. (C) T8 staining shows that now the majority of stained lymphocytes are in the epithelium; few cells in the lamina propria are stained. (See Chapter 3, section on epithelial lymphocytes, for discussion.)

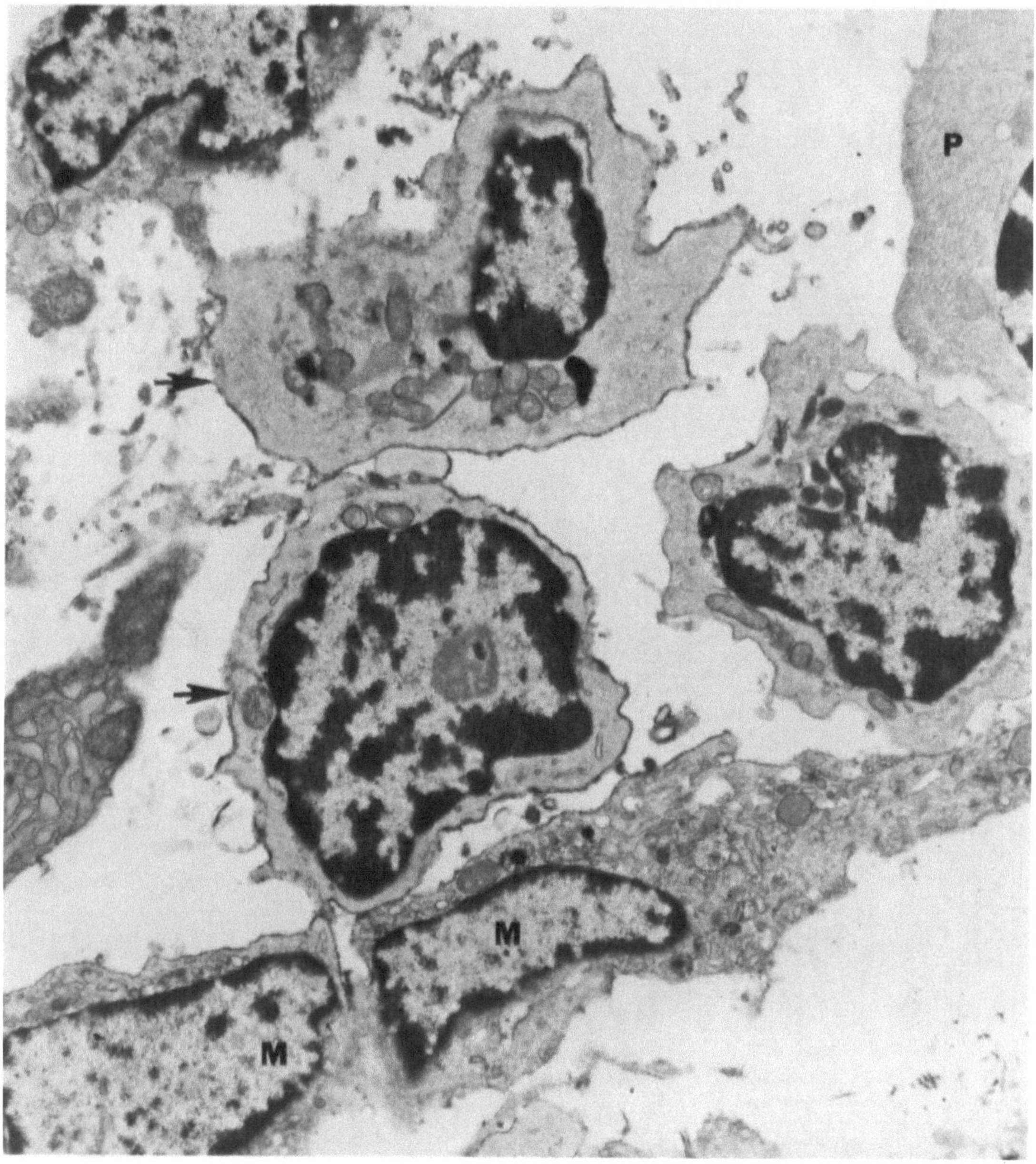

Figure 4-2. Electron micrograph of portion of ileal lamina propria in Crohn's disease after processing for immunoperoxidase localization of monoclonal antibody to helper T cells (T4). Note that there is surface staining (arrows) of two of three lymphocytes. There is no staining of the plasma cell (P) and minimal staining of the surface of two macrophages (M) immediately adjacent to the stained lymphocyte in the center. The slight macrophage staining is considered to be a diffusion artifact. × 10,500

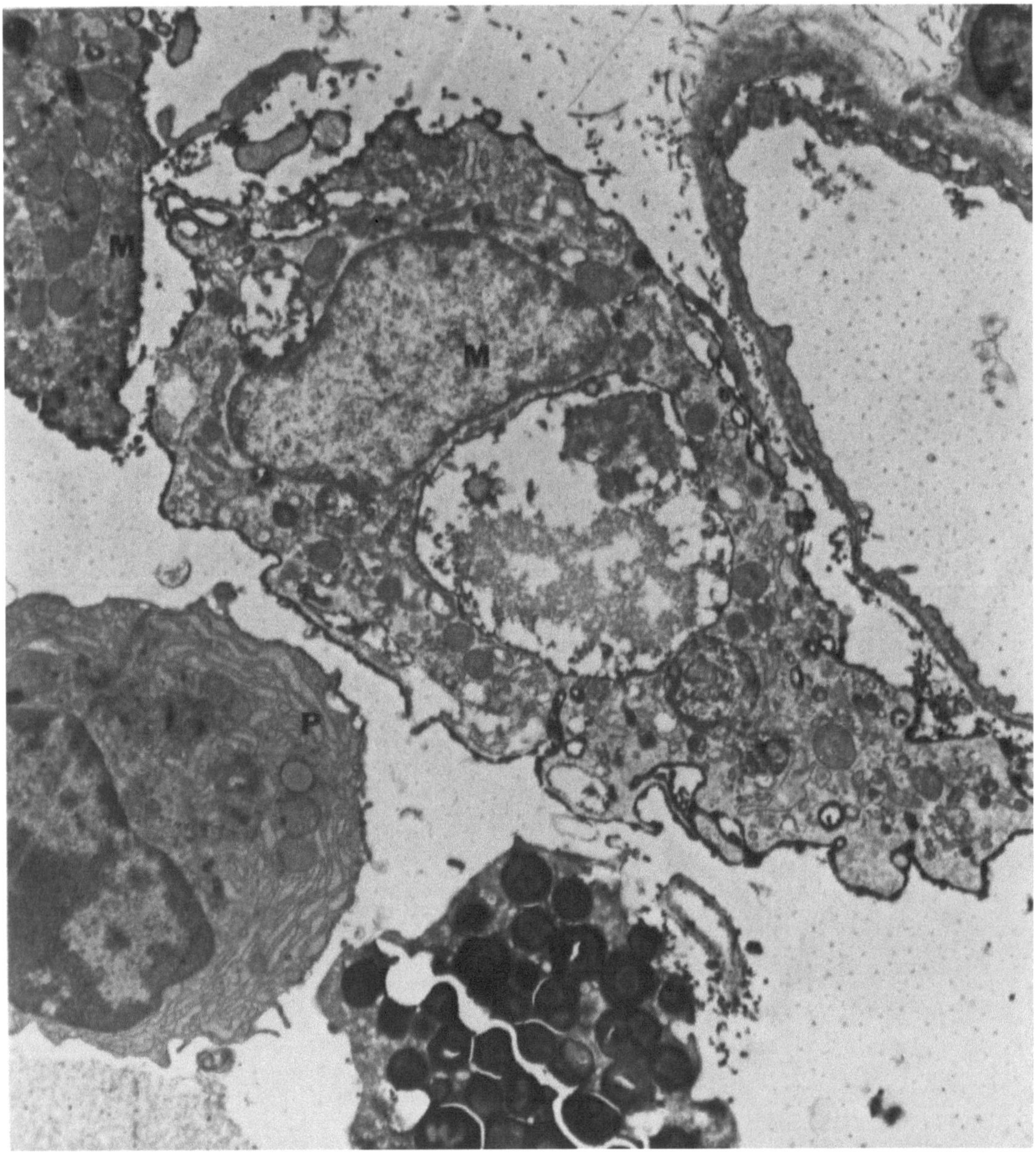

Figure 4-3. Electron micrograph of portion of normal colonic lamina propria after immunoperoxidase staining with a monoclonal antibody to HLA-DR antigen (I2). Note intense surface staining of the two macrophages (M) and minimal artifactual staining of the plasma cell (P). Note intense peroxidase (endogenous) "staining" of the eosinophil granules. × 10,500

immunoelectron microscopy, tissues (1×1×1 mm) are fixed in Nakane's[9] fixative consisting of 2% paraformaldehyde, 0.075M lysine, and 0.01M sodium periodate in 0.0375M phosphate buffer, pH 7.4. Although there are many similarities in each of the techniques, significant differences in processing of the tissues requires that details of each of the procedures be given.

Paraffin-Embedded, Routinely (Bouin) Fixed Tissues

This method gives excellent results when polyclonal antibodies to cytoplasmic antigens are used but is less successful with monoclonal antibodies, especially those directed toward surface antigens.

1. Dewax with xylene, 10 minutes, followed by acetone, 10 minutes.[8] Do not permit sections to dry at any time after dewaxing.
2. Wash three minutes in PBS.
3. React with 3,3′ diaminobenzidine (DAB), 0.5 mg/ml, in PBS (phosphate buffered saline) containing 0.015% H_2O_2, for 5 to 10 minutes. The DAB should be made up fresh, filtered, and used within one hour. Otherwise there may be nonspecific binding to tissues and generation of DAB autooxidation products that can complex nonspecifically to proteins.[10] The reaction with DAB blocks endogenous peroxidase or stains it a contrasting color.[11]
4. Incubate in diluted normal serum of animal in which primary antibody is derived when using biotinylated (two-step procedure) primary antibody.[12] Use 5% serum in 2% fetal calf serum in PBS for 20 to 30 minutes.
5. Shake off serum and add directly the diluted primary antibody to the antigen being sought. Incubations of 15 to 30 minutes are usually adequate, but longer intervals may be required.
6. Wash in PBS, three times, two minutes each time.
7. In three-step procedure, incubate in diluted normal serum of animal in which secondary antibody is derived (5% serum in 2% fetal calf serum in PBS for 20–30 minutes) and then incubate in biotinylated anti-IgG for 15 to 30 minutes, usually at a 1:200 dilution. Then wash in PBS, two times, three minutes each time.
8. (optional) Treat with 0.1M sodium azide, 1% hydrogen peroxide in PBS, for 20 to 30 minutes to inhibit endogenous peroxidase.[13] Wash in PBS for 20 minutes if this step is utilized.
9. Incubate in ABC reagent (Vectastain, Vector Laboratories, Inc.) for 15 to 30 minutes.
10. Rinse in PBS, two changes, three minutes each time.
11. Develop the color reaction using 3-amino 9-ethylcarbazole (AEC), 0.2 mg/ml with 0.015% H_2O_2 in acetate buffer, for 3 to 5 minutes. The solution is prepared by dissolving 2 mg AEC in 0.5 ml N,N-dimethylformamide, to which 9.5 ml acetate buffer, 0.05M, pH 5.0, is added and mixed. Just before use, 1 drop (approximately 0.05 ml) of 3% H_2O_2 is added.
12. Wash in water two times, five minutes each time.
13. Counterstain with hematoxylin, 15 seconds, or with 1% methylene blue, 5 seconds.
14. Dehydrate, clear, and mount.

Comments

The addition of acetone to the dewaxing procedure clearly enhances immunostaining in freeze-dried paraffin-embedded tissues[8] and seems to be useful for routinely fixed paraffin-embedded tissues as well. The two-step procedure utiliz ing biotinylated primary antibodies is superior to the three-step procedure using a biotinylated secondary antibody.[12] The Vectastain ABC kit gives superior results to those obtained with avidin-horseradish peroxidase. When the primary antibody is omitted, diffuse cytoplasmic staining of many mononuclear cells is found in the submucosa, lamina propria, and occasionally in the epithelium. This staining is not blocked by incubation in azide and probably represents nonspecific (? specific) affinity of avidin-HRP for mast cell cytoplasm and/or granules. The cells stained are mononuclear; the nucleus is located centrally and is rather dense in appearance without the "cartwheel" structure of plasma cells. Faint brown endogenous peroxidase staining can be seen in the cytoplasm of polynucleated cells with the morphology of neutrophils and/or eosinophils. Plasma cells are not stained in the absence of primary antibody. Plasma cells are not found in the epithelium; mast cells are occasional occupants of the epithelium.[14] The surprising feature of this observation is the remarkable number of "mast cells" found within the lamina propria and epithelium. It has long been known that mast cells are easily detected in the submucosa, but they have been (apparently) less easily detected in the lamina propria and epithelium.[15] This greatly complicates the detection of small numbers of positively stained plasma cells (e.g., for IgG) in the lamina propria because of the uncertainty of distinguishing small numbers of positively stained plasma cells from equally prominent (or even more numerous) nonspecifically stained mast cells. Once both sets of cells are stained with DAB they are not easily distinguished, because cellular details are often obscured by the DAB.

Trypsin treatment of the deparaffinized sections prior to immunostaining appears to "unmask" antigens and give more consistent results.[16] The sections, after deparaffinization with cold (4°C) xylene, two times, five minutes each time, and rehydration are incubated at 37°C for two hours in Coplin jars containing 0.1% trypsin (type III, Sigma Chemical Co.) and 0.1% $CaCl_2$ in distilled water (pH adjusted to 7.8 with 0.1N NaOH). Sections are then washed in distilled water five times and placed in PBS at 4°C overnight.[16] Immunostaining is then done. This procedure has improved the quality of detection of cIg and has revealed the presence of sIg when the section is stained for IgA.

Frozen Sections

We follow the methods outlined by Warnke and Levy[4] but with some modifications. Cryostat sections are dipped in acetone at room temperature for less than five seconds and then stored at −20°C until stained (preferably within 30 days). Just before staining the slides are fixed in acetone at 4°C for 10 minutes.

1. We do not rehydrate the sections[8] but immediately incubate them with appropriately diluted primary antibody (range of 1–20 μg/ml of monoclonal antibody) in 2% fetal calf serum in PBS[17] for 15 minutes. A control done with 5% normal serum or immunoglobulin of primary antibody species should be a part of every series.
2. Wash in PBS for three minutes. The PBS recommended by Wood and Warnke[17] consists of 0.12M NaCl, 0.01M $NaH_2PO_4 \cdot H_2O$, 0.04M K_2HPO_4, and 4.9×10^{-5}M merthiolate, and has a pH range of 7.2 to 7.4. These authors indicated that they obtain superior results using this modified PBS, particularly during the development of the DAB reaction product.
3. If background staining is a problem, incubate the sections for 20 minutes in 5% normal serum (of the secondary antibody species) in 2% FCS/PBS (FCS

= fetal calf serum). Tap excess serum off the slide, leaving a thin layer coating the tissue.

4. Incubate in biotinylated IgG directed toward the primary antibody species for 15 minutes, a 1:100 to 1:200 dilution (generally 7.5–15 μg/ml, although some investigators use 25–30 μg/ml[18]).
5. If endogenous peroxidase is a problem, incubate (optional) for 20 minutes in 0.1M sodium azide in 1.0% H_2O_2 at this point. We prefer to prepare sections routinely with and without azide when staining the gut mucosa.
6. Wash in PBS, two times, three minutes each time (or two times, 10 minutes each time, if azide is used). One may use the Vectastain-ABC method at this point[5] but we do not find it superior to the use of avidin-HRP when preparing frozen sections.
7. Incubate with avidin-HRP, 25 μg/ml, for 15 minutes. (Van Voorhis et al.[18] use 50 μg/ml when using avidin-fluorescein.)
8. Wash in PBS, two times, three minutes each time.
9. Develop the color reaction using 3-amino 9-ethylcarbazole (AEC), 0.2 mg/ml with 0.015% H_2O_2 in acetate buffer, for three to five minutes. The solution is prepared by dissolving 2 mg AEC in 0.5 ml N,N-dimethylformamide, to which 9.5 ml acetate buffer, 0.05M, pH 5.0, is added and mixed. Just before use, 1 drop (approximately 0.05 ml) of 3% H_2O_2 is added. (The color reaction may also be developed using DAB, 0.5 mg/ml in 0.015% H_2O_2 using the Wood and Warnke PBS [see step 2].)
10. Wash two times, three minutes each time, in PBS or distilled water.
11. Postfix in 4% formaldehyde for five minutes. Wash in water for three minutes.
12. Counterstain in Mayer hematoxylin for three to four minutes.
13. Coverslip specimens developed in AEC with Gelvatol,* pH 7.2. AEC is lipid soluble and thus cannot be mounted in permount. Forty grams of Gelvatol is placed in 160 ml cold PBS, pH 7.2, and heated in a water bath to 70°C until dissolved. When cool, 80 ml of glycerol is added. Add several drops of 2N NaOH to return the pH to 7.2. Place slides in refrigerator overnight to polymerize the Gelvatol. AEC fades rapidly in other water-soluble mounting media; Gelvatol significantly prolongs the useful life of the stained specimens (6–12 months), although we sometimes have rapid fading of the stained specimens, even when using Gelvatol. It is helpful to store the stained sections at 4°C if fading is a problem. Nail polish around the coverslip may prolong slide life for three or more years.

Comments

DAB may be utilized to develop the color reaction in frozen sections, but it often gives less contrast than does AEC. However, we find that DAB gives a more consistent color reaction compared with AEC. We incubate with 0.5 mg/ml DAB in 0.015% H_2O_2 (Warnke prefers 1–3 mg/ml DAB in 0.01% H_2O_2 [4,17]).

Following development of the color reaction with DAB or AEC, incubation in 0.5% copper sulfate in normal saline may enhance the contrast, although we have not found this to be the case. Use of 1% methylene blue in distilled water

---

*One-pound samples of Gelvatol 20/30 can be obtained from Monsanto Company, Chicago.

as the counterstain is a useful alternative to Mayer's hematoxylin. We have found that most methods generally recommended for inhibiting endogenous peroxidase (H_2O_2 in methanol, periodic acid-sodium borohydride,[19] and 0.1% phenylhydrazine[20]) inactivate the antigenic sites recognized by monoclonal antibodies and result in diffusion of the stain. We find incubation of sections with 0.1M azide in 1.0% H_2O_2 to be most useful, and we prepare all sections to be stained with monoclonal antibodies in the presence and absence of this step. Fc receptor binding is not a cause of nonspecific staining when fixed tissue sections are used.[6]

One technical problem is that even freshly prepared solutions rapidly become contaminated with bacteria. Azide should not be used to inhibit bacterial growth because it inhibits antigen-antibody reactions. We routinely add ethylmercurithiosalicylic acid sodium salt (merthiolate), 0.002%, as a preservative to all solutions and have no problem with bacterial contamination. The shelf life of most monoclonal antibodies, once reconstituted, is 120 days (possibly longer) when stored at −20°C and 30 days when stored at 4°C. Some monoclonal antibodies may have a shelf life of greater than two years when stored at −70°C, depending on the protein concentration and nature of diluent. We keep our working solutions (final dilution optimal for staining) usually no more than 24 hours at 4°C. Most fixatives (glutaraldehyde 3% in PBS, paraformaldehyde 3% in PBS, formalin) abolish staining reactions with monoclonal antibodies, even with only five minutes of exposure, and hence the need for frozen sections. Fixation of the sections in formalin just before counterstaining results in better morphological preservation.

Endogenous avidin binding activity (EABA) has been shown to be a problem in tissues relatively rich in biotin-containing enzymes. These tissues include liver, adipose tissue, and kidney.[17] We have not found EABA to be a problem in intestinal tissues. If there is concern in this regard, tissue sections should be incubated with avidin, 1 mg/ml in PBS (to block all endogenous biotin) for 20 minutes followed by biotin, 0.1 mg/ml in PBS (to saturate all avidin-binding sites for biotin) for 20 minutes prior to specific staining.[17,21]

Failure of frozen sections to adhere to the microscopic slides may be a problem and the use of gelatin, ammoniated egg albumin, Elmer's glue (1 part glue, 2 parts water), and Sobo glue (1 part glue, 3 parts water) have been recommended to correct this.[3,16,19] Use of polylysine, 0.1%, to coat the microscopic slides has been helpful in promoting adherence of frozen sections to the slides. We find that coating of microscopic slides with egg albumin alone gives excellent adherence. Alternatively, Tissue-Grip,* containing gelatin and potassium dichromate, also gives excellent adherence.

Freeze-Dried Paraffin-embedded (FDPE) Sections

Antigens recognized by monoclonal antibodies do not survive conventional tissue processing and thus are usually identified in frozen sections. This antigenic denaturation appears to result from exposure of tissues to aldehyde-containing fixatives and alcohol during conventional processing.[8] Heating and exposure to paraffin do not seem to result in denaturation of antigens, and use

* Available from Fisher Scientific Company.

of FDPE sections for immunohistological staining with monoclonal antibodies has been shown to give results equal or superior to those obtained using frozen sections.[8] The antigens in FDPE tissues appear to be stable, permitting long-term storage at room temperature. Further, sections of FDPE tissues may be fixed with formalin without loss of antigenicity.

Our methods are as follows.

1. Place freshly obtained mucosal specimens, generally 1- to 2-mm cubes (Stein et al.[8] use tissue samples as large as 1×1×0.5 cm) in airtight plastic capsules (Beem #00) containing small amounts of saline and snap frozen in isopentane (2-methylbutane) cooled in liquid nitrogen, and store at −70°C until freeze dried.
2. Place the frozen specimen on a precooled plate of an Edwards-Pearce tissue dryer* and freeze dry at −45°C, 10^{-2} torr in the presence of phosphorus pentoxide for 24 hours. Allow the tissue to warm to room temperature, bring to atmospheric pressure, and then embed in paraffin in a vacuum oven for 30 minutes at 55 to 56°C.
3. Cut 4- to 6-μm sections at various times at room temperature. It is preferable to stain the sections immediately, although they maintain antigenic stability for one week at room temperature, possibly longer at 4°C, and are stable for at least 18 weeks at −20°C.[8]
4. Just before staining, dewax by immersing the sections for 10 minutes in xylene followed by 10 minutes in acetone. Place the sections in PBS or in tris-buffered saline (TBS). Do not permit the slides to dry at any stage after dewaxing.
5. Stain with monoclonal antibodies as previously described for frozen sections. However, much higher concentrations of antibody are generally needed. For example, we obtain excellent results staining frozen sections for T cells using the monoclonal antibody, OK-T11, at dilutions of 1:1280, whereas we find a 1:80 dilution gives similar results in FDPE sections. Also, incubation with primary antibody for 30 minutes to one hour may be necessary,[8] although we have obtained good results after only 15-minute incubations.
6. Stain using the two-stage or three-stage avidin-biotin system used for frozen sections. We prefer to contrast with AEC and to counterstain with Mayer's hematoxylin. The alkaline phosphatase immunoenzyme technique[8] is not suitable for the intestine because of the intestine's high endogenous content of alkaline phosphatase that is not readily inhibited.[22]

Comments

Stein et al.[8] reported that brief fixation of FDPE sections, for example, 10 seconds in Bouin's solution, followed by incubation in buffer prior to incubation with primary antibody resulted in improved morphology without loss of antigenicity. We have not tried this modification.

The FDPE technique clearly results in improved morphological results without loss of antigenic reactivity. There tends to be less diffusion artifact and less background staining. Paraffin blocks can be stored for months (years?) at room temperature and still be suitable for sectioning and staining with monoclonal

---

*Edwards High Vacuum Inc., 3279 Grand Island Blvd, Grand Island, NY 14072.

antibodies. Nemes, Thomazy, and Szeifert[23] have found that cytoplasmic immunoglobulin is inconsistently preserved in FDPE sections while surface immunoglobulin is well preserved. Methods are also available for light microscopy localization of immunoglobulins in sections embedded in epoxy resins.[24]

Materials and Methods: Electron Microscopy

Fixation for electron microscopy immunohistochemistry often requires painstaking trial and error in determining the most appropriate fixative for the antigen being investigated. Generally, Nakane's PLP containing 2% paraformaldehyde has been most effective for demonstration of surface immunoglobulins defined by monoclonal antibodies,[9] and we have found a combination of the methods described by Cerf-Bensussan et al.[6] and by Kurosaka and Ziff[7] to be most useful. Because the techniques differ considerably, each will be described and the modifications used in our laboratory will then be detailed.

Cerf-Bensussan Method

1. Fix tissues (no larger than 1×2×4 mm) in PLP containing 2% paraformaldehyde for 15 minutes at 4°C.[9] See later section for PLP recipe.
2. Wash for 24 hours at 4°C in phosphate-buffered saline, 0.1M, pH 7.4, containing 0.1M (3.4%) sucrose.
3. Orient and then embed the tissue in 5% agar* at 37°C. Cool the agar to room temperature and cut 125- to 150-μm sections using a Sorvall TC-2 tissue sectioner† or a vibratome.‡ Cerf-Bensussan et al. use 2.5% agar; we find that 5% agar gives better results.
4. Incubate the sections for 24 hours at room temperature with primary antibody, appropriately diluted, in a multiwell plate on an Orbital shaker§ at setting "3." Fifty microliters of antibody solution is generally sufficient when using Falcon 3040 Microtest II tissue culture plates (multiwell) in a humid atmosphere.
5. Wash in three changes of PBS over 30 to 45 minutes.
6. Incubate in biotinylated anti-mouse IgG, 1:100 dilution (15 μg/ml) in 2% FCS in PBS for two hours.
7. Wash in three changes of PBS over 30 to 45 minutes.
8. Incubate in avidin-HRP at 25 μg/ml in 2% FCS in PBS for two hours. Vectastain ABC may be utilized at this point, rather than the avidin-HRP.
9. Wash in three changes of PBS over 30 to 45 minutes.
10. Fix for 30 minutes at room temperature in 2% glutaraldehyde, 0.1M sucrose, 0.004 M $CaCl_2$ in 0.075M sodium cacodylate, pH 7.3.
11. Wash in PBS for 30 minutes.
12. Incubate in 180 μl DAB, 0.5 mg/ml in PBS, pH 7.6, for 30 minutes, followed by addition of 20 μl 0.3% H_2O_2 and an additional 30 minutes of incubation (final concentration of H_2O_2 is 0.03%).

---

*Difco Labs, P.O. Box 1058, Detroit, MI 48232, USA.
†Sorvall, Inc. (now DuPont Co. Inc. Instrument Products), Newtown, CN 06470, USA.
‡Lancer-Brunswick, St. Louis, MO 63103, USA.
§Bellco Glass, Inc., Vineland, NJ, USA.

13. Wash in PBS, 30 minutes to overnight.
14. Fix in 1% osmium, 1.5% potassium ferrocyanide in water, for one hour at room temperature.
15. Dehydrate in graded ethanol solutions and propylene oxide; embed in Epon 812 using Beem #00 plastic capsules. Dehydration must be done in glass chambers because propylene oxide reacts with the plastic in the Microtest II tissue culture plates.
16. Processing hereafter is routine. We obtain thick sections on all blocks, each of which is stained with toluidine blue or alkaline Giemsa for light microscopy study. Thin sections are then prepared from the best oriented samples and are examined electron microscopy both unstained and stained with uranyl magnesium acetate and with lead citrate.

Comments

Preservation of ultrastructure appears to be quite good in specimens embedded in agar,[25] although the consistency of penetration of the various immunoreagents is not always reliable. Also the prolonged incubation times at room temperature permits bacterial overgrowth, so that it is important to add merthiolate (ethylmercurithiosalicylic acid) 0.002% to all solutions.

Kurosaka and Ziff Method

This method of processing for the immunoperoxidase localization of monoclonal antibodies permits better penetration of the immunoreactants but results in poor morphology. The method is as follows:

1. Fix tissues up to 5 mm^3 in size at 4°C in 2% PLP, changing the PLP two times during the four hours.
2. Wash two times, five minutes each time, and then immerse overnight in PBS (0.01M phosphate buffer containing 0.15M sodium chloride, pH 7.6) containing 4×10^{-5} M digitonin* and 10% sucrose at 4°C.
3. Wash three hours in PBS plus 15% sucrose at 4°C.
4. Wash three hours in PBS plus 20% sucrose at 4°C.
5. Wash three hours or longer in PBS plus 5% glycerol and 25% sucrose at 4°C.
6. Embed in OCT compound in Beem #00 plastic capsules and snap freeze in liquid nitrogen. Store at −70°C (Kurosaka and Ziff recommend −80°C) until sectioned.
7. Cut 8- to 12-μm frozen sections (Kurosaka and Ziff recommend 6–8-μm sections) in a cryostat at −20 to −30°C and mount on Tissue Grip or double-coated glass slides (see step 25).
8. Let air dry for 30 minutes and then wash in chilled PBS for five minutes.
9. Incubate in 100 to 200 μl of appropriately diluted primary monoclonal antibody (usually 1:10 to 1:40) for 30 to 60 minutes. This, and all subsequent steps, are at room temperature.
10. Wash in PBS with two to three changes over 15 to 20 minutes.
11. Incubate in 5% normal serum of the secondary antibody species for 20 to 30 minutes to reduce background staining. Tap off excess serum leaving a thin layer coating the tissue.

*Sigma Chemical Co., P.O. Box 14508, St. Louis, MO 63178-9916, USA.

12. Incubate in antimouse biotinylated IgG (or IgM) for 20 to 60 minutes.
13. Wash as in step 10.
14. Incubate in avidin-HRP (or Vectastain ABC Reagent) for 45 to 90 minutes.
15. Wash as in step 10.
16. Fix in 2% glutaraldehyde containing 0.1M sucrose, 0.004M $CaCl_2$ in 0.075M cacodylate buffer, pH 7.3, for 30 minutes. (Kurosaka and Ziff do *not* fix in glutaraldehyde and omit this step.)
17. Wash as in step 10.
18. Incubate in incomplete Karnovsky's solution (0.2 mg DAB/ml, in 0.05M Tris-HCl, pH 7.6) for 30 minutes.
19. Transfer to complete Karnovsky's solution (containing 0.005% H_2O_2) for 2 to 10 minutes, depending on intensity of staining.
20. Wash in PBS for five minutes.
21. Fix in 1% osmium tetroxide in 0.2M cacodylate buffer, pH 7.4, for one hour.
22. Wash in PBS for five minutes or longer.
23. At this point the sections may be coverslipped using 90% glycerin in PBS, and photographed for a permanent record. We prefer to prepare two sections of each tissue sample on the same glass slide, one that is coverslipped permanently using Epon 812 and one that is dehydrated and embedded in Epon 812. The extra section(s) also provide insurance against accidental loss of one of the sections.
24. Dehydrate in graded alcohols to 100% followed by propylene oxide. While the sections are still wet (quickly) insert plastic capsules filled with Epon 812 over them and polymerize for 24 hours or longer at 60°C.
25. After polymerization of the Epon 812, remove the section now embedded in epon from the slide by (a) heating the slide on a hot plate at 90°C for 30 to 60 seconds and quickly snapping the epon from the slide, (b) cooling the slide in liquid nitrogen, or (c) placing the slides at −20°C until the embedded sections retract from the glass slides. Some experience is required to develop facility with these simple maneuvers. We have obtained best results when the slides were doubled coated with 6% gelatin in water containing 20% absolute alcohol followed by 0.5% egg albumin in 0.1% ammonium hydroxide.
26. Choose fields for study by electron microscopy by examination of the coverslipped section at light microscopy and then trimming the adjacent epon-embedded section using a dissecting microscope. Often sufficient details can be observed in the epoxy-embedded section to permit trimming it directly while observing it in the dissecting microscope. Observe the sections obtained with and without counterstaining (uranyl magnesium acetate and lead citrate).

Comments

As stated earlier, we have obtained good results using both the Cerf-Bensussan et al. and the Kurosaka and Ziff methods, but have found a combination of the two methods avoids some of the disadvantages of both, that is, the longer time of fixation and the cumbersome postfixation washes in the latter method and the long duration of antibody incubations in the former method. In the combination, the tissues are fixed and washed according to Cerf-Bensussan et al. and then embedded in OCT and processed thereafter according to Kurosaka and Ziff.

We also find it helpful to fix the 12-μm frozen sections in acetone for 10 minutes prior to the antibody incubations.

References

1. Hirata I, Berrebi G, Austin LL, et al: Immunohistological characterization of intraepithelial and lamina propria lymphocytes in control ileum and colon and in inflammatory bowel disease. *Dig Dis Sci* **31**:593–603, 1986.
2. Hirata I, Austin LL, Blackwell WH, et al: Immunoelectron microscopic localization of HLA-DR antigen in control small intestine and colon and in inflammatory bowel disease. *Dig Dis Sci* **31**:1317–1330, 1986.
3. Keren DF, Appelman HD, Dobbins WO III, et al: Correlation of histopathologic evidence of disease activity with the immunoglobulin containing cells in the colon of patients with inflammatory bowel disease. *Hum Pathol* **15**:757–763, 1984.
4. Warnke R, Levy D: Detection of T and B cell antigens with hybridoma monoclonal antibodies. *J Histochem Cytochem* **28**:771–776, 1979.
5. Hsu S-M, Raine L, Fanger H: Use of avidin-biotin-peroxidase complex (ABC) in immunoperoxidase techniques: A comparison between ABC and unlabelled antibody (PAP) procedures. *J Histochem Cytochem* **29**:577–580, 1981.
6. Cerf-Bensussan N, Schneeberger EE, Bahn AK: Immunohistologic and immunoelectron microscopic characterization of the mucosal lymphocytes of human small intestine by the use of monoclonal antibodies. *J Immunol* **130**:2615–2622, 1983.
7. Kurosaka M, Ziff M: Immunoelectron microscopic study of the distribution of T cell subsets in rheumatoid synovium. *J Exp Med* **158**:1191–1210, 1983.
8. Stein H, Gatter K, Asbahn H, et al: Methods in laboratory investigation: Use of freeze-dried paraffin-embedded sections for immunohistologic staining with monoclonal antibodies. *Lab Invest* **52**:676–683, 1985.
9. McLean IW, Nakane PK: Periodate-lysine-paraformaldehyde fixative: A new fixative for immunoelectron microscopy. *J Histochem Cytochem* **22**:1077–1083, 1974.
10. Farr OG, Nakane PK: Immuno-histochemistry with enzyme labeled antibodies: A brief review. *J Immunol Meth* **47**:129–144, 1981.
11. Taylor CR, Burns J: The demonstration of plasma cells and other immunoglobulin-containing cells in formalin-fixed, paraffin-embedded tissues using peroxidase-labelled antibody. *J Clin Pathol* **27**:14–20, 1974.
12. Hsu S-M, Cossman J, Jaffe ES: A comparison of ABC unlabeled antibody and conjugated immunohistochemical methods with monoclonal and polyclonal antibodies: An examination of germinal center tonsils. *Am J Clin Pathol* **80**:429–435, 1983.
13. Zeromski J, Perlmann P, Lagercrantz R, et al: Immunological studies in ulcerative colitis. *Clin Exp Immunol* **7**:463–467, 1970.
14. Austin LL, Dobbins WO III: Intraepithelial leukocytes of the intestinal mucosa in normal man and in Whipple's disease: A light and electron microscopic study. *Dig Dis Sci* **27**:311–320, 1982.
15. Dobbins WO III, Tomasini JT, Rollins EL: Electron and light microscopic identification of the mast cell of the gastrointestinal tract. *Gastroenterology* **56**:268–279, 1969.
16. Qualman SJ, Keren DF: Immunofluorescence of deparaffinized, trypsin-treated renal tissues: Preservation of antigens as an adjunct to diagnosis of disease. *Lab Invest* **41**:483–489, 1979.
17. Wood GS, Warnke R: Suppression of endogenous avidin-binding activity in tissues and its relevance to biotin-avidin detections systems. *J Histochem Cytochem* **29**: 1196–1204, 1981.
18. Van Voorhis WC, Kaplan G, Sarno EN, et al: The cutaneous infiltrates of leprosy: Cellular characteristics and the predominant T-cell phenotypes. *N Engl J Med* **307**: 1593–1597, 1982.

19. Isobe Y, Chen ST, Nakane PK, et al: Studies on translocation of immunoglobulins across intestinal epithelium. I. Improvements in the peroxidase-labeled antibody method for application to study of human intestinal mucosa. *Acta Histochem Cytochem* **10**:161–171, 1977.
20. Straus W: Peroxidase procedures: Technical problems encountered during their application. *J Histochem Cytochem* **27**:1349–1351, 1979.
21. Colucci G, Colombo M, Del Ninno E, et al: In situ characterization by mononuclear cell infiltrate in chronic active hepatitis. *Gastroenterology* **85**:1138–1145, 1983.
22. Ponder BA, Wilkinson MM: Inhibition of endogenous tissue alkaline phosphatase with the use of alkaline phosphatase conjugates in immunohistochemistry. *J Histochem Cytochem* **29**:981–984, 1981.
23. Nemes Z, Thomazy V, Szeifert G: Demonstration of light chain monotypia in B-cell non-Hodgkins' lymphomas using unfixed freeze-dried and formalin-fixed trypsinized paraffin sections. *J Clin Pathol* **36**:883–893, 1983.
24. Rodering CB, Erlandsen SL, Coulter HE, et al: Immunohistochemical localization of IgA antigens in sections embedded in epoxy resin. *J Histochem Cytochem* **28**:199–205, 1980.
25. van Duinen SG, Mauw BJ, de Graff-Reitsma CB, et al: Methods in laboratory investigation: Immunoelectron microscopic methods for demonstration of antigen on normal human melanocytes and other epidermal cells. *Lab Invest* **50**:733–741, 1984.

5
The Abnormal Biopsy

Recognizing Artifacts

The importance of recognition of histological artifacts that can be confused with pathological abnormalities cannot be overemphasized. These artifacts may be induced by cutting or suction trauma during removal of the biopsy specimen, by the presence of Brunner glands or of lymphoid hyperplasia, or by the failure to orient biopsy specimens so that the sections are obtained parallel to crypts and villi (Figure 5-1).[1,2] How to recognize artifacts in light microscopy sections has been discussed in Chapter 2. It is discussed further at the end of this chapter in the section dealing with the normal histology of the duodenal bulb.

Early reports from Southeast Asia indicated that there was a characteristic intestinal lesion in cholera. When control biopsy specimens were obtained from presumably normal Southeast Asians, all were found to have "intestinal lesions" when their specimens were compared with biopsy specimens obtained from North Americans. Similarly, reports of a characteristic intestinal lesion in patients with tropical sprue in India do not bear up when their biopsy specimens are compared with biopsy specimens from Indians without tropical sprue. It is easily shown that large numbers of asymptomatic subjects who reside in areas where tropical sprue is endemic have, on intestinal biopsy, histological changes indistinguishable from those of "tropical sprue." Thus a mild intestinal lesion compatible with the diagnosis of tropical sprue in North Americans must be considered within the range of normal in India.

For appropriate analysis of electron micrographs obtained in disease states of the intestine, a brief introduction concerning the problems of light microscopy interpretation of biopsy specimens in intestinal disease is necessary. The majority of intestinal mucosal biopsy specimens showing minor or nonspecific changes are of no positive diagnostic importance. (This statement is equally true for electron microscopy.) Light microscopy sections of biopsy specimens are best classified histologically as being normal, nonspecifically abnormal, or abnormal. The abnormal category means that a characteristic and sometimes diagnostic lesion is present, such as that of celiac sprue or of Whipple's disease. The lesion seen in celiac sprue can serve as a standard of comparison for intestinal mucosal lesions

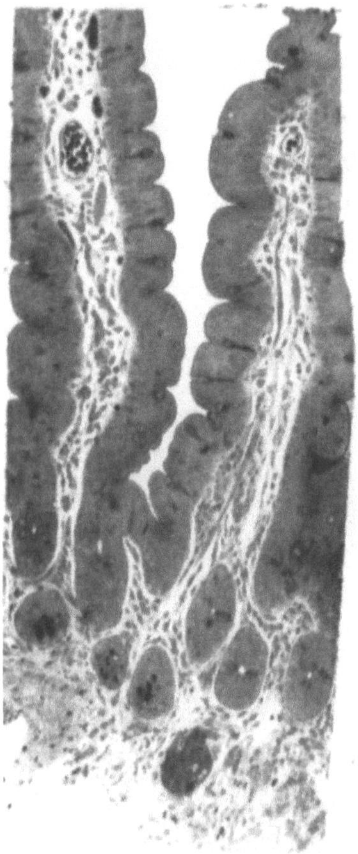

Figure 5-1. Light microscopy photograph of thick section of epoxy-embedded biopsy specimen showing that portion of the specimen that has been trimmed for thin sectioning. Note that the full length of two villi and the associated crypts are included in the section. Toluidine blue, × 125. (From Dobbins,[138] reprinted by permission of John Wiley & Sons, Inc. Copyright © 1978, John Wiley & Sons, Inc.)

observed in other disease states. Biopsy specimens with a lesion of celiac sprue are generally graded as showing mild, moderate, or severe alterations in villous structure. The severe lesion shows virtual absence of villi, elongated crypts, increased cell content of the lamina propria, and surface epithelial abnormalities. Lesser degrees of abnormality are characterized by a more normal surface epithelium and by better formed villi.

With appreciation of the necessity of appropriate orientation, of obtaining suitable controls from the population being studied, and of the nonspecificity of many of the changes reported at the light microscopy level, consideration can now be given to the artifacts found at electron microscopy and which should not be confused with abnormal states.

Figure 5-2 illustrates the usual appearance of lysosomes present at the apex of normal intestinal absorptive cells. These acid hydrolase-containing organelles consist of heterogeneous dense bodies, autophagic vacuoles, and numerous multivesicular bodies. Cells with a similar appearance have been reported as showing an abnormal increase in number of epithelial cell lysosomes, an interpretation that may be inappropriate.[3]

Figure 5-3 illustrates possibly the maximal range of normal lysosomal content of absorptive cells. One absorptive cell is intact, whereas the adjacent one shows dilatation of the endoplasmic reticulum, dilated mitochondria with disrupted cristae, and numerous dense bodies within the mitochondria. This

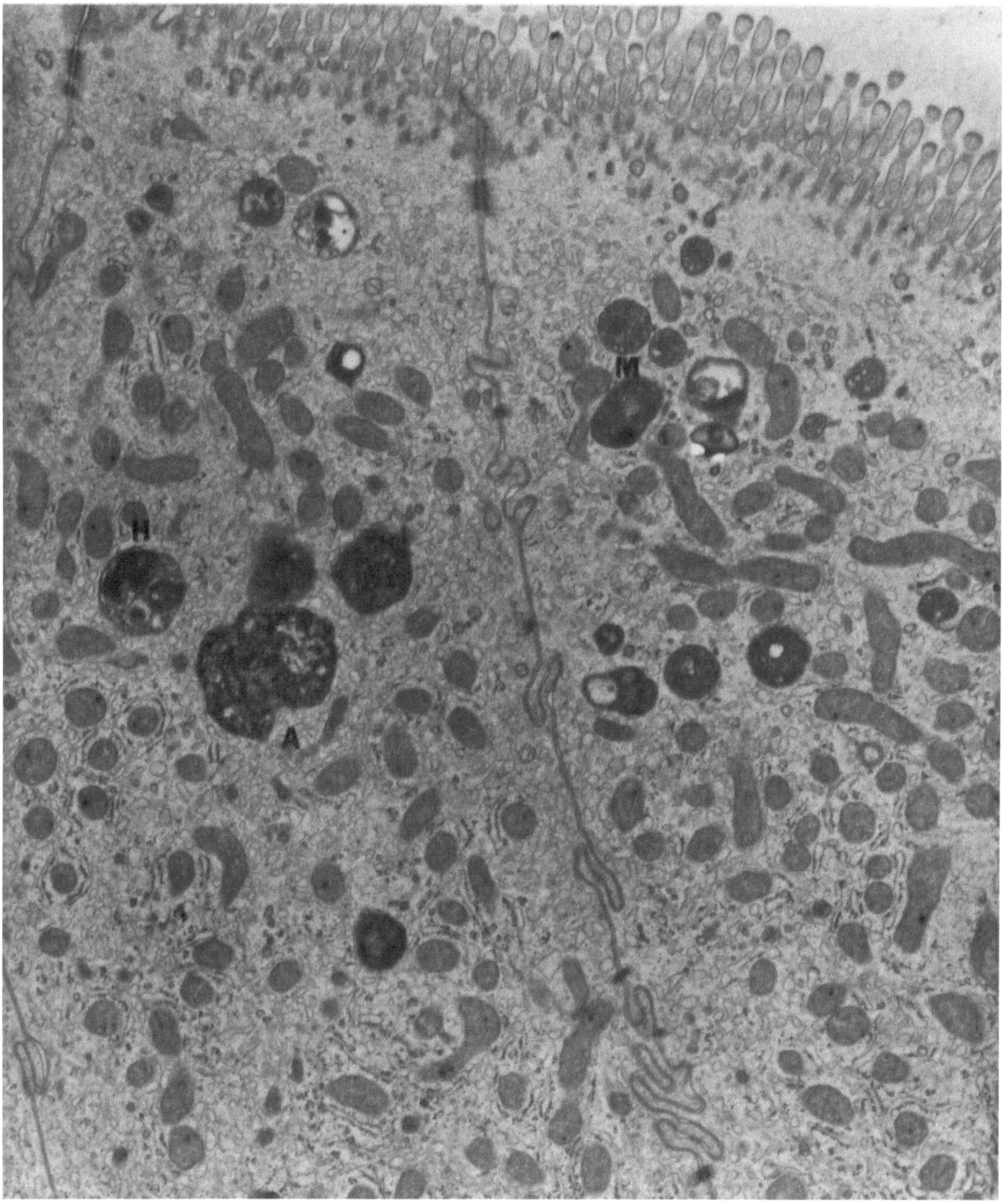

Figure 5-2. Electron micrograph of epithelial cells in normal jejunal biopsy specimen of a patient with ulcerative colitis. It illustrates the range in appearance of lysosomes often seen in normal intestinal epithelial cells. The lysosomes consist of heterogeneous dense bodies (H), autophagic vacuoles (A), and multivesicular bodies (M). × 16,000. (From Dobbins,[138] reprinted by permission of John Wiley & Sons, Inc. Copyright © 1978, John Wiley & Sons, Inc.)

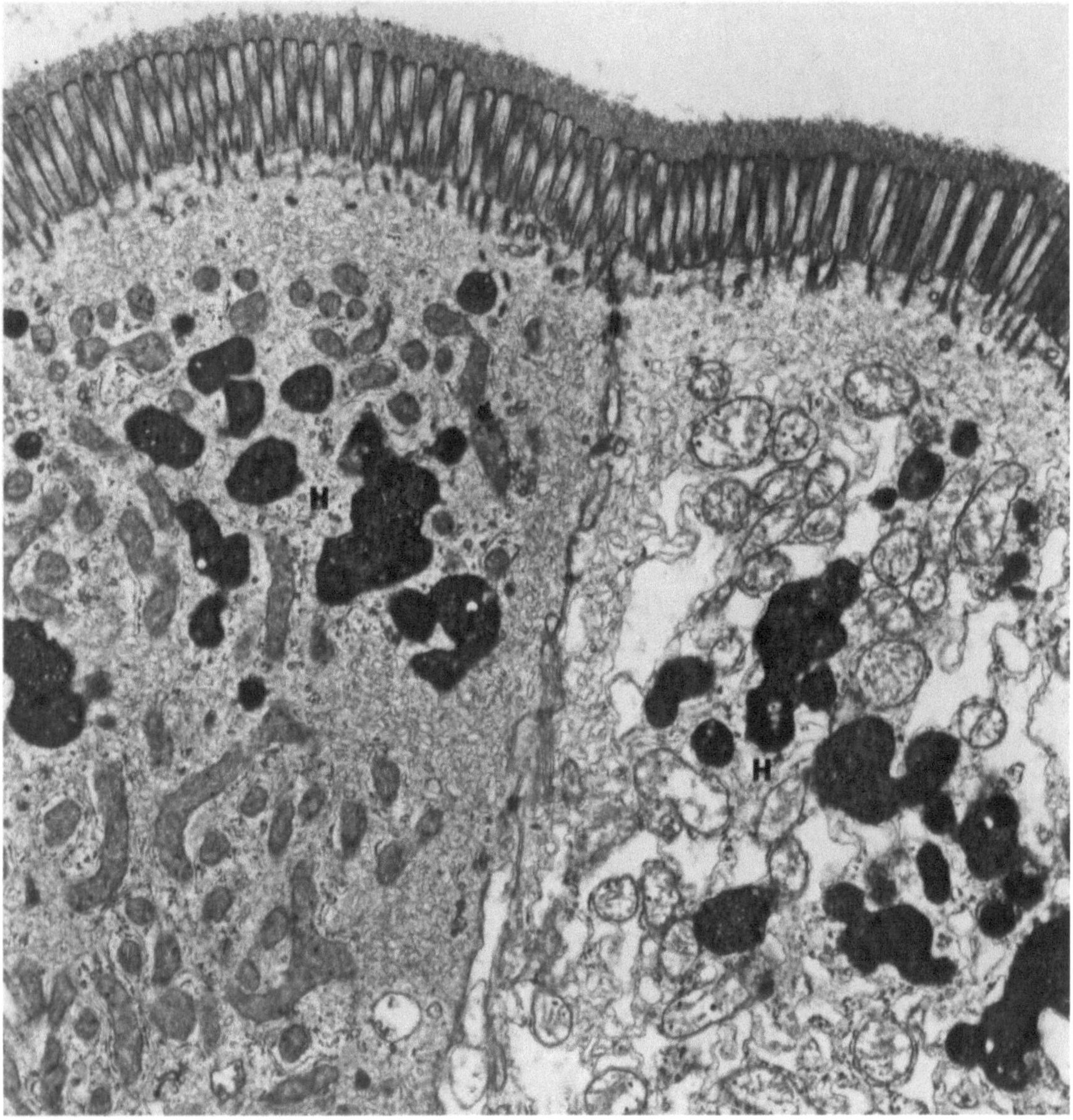

Figure 5-3. Electron micrograph of intestinal absorptive cells from biopsy specimen of a normal individual. Note the numerous heterogeneous dense bodies (H) in both the normal cell on the left and the senescent cell on the right. Note normal microvilli and surface coat over both cells. × 15,400. (From Dobbins,[138] reprinted by permission of John Wiley & Sons, Inc. Copyright © 1978, John Wiley & Sons, Inc.)

represents a senescent absorptive cell, a frequent finding at the apex of normal intestinal villi. The normal cell turnover time for intestinal epithelial cells in humans ranges from three to five days. Intestinal absorptive cells and goblet cells arise from mitoses of undifferentiated crypt cells, migrate up the side of intestinal villi during the period of three to five days, and are lost at the apex of the villus, the cell extrusion zone. Thus, senescent cells are often present at the apex of villi.[4] Similar-appearing cells at the mid villus or villous crypt would represent a pathological change; such changes have frequently been described

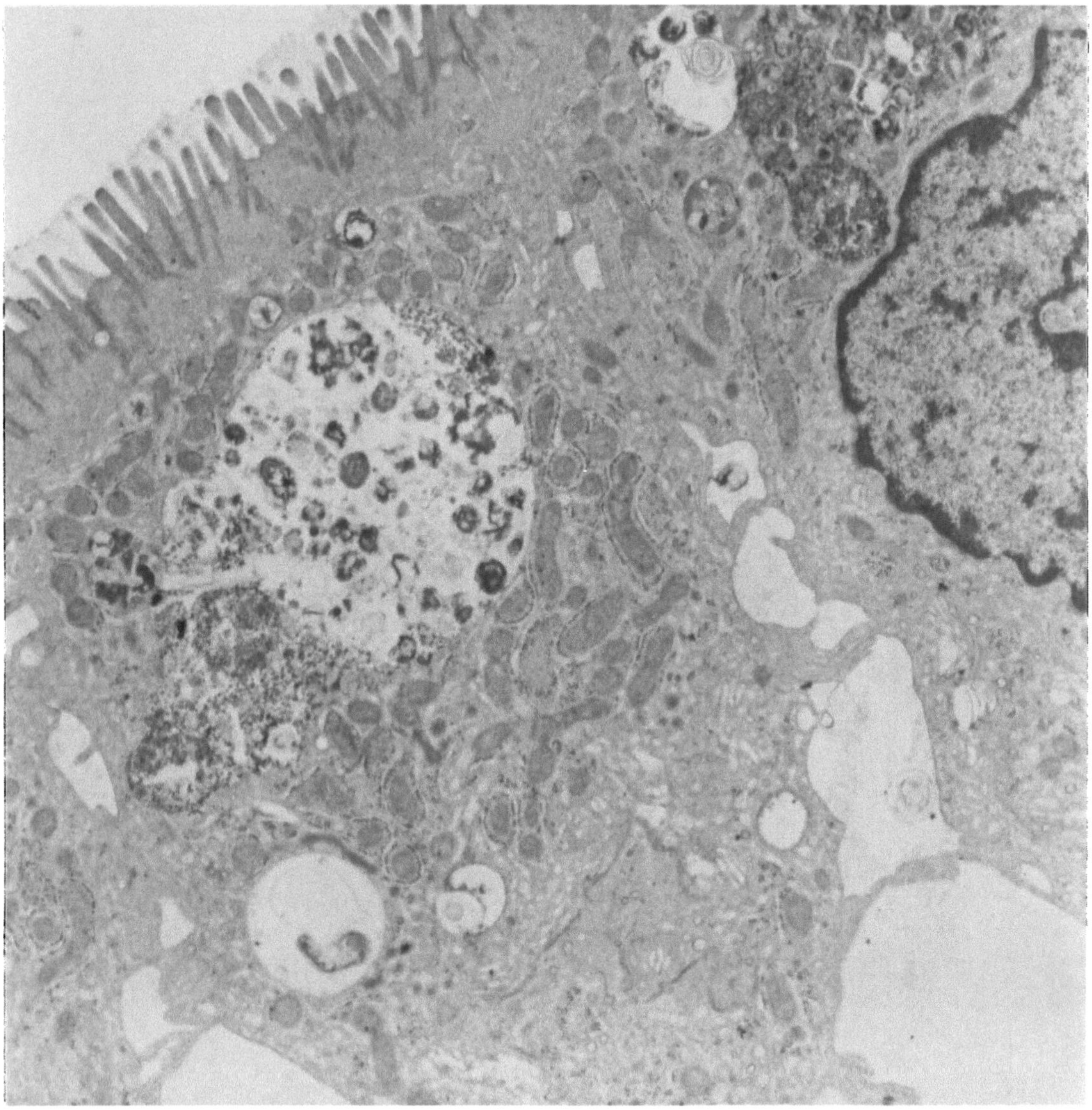

Figure 5-4. Electron micrograph showing large lysosomal inclusions in intestinal epithelial cells of a patient with Zollinger-Ellison syndrome. × 15,400. (From Dobbins,[138] reprinted by permission of John Wiley & Sons, Inc. Copyright © 1978, John Wiley & Sons, Inc.)

following radiation injury to the intestine or following use of antimetabolic agents for treatment of cancer.[5–7]

Figure 5-4 shows an abnormally increased lysosomal content within intestinal absorptive cells from a patient with Zollinger-Ellison syndrome. This change is quite nonspecific. Presumably the increased number of lysosomes reflects a response of absorptive cells to injury induced by the incredibly high gastric acid output that occurs in this syndrome.[8]

Figure 5-5, a low-magnification electron micrograph, demonstrates a variety of "degenerative" changes seen in intestinal absorptive cells at the apex of a villus in a patient with liver cirrhosis and chronic diarrhea. Such advanced

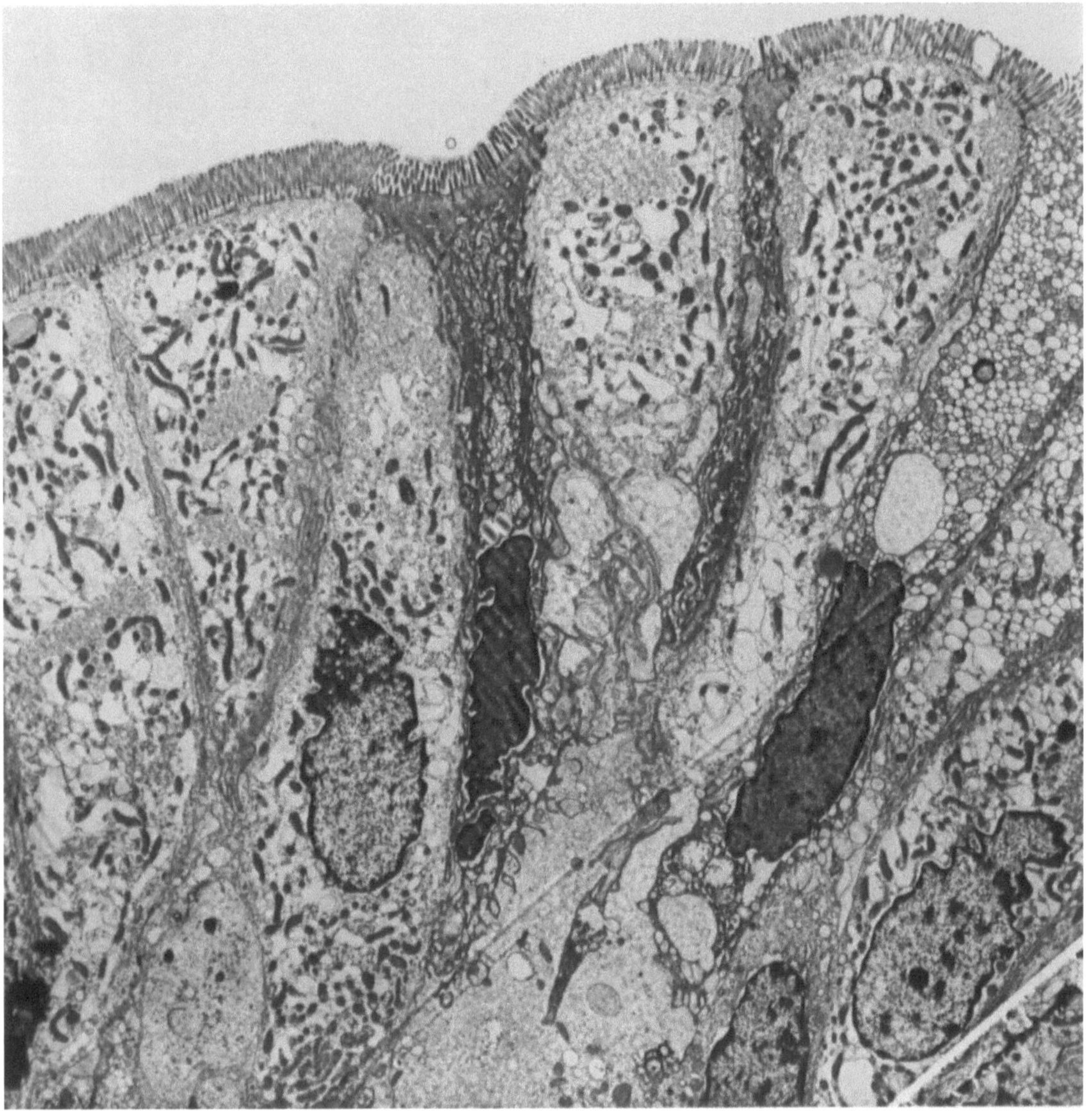

Figure 5-5. Electron micrograph showing epithelial cells in the cell extrusion zone of the villous tip of an intestinal biopsy specimen of a patient with cirrhosis of the liver. See text for details. × 3,400

changes may well represent an abnormality but should be considered the extreme range of normal for cells present at the apex of villi. Portela-Gomes, Martins, and Correia[9] have suggested that these changes (dilatation of endoplasmic reticulum, altered appearance of mitochondria, increased and decreased density of cytoplasmic matrix, rounded and irregular distribution of microvilli) are characteristic of intestinal absorptive cells of cirrhotics. Similar changes have been described following acute ethanol ingestion and are nonspecific.[10]

Figure 5-6 shows the striking dilatation of the intercellular spaces of intestinal epithelial cells normally found at the apex of villi, especially during active water and electrolyte absorption.[11] A comparable degree of dilatation of the intercellular spaces in the midvillus and villous crypt is not ordinarily seen.

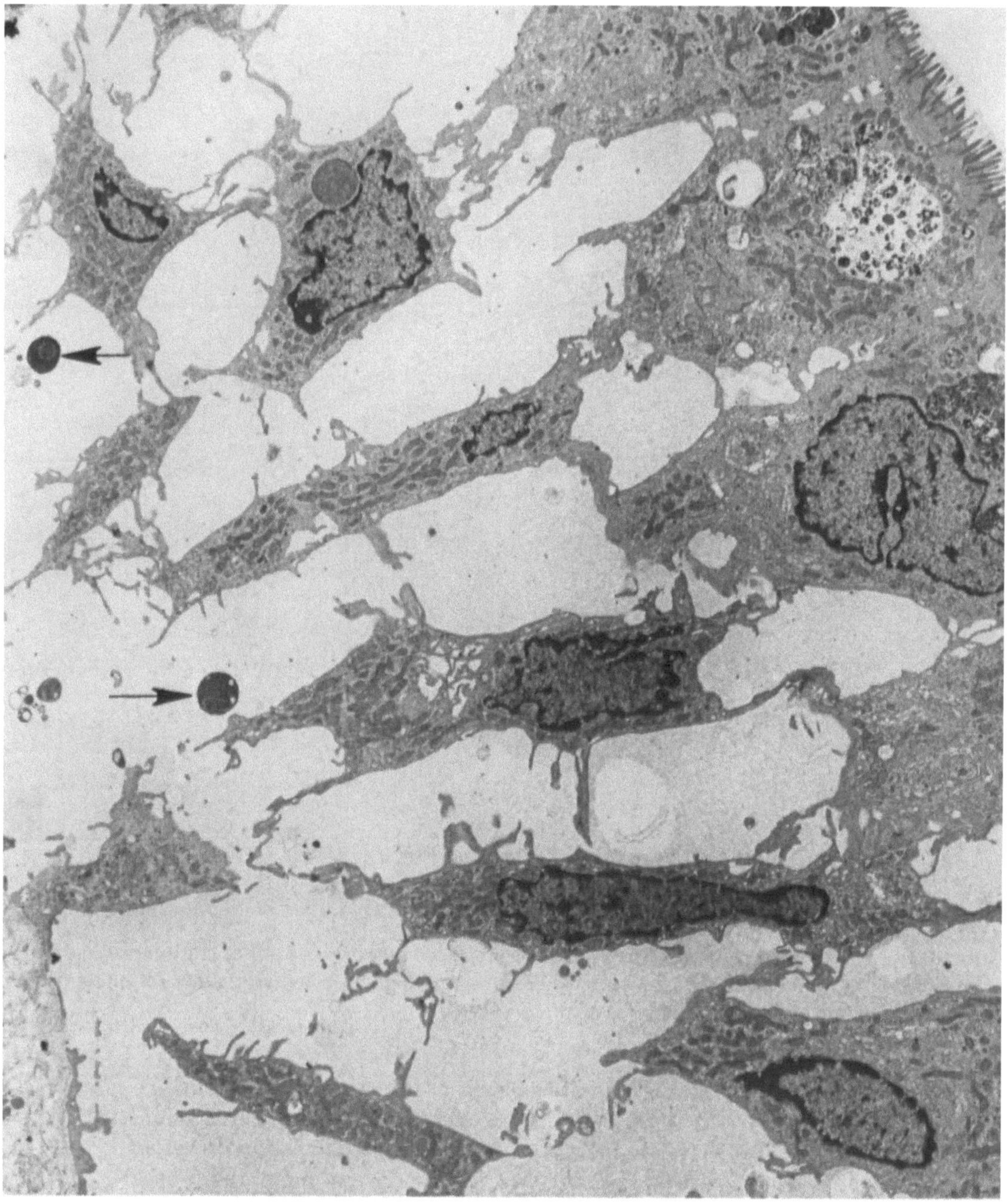

Figure 5-6. Electron micrograph showing dilated intercellular spaces between epithelial cells (of a patient with Zollinger-Ellison syndrome) usually present at the upper portion of villi, particularly in the absorptive state. The small dense bodies (arrows) are seen occasionally in intercellular spaces and are presumably "cellular debris." × 4,800

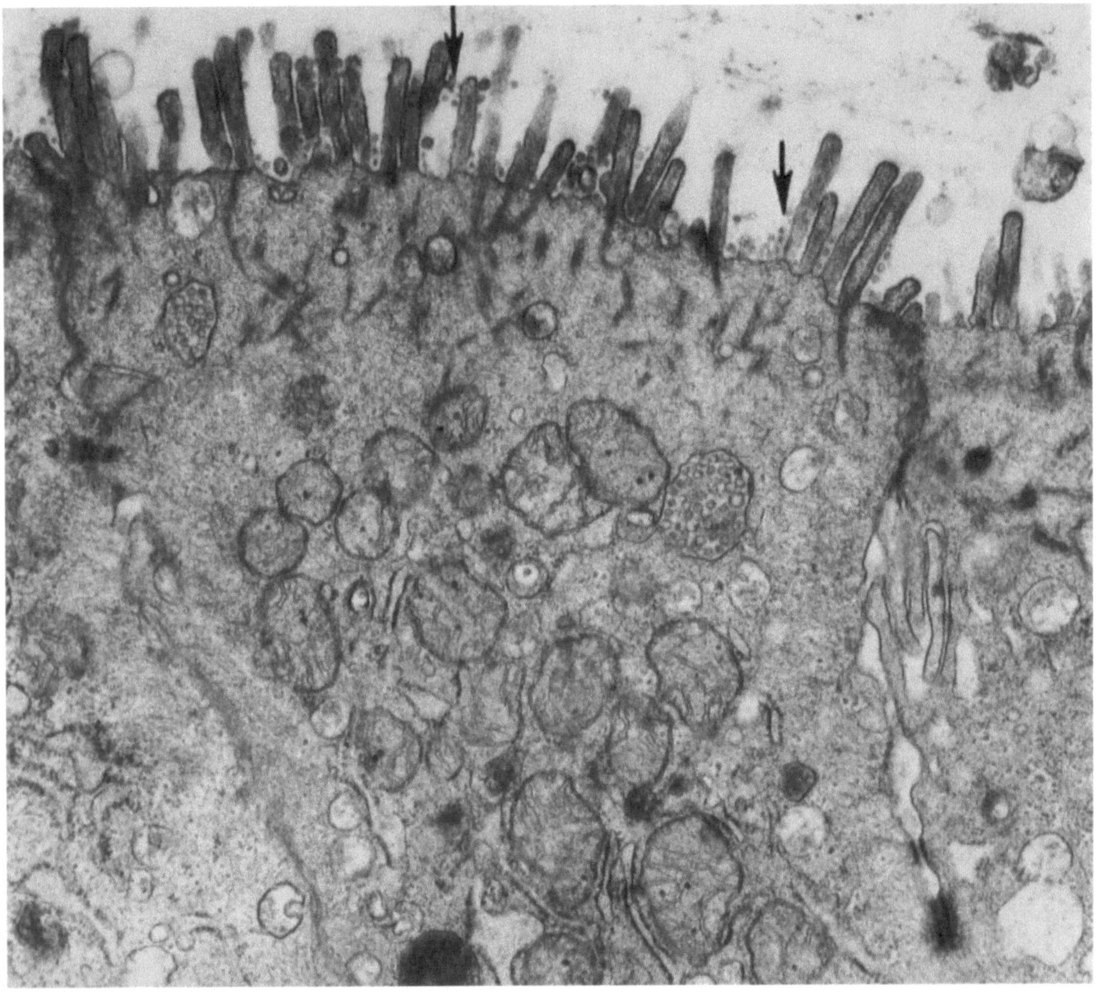

Figure 5-7. Electron micrograph showing numerous rounded membrane profiles or glycocalyceal bodies (arrows) present around microvilli of intestinal absorptive cells. These bodies are more often seen about microvilli of colonic epithelial cells. × 33,600

Figure 5-7, an electron micrograph of the apex of intestinal epithelial cells from a patient with treated Whipple's disease, shows numerous round membrane profiles or vesicles between and above microvilli of the absorptive cells. Such profiles have been confused with viruses but are normal membrane remnants frequently present over microvilli of colonic and intestinal absorptive cells. These spherical bodies have also been called glycocalyceal bodies and have been considered to be "real structures," not just shedding of cell membrane,[12] an interpretation I consider inaccurate. These membrane remnants were once thought to be specific for colonic epithelial cells (and have served as an "ultrastructural marker" for colonic adenocarcinoma) but are also found in respiratory and intestinal epithelium. These profiles presumably represent remnants of degenerating microvilli and are more prominent in the colon than in the small intestine. Figure 5-8 shows numerous cytoplasmic protrusions and apparent loss of microvilli, often seen at the apex of undifferentiated crypt cells and

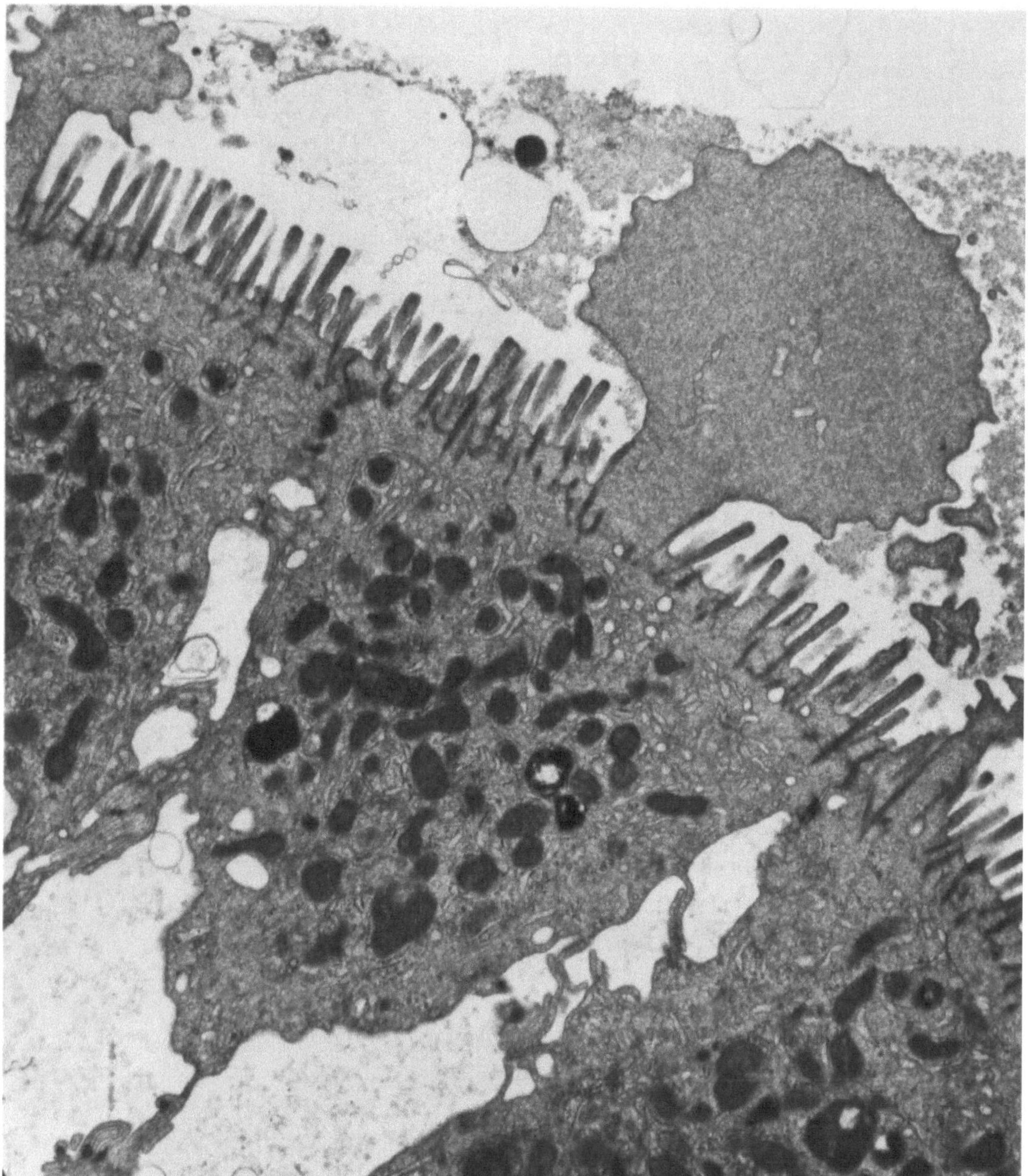

Figure 5-8. Electron micrograph of epithelial cells of intestinal biopsy specimen of a patient with chronic diarrhea and hypoalbuminemia. Note remarkable cytoplasmic protrusions from the apex of the epithelial cells. × 14,500

sometimes of absorptive cells. This may be a fixation artifact due to poor penetration of the fixative into crypts.[13] Figure 5-9 illustrates the interlocking of basal processes of normal intestinal epithelial cells, a finding considered, without specificity, to be a manifestation of adenocarcinoma of the colon.[14]

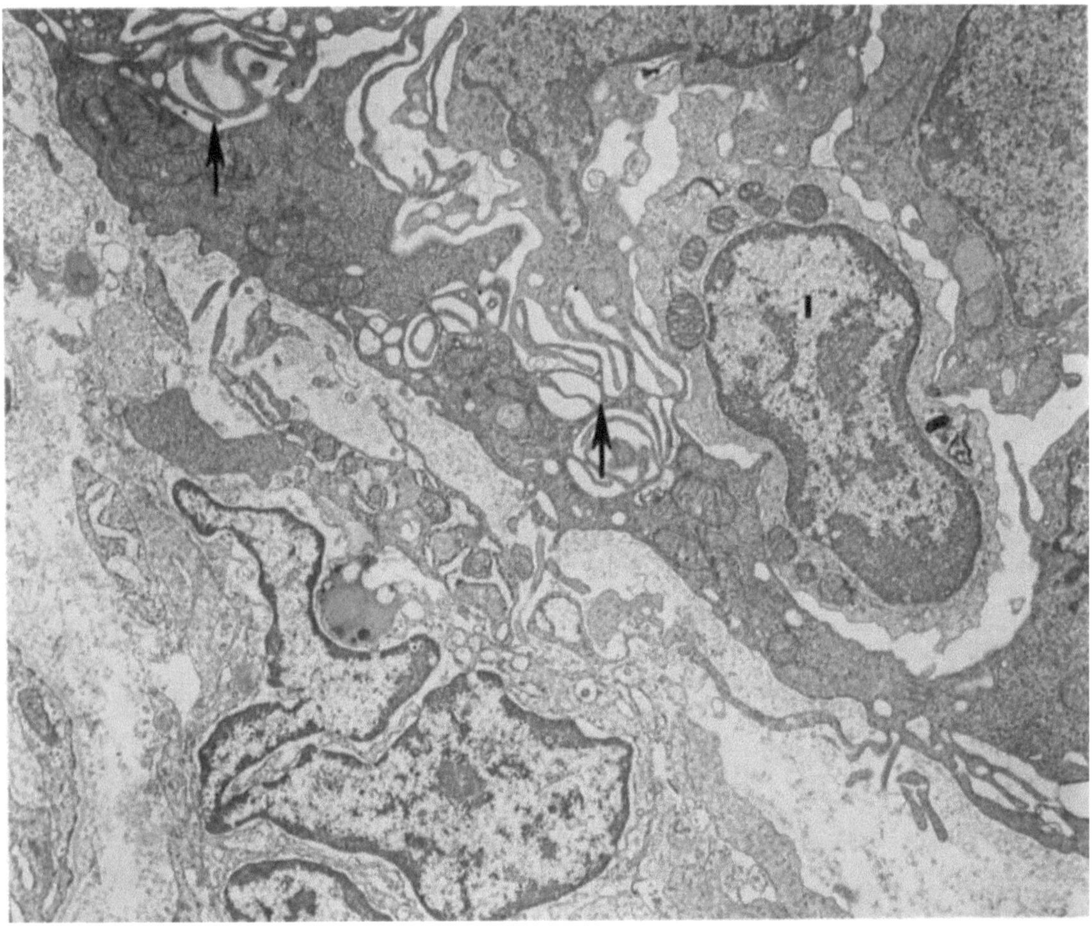

Figure 5-9. Electron micrograph showing extensive interlocking and interdigitation of cytoplasmic processes (arrows) of adjacent epithelial cells. Intraepithelial lymphocyte (I). × 12,500

Interpreting the Abnormal Biopsy

Overview

Villous architecture is altered in most mucosal disorders of the intestine. Thus, the severity of a mucosal abnormality should be graded first according to changes in villous architecture. Abnormal specimens should be graded as showing a mild, moderate, or severe abnormality. There is excellent agreement among experienced observers regarding the interpretation of moderate and severe abnormalities when biopsy specimens are read in a coded fashion. It may be more difficult to distinguish consistently a mild abnormality from a normal specimen.[15]

1. The mild villous abnormality. In this case many villi are branched, broadened, or fused above the crypt region. Other villi may appear normal. The surface epithelium is often abnormal, with a loss of nuclear polarity and an increase in intraepithelial lymphocytes (Figure 5-10). Sometimes the epi-

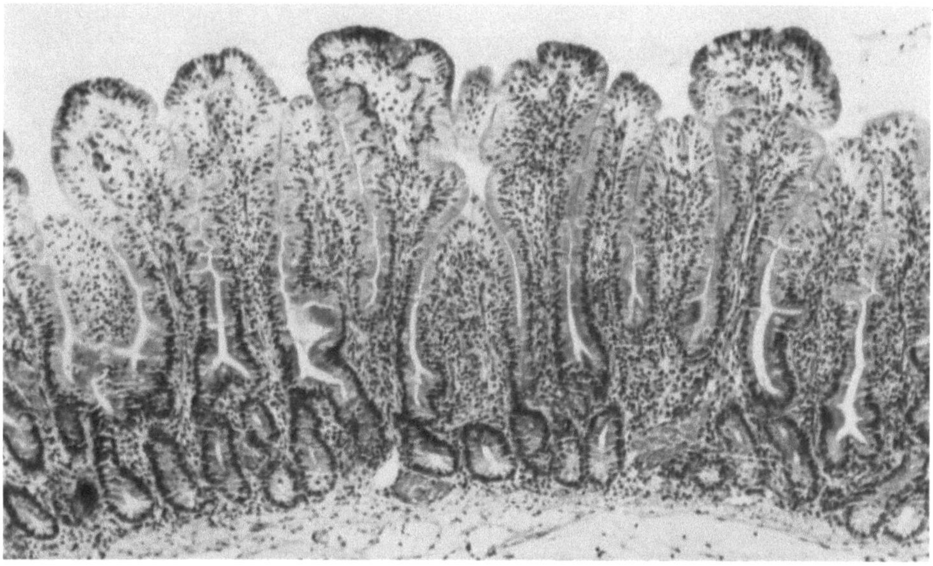

Figure 5-10. Mild villous abnormality of proximal jejunal mucosa showing broadened villi, loss of nuclear polarity of the surface epithelium, increased numbers of intraepithelial lymphocytes, and a mild increase in the cellular content of the lamina propria. Some of the villi are branched. H & E stain, × 100. (Reproduced from Dobbins,[16] by permission of Churchill Livingstone, Inc., New York.)

thelial changes will be more striking than the changes in villous structure. Mitotic figures, which are normally present only in the crypts, may be more prominent and may be found just above the crypt level. The number of round cells and of acute inflammatory cells may be increased in the lamina propria. However, a slight increase in number of cells within the lamina propria in the absence of villous changes or changes in the surface epithelium is more likely a normal variant.

2. The moderate lesion. This lesion is characterized by broadened and shortened villi (Figure 5-11). The surface epithelium over the tips of villi may be cuboidal and the round cell content of the lamina propria will be unequivocally increased. Intraepithelial lymphocytes appear to be greatly increased in number.
3. The severe lesion. This lesion is easily recognized because villi are almost completely absent (Figure 5-12). The cellular content of the lamina propria is obviously markedly increased and intraepithelial lymphocytes are generally quite numerous.

The foregoing classification is based on assessment of villous architecture alone.[1,16] Other changes may be found in the epithelium regardless of the degree of villous change. For example, the surface epithelium may be severely vacuolated, the vacuoles generally representing cell injury or collections of fat. Frank macrocytosis is sometimes a feature of the surface epithelium. The presence of more than occasional polymorphonuclear (PMN) leukocytes within the lamina propria is generally an abnormality if hemorrhagic artifact has been excluded. Clearly excessive numbers of PMN leukocytes may justify the pathological

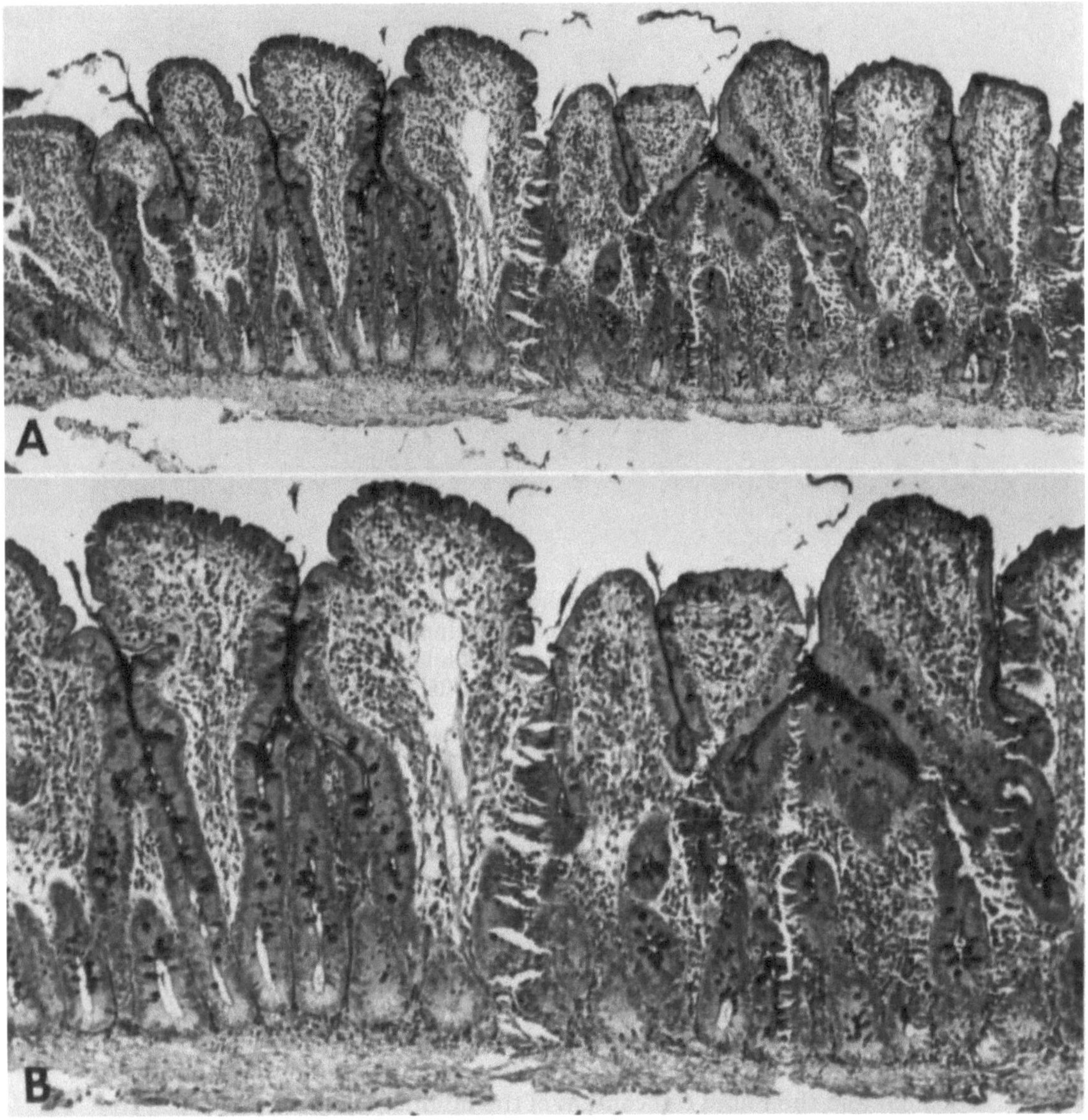

Figure 5-11. Moderate villous abnormality of proximal jejunal mucosa in a North American patient with tropical sprue acquired in the Philippines. (A) The villi are broadened and there is a marked increase of cellular content within the lamina propria. The epithelium is a bit flattened, with an increased number of intraepithelial lymphocytes. The crypts are slightly lengthened. PAS and hematoxylin stain, × 80. (B) Higher magnification of portion of section illustrated in (A) showing more clearly the epithelial changes and increased numbers of intraepithelial lymphocytes. PAS and hematoxylin stain, × 132. (A and B reproduced from Dobbins,[16] by permission of Churchill Livingstone, Inc., New York.)

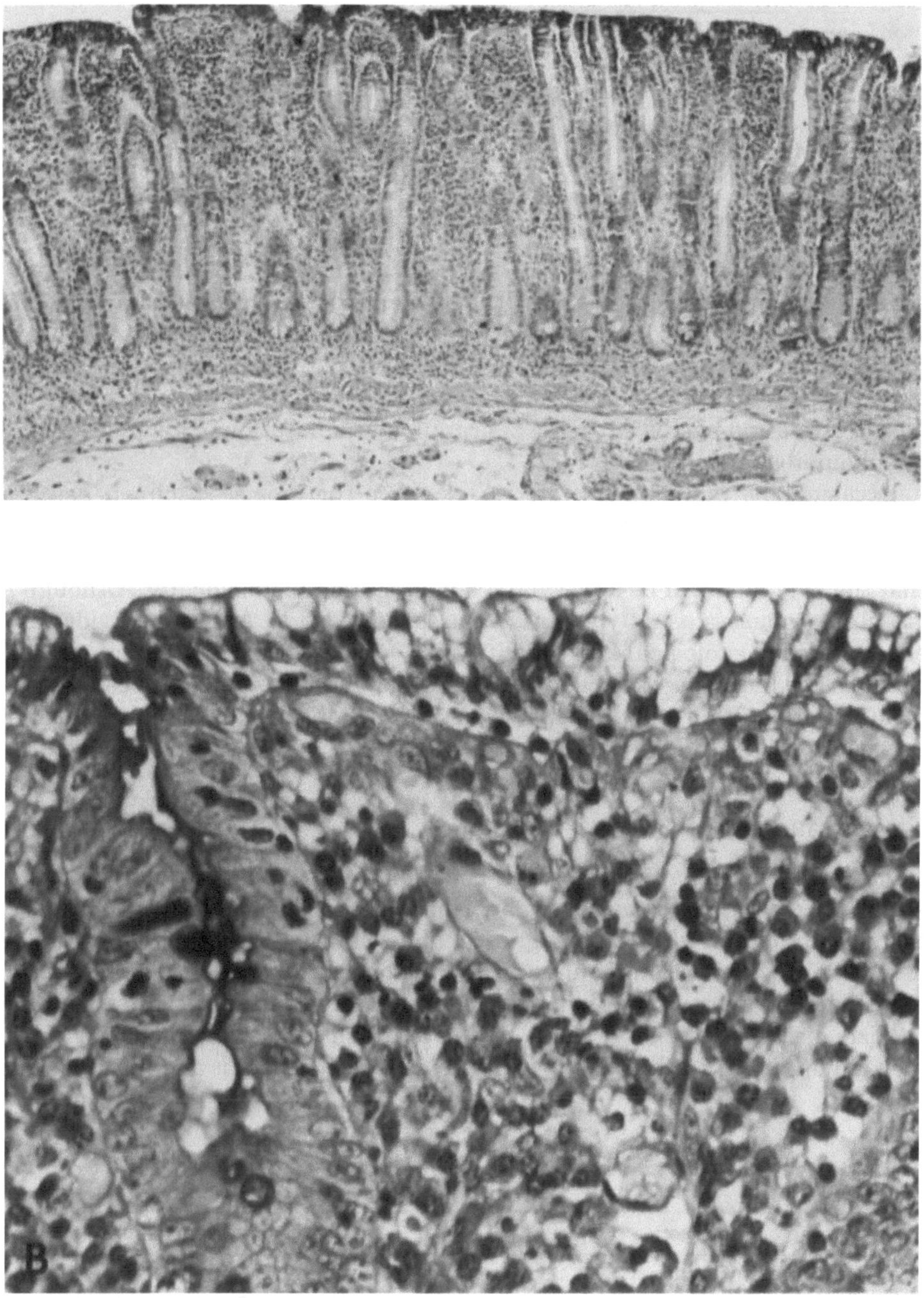

Figure 5-12. Characteristic lesions seen in untreated celiac sprue. (A) Severe villous abnormality. Note absence of villi, flattened surface epithelium, lengthened crypts, and increased cellular content of the lamina propria. H & E stain, × 132. (B) Severe epithelial cell injury. Note marked vacuolization of the surface epithelial cells and the marked plasma cell infiltrate in the lamina propria. The crypt epithelial cells are more normal in appearance, though cuboidal in shape. H & E stain, × 500. (A and B reproduced from Dobbins,[16] by permission of Churchill Livingstone, Inc., New York.)

Table 5-1. Rubin Classification of Small-Bowel Specimens

Villi	Conclusion
Flat	Nonspecific
	Diagnostic
Variably abnormal	Nonspecific
	Diagnostic
Normal	Diagnostic

From ref. 16, by permission. Data from Perera et al.[1]

diagnosis of acute inflammation. Other inflammatory changes that are occasionally observed include crypt abscesses, microulcerations, and the presence of granulomas. These changes are more likely to be detected if many sections are examined. Changes that may be observed in the submucosa include changes in blood vessels and nerves, presence of granulomas, dilatation of lymphatics, and lymphomatous infiltration.

Occasionally, individuals prefer to examine the biopsy specimen using a hand lens or a dissecting microscope before the specimen is processed. This may be worthwhile, because observation of a grossly flat mucosa leaves little doubt that the specimen will be abnormal when processed histologically. Although a grossly flat mucosa usually means celiac sprue in North America, a wide variety of specific histological abnormalities can only be revealed after careful sectioning. The hand lens (the poor person's dissecting microscope) is best used as an aid in orientation of the specimen, so that sections may be obtained parallel to crypts and villi.

Classification

Rubin[1] classified the abnormal small-bowel biopsy as (1) flat biopsy, either nonspecific or diagnostic; (2) variably abnormal villi, either nonspecific or diagnostic; and (3) normal villi, but diagnostic abnormality present (Table 5-1). The definitive diagnosis often requires a finding not only of an abnormal biopsy but of a second diagnostic criterion, usually one relating to specific therapy. Table 5-2 lists diseases associated with a flat (severe lesion) small-bowel biopsy specimen but one in which the histological changes are nonspecific and the diagnosis

Table 5-2. Conditions Associated with Flat Nonspecific (Nondiagnostic) Small-Bowel Biopsy Specimens

Disease or condition	Treatment
Symptomatic celiac sprue	Gluten-free diet
Other protein injury	Diet free of soy, chicken, milk
Tropical sprue (some forms)	Antibiotic plus folic acid
Childhood kwashiorkor	Diet with adequate protein
Refractory sprue	None known
Familial enteropathy	None known
Drug-induced injury (rare)	Withdrawal of drug
IgA deficiency (rare)	None known
Infectious gastroenteritis (rare)	None – illness is self-limited
Stasis syndrome (some forms)	Antibiotics: surgery

From ref. 16, by permission.

Table 5-3. Conditions Associated with Flat (Diagnostic) Small-Bowel Biopsy Specimens

Disease	Diagnostic histology
Collagenous sprue	Broad band of collagen below surface epithelium
Late-onset immunodeficiency (common variable hypogammaglobulinemia)	Virtually absent lamina propria plasma cells; nodular lymphoid hyperplasia
Whipple's disease (some forms)	Typical macrophages in lamina propria
Eosinophilic gastroenteritis (some forms)	Eosinophilic infiltrate
Primary intestinal lymphoma (some forms) (immunoproliferative intestinal disease)	Malignant lymphoid cells in lamina propria

From ref. 16, by permission.

hinges on a therapeutic response. Table 5-3 lists diseases associated with a flat specimen in which the histological changes are diagnostic. Table 5-4 lists diseases associated with specimens with variably abnormal villi (mild to moderate lesions) but in which the histological changes are nonspecific, again a treatment response being required to establish a diagnosis. Table 5-5 lists causes of variably abnormal villi when the histological changes are diagnostic. Table 5-6 lists diseases associated with normal villi but in which the diagnosis can be made histologically. Finally, Table 5-7 lists a variety of conditions in which "abnormal" histological changes have been reported in "normal" small-bowel biopsies. In this case the biopsies were probably normal because such "abnormalities" have not been confirmed in laboratories that process and interpret specimens using the careful methods of processing recommended by Rubin.[1,16]

In the discussion that follows, lesions are described as either diffuse or patchy in distribution. In the first case, every biopsy specimen taken at the duodenojejunal junction is abnormal. In the second, some specimens will show histological changes and others will not. A lesion described as severe always lacks villi. On the other hand, a lesion described as variable exhibits a spectrum of changes ranging from normal to severe either in the same patient or in different patients with the same disease.

Table 5-4. Conditions Associated with Variable Abnormal (Nonspecific) Villi in Small-Bowel Biopsy Specimens

Disease or condition	Treatment
Subclinical celiac sprue (25% dermatitis herpetiformis)	Gluten-free diet
Infectious gastroenteritis	None; spontaneous cure
Stasis syndromes	Antibiotics
Tropical sprue (some forms)	Antibiotics plus folic acid
Geographical variation	Environmental improvement
Zollinger-Ellison syndrome (gastrinoma)	Cimetidine
Graft-vs.-host reaction	Steroids; antithymocyte globulin
Chronic idiopathic pseudoobstruction and scleroderma	None known
Sarcoidosis	None known
Drug-induced injury	Withdrawal of drug

From ref. 16, by permission.

Table 5-5. Conditions Associated with Variably Abnormal (Diagnostic) Villi in Small-Bowel Biopsy Specimens

Disease or condition	Diagnostic histology
Whipple's disease	Typical macrophages filling lamina propria
Mycobacterium avium	Macrophages containing acid-fast bacilli
Eosinophilic gastroenteritis	Clumps of eosinophils
Primary intestinal lymphoma	Malignant lymphoid cells in lamina propria
Parasitic disease	*Strongyloides*, *Giardia*, *Coccidia*, *Microsporidia*, *Capillaria*, schistosomiasis, leishmaniasis
Fungal disease	*Histoplasma*, *Candida*, *Cryptococcus*
Viral disease	Cytomegalovirus, herpes, rotavirus, Norwalk agent
Macroglobulinemia	Hyaline masses in lamina propria
Lymphangiectasia (primary or secondary)	Dilated lymphatics in lamina propria
Late-onset immunodeficiency (common variable hypogammaglobulinemia)	Virtually absent plasma cells; nodular lymphoid hyperplasia
Severe B_{12} or folate deficiency	Epithelial macrocytosis
Irradiation, chemotherapy	Deficiency of mitoses

From ref. 16, by permission.

Table 5-6. Conditions Associated with Normal Villi but Diagnostic Histological Change

Disease or condition	Diagnostic histology
Abetalipoproteinemia	Fasting absorptive cells loaded with fat
Crohn's disease	Noncaseating granulomas
X-linked immunodeficiency	Virtually absent lamina propria plasma cells
Lipid storage diseases	Vacuolated ganglion cells, capillaries, and macrophages
Amyloidosis	Congo-red-positive material in capillaries/lamina propria
Chronic granulomatous disease	Pigmented, vacuolated macrophages in lamina propria
Melanosis intestini	Densely pigmented macrophages in lamina propria
Lipofuscinosis	Pigment containing macrophages
Mastocytosis	Mast cell infiltrate

From ref. 16, by permission.

Table 5-7. Conditions in which "Abnormal" Histological Changes Have Been Reported in Normal Duodenojejunal Biopsies

Dermatoses other than dermatitis herpetiformis
Pancreatitis
Alcoholism
Cirrhosis
Hepatitis
Iron deficiency anemia
Ulcerative colitis
Postgastrectomy without bacterial overgrowth
Malignancy not involving gut
Cholera*
Hookworm disease*
Primary disaccharidase deficiency

*Normal by local geographical standards, that is, same as in uninfected controls in the same environment.
From ref. 16, by permission.

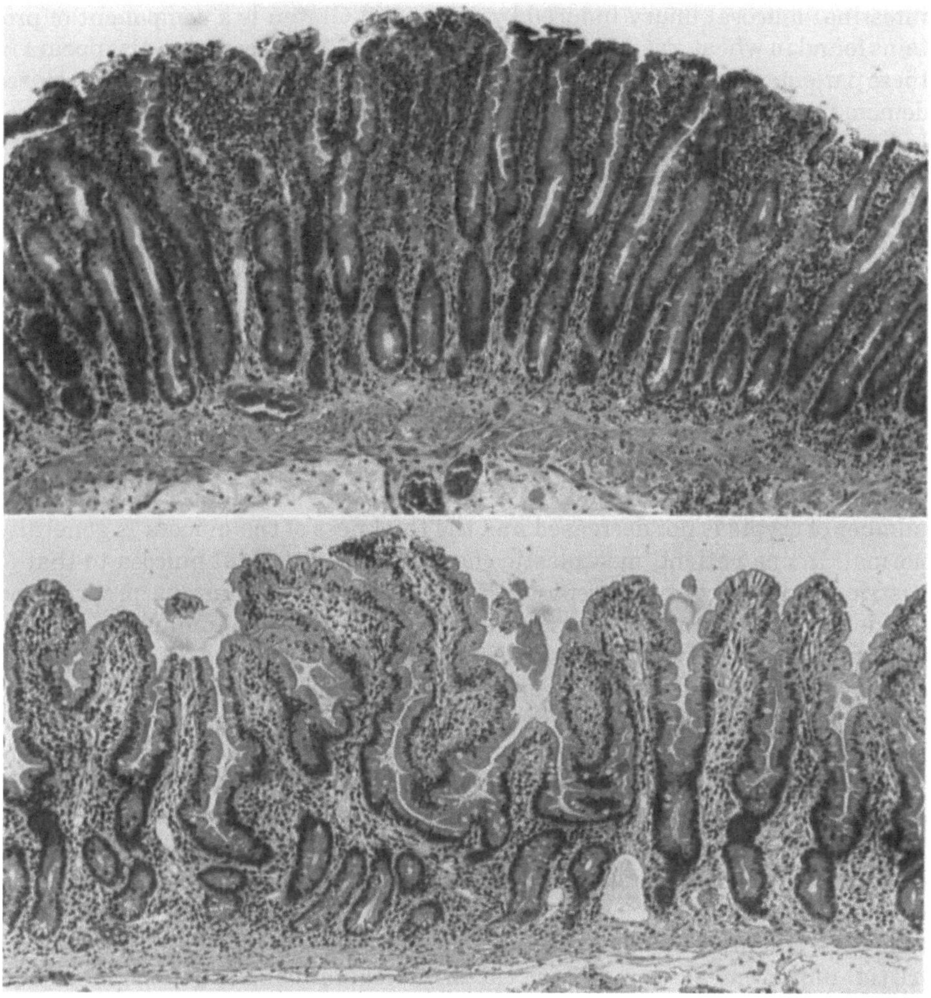

Figure 5-13. Reversible abnormality in celiac sprue. (a) Severe abnormality in proximal jejunum of a patient with untreated celiac sprue. (b) Moderate abnormality indicating only partial improvement in the mucosa of the same patient after five months of a gluten-free diet. H & E stain, × 100.

Diseases Associated with a Flat Small-Bowel Biopsy Specimen

Celiac Sprue: Gluten-Induced Injury

Patients with untreated celiac sprue, who present with overt malabsorption or with various deficiencies secondary to malabsorption, show a severe abnormality of the villous architecture in the proximal jejunal mucosa[15] (Figure 5-13). Almost all North Americans with this intestinal lesion turn out to have celiac sprue. The intestinal lesion is characteristic but not diagnostic, because it is seen in several other diseases that are rare in North America. Celiac sprue appears to be the result of a genetically determined defect in intestinal absorption, the pathogenesis being related to intestinal mucosal injury induced by gluten.[17,18] Gluten is a component of proteins found in wheat, rye, barley, and possibly oats. Malabsorption disappears in these patients when gluten is removed from the diet. Thus the diagnosis requires demonstration of the presence of the characteristic severe intestinal mucosal lesion as well as documentation of a clinical response to a gluten-free diet.

The severe mucosal lesion of the untreated patient generally shows absence of villi. The surface epithelium may be cuboidal, vacuolated, and basophilic, and contains many intraepithelial lymphocytes. Occasional PMN leukocytes may be seen in the epithelium. In some patients the surface epithelium may be elongated, as evidenced by the presence of the flat lesion. Crypts are elongated, as evidenced by the presence of scattered mitoses extending from the base of the crypts almost to the surface of the specimen. Crypt epithelial cells tend to be normal in appearance.

The cell content of the lamina propria is obviously increased. Plasma cells predominate; lymphocytes, eosinophils, and PMN leukocytes are increased in number but to a lesser extent. Crypt abscesses may be seen occasionally. The number of crypts is not decreased and the thickness of the mucosa is generally normal. In one patient, metaplastic change of the duodenal mucosa to that of gastric fundal glands was reported.[19] Infiltration of the lamina propria and epithelium with lymphocytes suggests that the lesion is immune mediated, presumably by a delayed hypersensitivity reaction.[20]

The mucosal lesion is more severe in the proximal intestine than in the distal intestine. This may relate to the fact that the proximal intestine is exposed to higher concentrations of gluten than is the distal intestine. In fact, in patients with little or no steatorrhea, only the most proximal jejunum is abnormal. However, in patients with severe steatorrhea the whole small bowel, including the ileum, tends to be severely involved.[21] In response to a gluten-free diet, the ileal lesion repairs itself much more rapidly than does the jejunal lesion. Several years or more may be required for the lesion at the duodenojejunal junction to return to normal (Figure 5-14). Usually the proximal jejunum shows evidence of only partial improvement, even though the distal jejunum and the ileum may show return to complete normality. Further, many patients fail to adhere to an absolutely gluten-free diet, and this may account for the continued evidence of injury in the proximal jejunum. The pathologist must remind the clinician that a persistent proximal mucosal abnormality does not necessarily reflect the state of the rest of the jejunum and is not a cause for concern if the patient is doing well clinically. If the patient can adhere to a completely gluten-free diet, the intestinal mucosa will almost certainly eventually return to normal. Few individuals are capable of following such a strict diet, however, particularly when they find that they have few clinical consequences from occasional gluten ingestion.

Electron microscopy studies of the intestinal mucosa in celiac sprue have shown that crypt epithelial cells (goblet, Paneth, enteroendocrine) are normal in appearance.[22,23] Undifferentiated crypt cells, characterized by a great abundance of free ribosomes and secretory vesicles and by poorly developed microvilli, are often present in the midvillus as well as in their normal location in the crypts. Major pathological changes are found only in the surface epithelial cells. The microvilli of surface epithelial cells are often shortened and may be irregular in size and shape (Figure 5-15). Mitochondria and the endoplasmic reticulum are dilated and irregular in shape, and there is a marked increase in number of

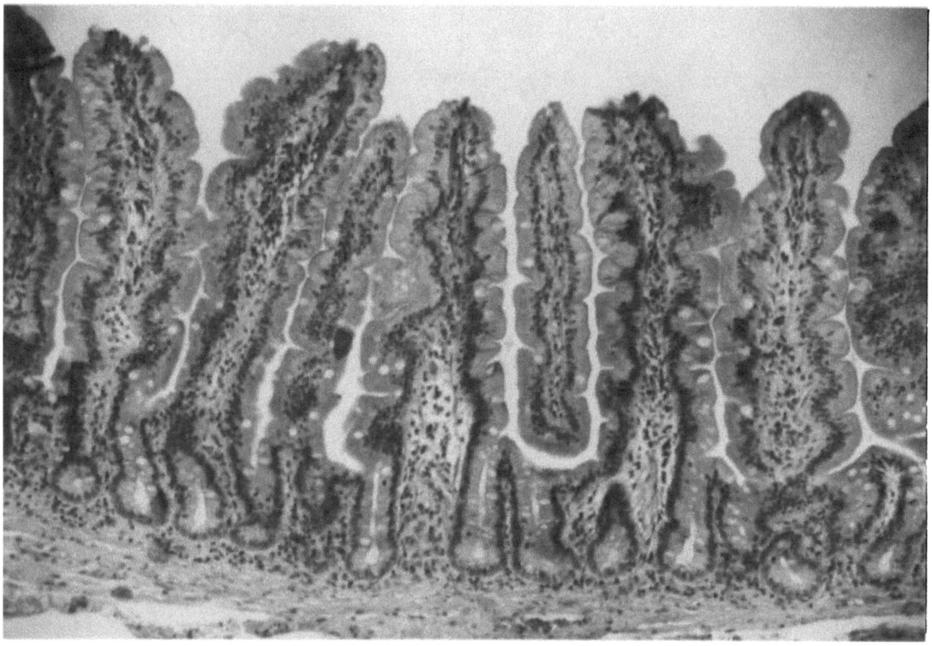

Figure 5-14. Appearance of normal proximal jejunal mucosa. H & E stain, × 175

lysosomes. The epithelial cells may be packed with huge lipid inclusions, which implies that absorbed fatty acids are unable to exit the cell. This "exit-blockade" of fatty acids and of chylomicrons (Figure 5-16) may be related in part to the thickened basal lamina that is uniformly present in the severe sprue lesion.[24] Similar ultrastructural changes have been described in the celiac sprue lesion that accompanies dermatitis herpetiformis.[25]

Freeze-fracture replicas of intestinal epithelial cells in celiac sprue have been examined for density of intramembranous proteins (IMPs) and for tight junction structure.[26] IMP density in absorptive cells in celiac sprue was 30% less than that of controls and the tight junctions had fewer strands and less depth than controls. The decrease in number of IMPs in celiac sprue correlated with decreased specific activity of brush border enzymes, while the appearance of tight junctions of absorptive cells in celiac sprue was similar to that of tight junctions of normal *crypt* epithelium. That is, the tight junctions were "leaky," and this presumably accounts for the increased intestinal permeability seen in celiac sprue.

The rare individual who fails to respond to a gluten-free diet may have "*refractory sprue*" or "*unclassified sprue*."[27] Some of these patients will recover with corticosteroid or cyclophosphamide therapy. Rarely, patients are refractory to all therapy and the disease pursues a relentless and generally fatal course. One individual was shown to require not only a gluten-free diet, but a diet free of eggs, chicken, and tuna.[27] Occasionally other lesions may be confused with the lesion of celiac sprue. In patients who fail to respond to a gluten-free diet, the original biopsy specimen should be reexamined to look for lesions of collagenous sprue, hypogammaglobulinemic (late-onset immunodeficiency) sprue, intestinal

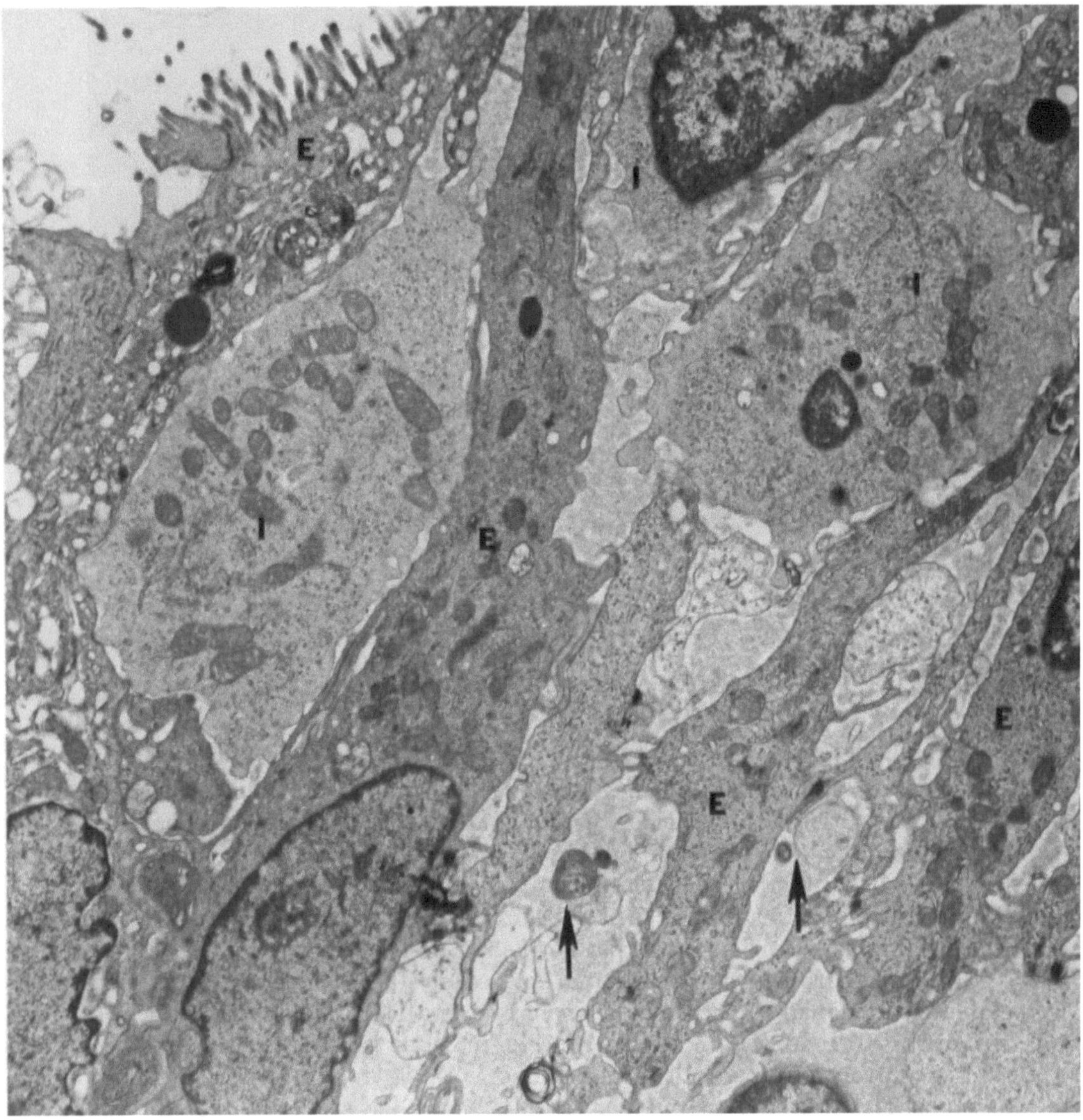

Figure 5-15. Electron micrograph showing appearance of epithelial cells in untreated celiac sprue. The microvilli are sparse to absent, intraepithelial lymphocytes (I) are prominent, epithelial cells (E) have dilated endoplasmic reticulum, and there is moderate "cellular debris" in the intercellular spaces (arrows). × 11,500

lymphoma, and treatable infections such as giardiasis or coccidiosis, which may have been missed on review of the original biopsy specimen. The unusual causes, at least in North America, of the severe nonspecific jejunal abnormality must be ruled out by the therapeutic response to appropriate treatment (Table 5-2). Additional biopsies may be needed to detect diseases with a patchy distribution such as collagenous sprue and intestinal lymphoma. Patients with celiac sprue have an increased incidence of abdominal lymphoma. Thus, lymphoma

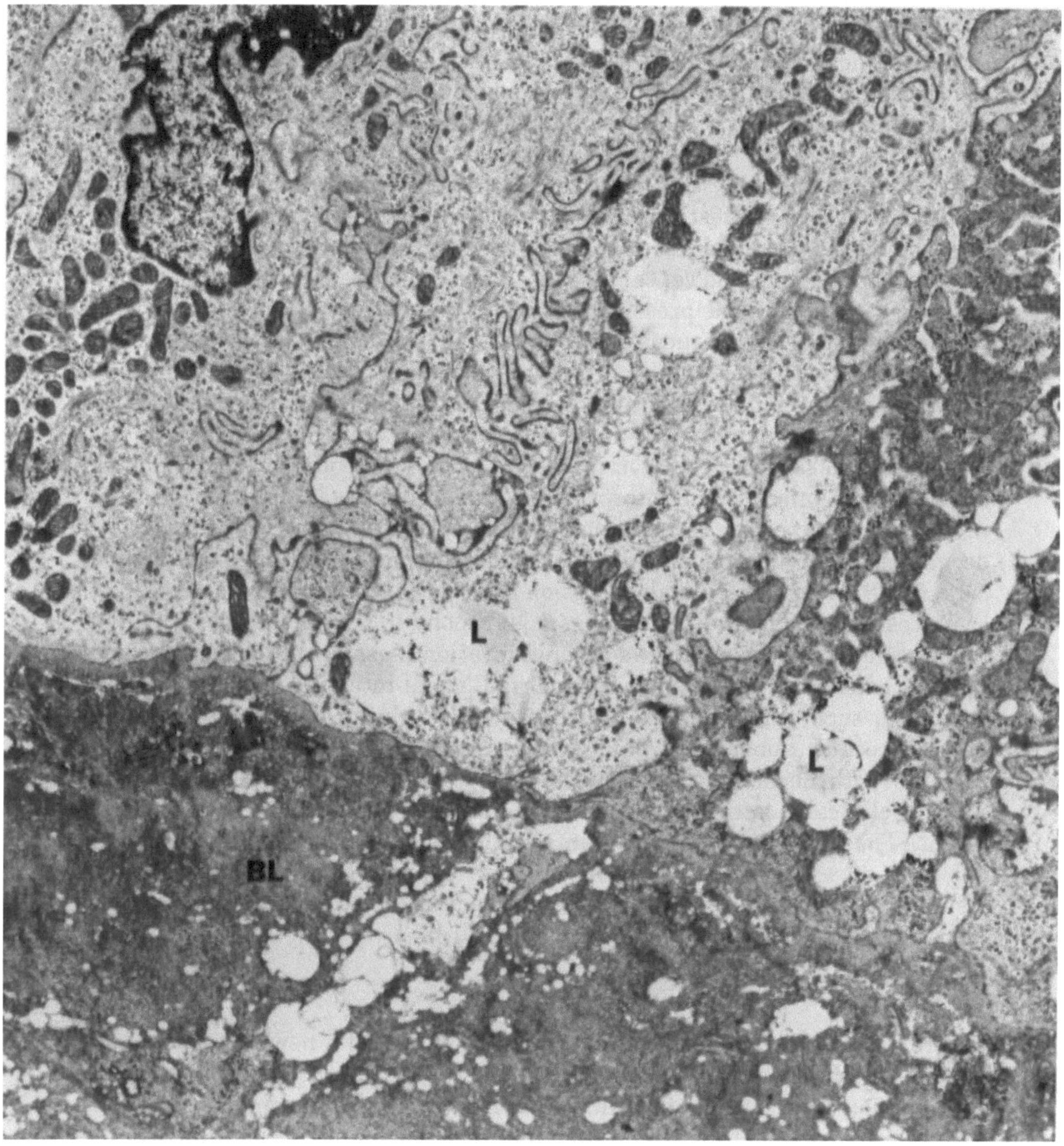

Figure 5-16. Electron micrograph showing base of epithelial cells above a greatly thickened "basal lamina" in a patient with celiac sprue. The basal lamina (BL) consists of numerous fibrils and collagen fibers. The clear areas in the basal lamina are fat droplets and chylomicrons. Note the lipid droplets (L) at the base of the epithelial cells. × 11,500

should be considered in patients who respond suboptimally to the initial gluten-free diet or who relapse while on the diet.

Diagnosis of celiac sprue in children is often complicated by the fact that the child has been placed on a gluten-free diet before an intestinal biopsy specimen has been obtained. Response to a gluten-free diet without prior demonstration of the characteristic intestinal lesion does *not* make the diagnosis of celiac sprue. The patient should still undergo biopsy examination, because an established diagnosis of celiac sprue requires a life-long gluten-free diet. Rubin found that most children "diagnosed" as having celiac sprue by a therapeutic trial of the gluten-free diet early in life proved not to have the disease when intestinal biopsy specimens were taken after institution of a high-gluten diet.[1] Some of these children probably had infectious gastroenteritis, which may induce a transient but severe intestinal mucosal abnormality in children (Figure 5-24).[28] Some children with established celiac sprue may abandon the gluten-free diet after responding to it and yet continue to do well. However, a repeat small-bowel biopsy will still show the presence of the characteristic jejunal lesion.

It may be that celiac sprue is more often a subclinical disorder than a disorder that presents with overt malabsorption. For example, family studies have revealed severe intestinal mucosal lesions in asymptomatic members.[17] Some patients may present with an isolated deficiency of absorption, such as iron deficiency anemia secondary to iron malabsorption, or osteomalacia secondary to calcium and vitamin D malabsorption. Biopsy specimens in these patients will reveal the characteristic severe lesion of celiac sprue. Two thirds of patients with dermatitis herpetiformis have the characteristic lesion of celiac sprue (Figures 5-12, 5-13).[21] Even though malabsorption is not apparent in these patients, both the intestinal lesion and the skin lesions respond to treatment with a gluten-free diet if the diet is strict and of long duration. A small number of patients with dermatitis herpetiformis have a normal small-intestine mucosa. However, when two such patients were placed on a high-gluten diet, their intestinal mucosa became abnormal.[29] Thus it appears that the majority of (if not all) patients with dermatitis herpetiformia have a latent form of celiac sprue, usually clinically latent but sometimes both morphologically and clinically latent. The vast majority of patients with dermatitis herpetiformis can be shown by means of direct immunofluorescence to have IgA deposits in the skin lesions.[30] Similar deposits of IgA have not been detected in intestinal mucosal biopsies of these patients.[30]

Protein Injury Other than Gluten

Severe gastrointestinal reactions to the introduction of a new protein into the diet of an infant are sometimes observed.[31] In one case, the normal proximal jejunal mucosa became acutely inflamed and lost its villi within 12 hours after soy protein ingestion by a susceptible infant.[31] This infant concomitantly developed shock, fever, and mild gastrointestinal bleeding. The lesion produced was indistinguishable from that seen in untreated celiac sprue. Similar reactions to various dietary proteins, especially those in milk, have been detected.[32] As noted previously, one adult needed not only a gluten-free diet, but a diet free of eggs, chicken, and tuna, for correction of the severe intestinal mucosal lesion.[27]

Refractory Sprue: Ulcerative Jejunoileitis

Patients with this condition have the characteristic severe lesion of untreated celiac sprue and do not respond to a strict gluten-free diet.[33] The diagnosis is one of exclusion; more than one, as yet unidentified, disease process may be represented. There appear to be two clinical categories of this syndrome, one in which malabsorption is not life threatening and one in which malabsorption

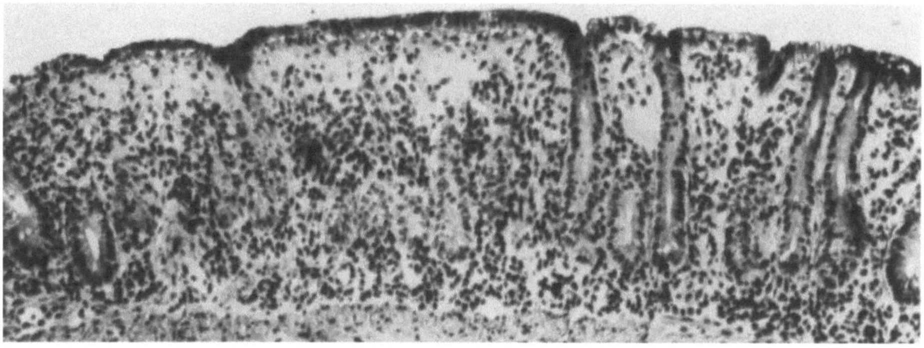

Figure 5-17. Collagenous sprue. Severe proximal jejunal abnormality in a patient with fatal malabsorption. Note the broad hyalin band of collagen separating the surface epithelium from the underlying lamina propria. H & E stain, × 184. (Reproduced from Dobbins,[16] by permission of Churchill Livingstone, Inc., New York. Original supplied by C.E. Rubin.)

runs a fulminant course. Patients in the latter group have often been reported as having "ulcerative jejunoileitis."[34] Occasionally the benign syndrome will evolve into a "malignant" syndrome, but long survival with variable disability is often the case. Patients in both categories may respond initially to a gluten-free diet, only later becoming unresponsive. Some of these patients eventually turn out to have abdominal lymphoma. Whether the lymphoma was a primary event or developed as a complication of preexisting celiac sprue generally cannot be determined.[33] Paneth cells may be absent in a few patients with unclassified sprue, but the significance of this observation is not at all clear.[1,16]

Collagenous Sprue

In patients with collagenous sprue there is a complete absence of villi, and all the other characteristic features of the severe lesion of untreated celiac sprue is found in intestinal biopsy specimens.[35] The unique and diagnostic feature, however, is a strikingly broad band of collagen found just below the epithelial basal lamina (Figure 5-17). Most patients with this rare lesion have a fulminant and generally fatal course. There is no known treatment for this syndrome. Some patients respond transiently to a gluten-free diet, and it is reasonable to recommend a prolonged gluten-free diet in all patients. Immunosuppressive agents (corticosteroids, cyclophosphamide) may be tried. Collagen deposition is not uniformly present early in the disease. Biopsy may be required at repeated intervals to establish the diagnosis.

Collagenous sprue may be an end-stage of unclassified sprue or even of celiac sprue that has become refractory to treatment. Because its course is different from that of celiac sprue, and because the long-term prognosis is grim, it is reasonable to classify this as a separate diagnostic entity until it can be more clearly defined. Bossart et al.[36] have suggested that collagenous sprue is nothing more than a variant of celiac sprue and that it does not deserve a separate classification. Rubin, however, has pointed out that Bossart did not demonstrate the presence of a broad band of collagen below the epithelium but only the expanded basement membrane complex seen in any flat lesion.

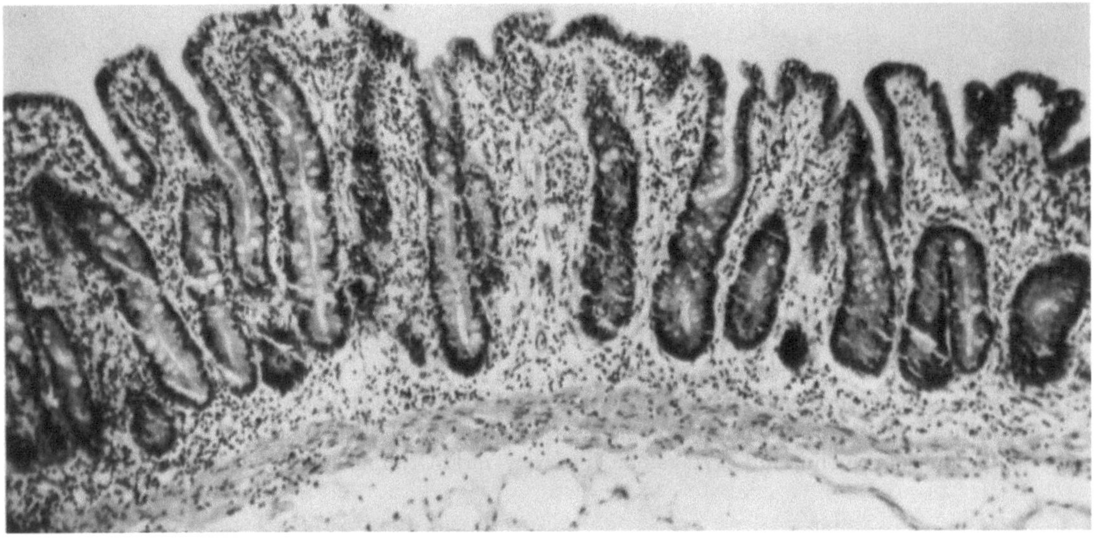

Figure 5-18. Severe jejunal mucosal lesion in a patient with late-onset immunodeficiency. Plasma cells are absent from the lamina propria. The patient had a severe deficiency of serum IgG, IgA, and IgM. H & E stain, × 132. (Reproduced from Dobbins,[16] by permission of Churchill Livingstone, Inc., New York.)

Immunodeficiency

There are two major *adult* immunodeficiency syndromes: (1) late-onset immunodeficiency, or common variable hypogammaglobulinemia (CVH), and (2) selective IgA deficiency.[37] CVH is represented by two major subgroups–one in which there is variable deficiency of immunoglobulins, especially IgA and IgM, and in which there may or may not be clinically significant T cell deficiency; and the other in which there is predominant T cell deficiency, antibody deficiency being less prominent. CVH is also characterized by two major, usually mutually exclusive, presentations–recurrent respiratory tract infection or chronic diarrhea and malabsorption (hypogammaglobulinemic sprue) with recurrent gastrointestinal giardiasis. The latter presentation is common. The key pathological finding is a marked reduction or even absence of plasma cells in the intestinal mucosa (Figure 5-18). Plasma cells are also absent, or markedly diminished in number, at other sites such as the bone marrow and spleen. Mucosal structure varies from normal in appearance to a flat lesion comparable to that seen in individuals with celiac sprue (Figure 5-18). There is no clinical or histological response to a gluten-free diet. *Giardia lamblia* is often found on careful search of serial sections; it may be more easily seen in Giemsa-stained smears of mucus that was adherent to the specimen.[38] Generally, the flat mucosa will return to a more normal appearance following successful treatment of giardiasis.[38,39] Alternatively, it has been shown that a relatively normal-appearing mucosa in a patient with CVH will progress over a period of years, in association with recurrent diarrheal illness, to a flat mucosal appearance.[40] This progression is usually, but not always, related to recurrent giardiasis. Nodular lymphoid hyperplasia (NLH), consisting of hyperplastic lymphoid follicles containing immature B cells, may be found throughout the gastrointestinal tract. These follicles create a characteristic and often diagnostic nodular appearance of the small bowel when examined using barium contrast media. Suction biopsy specimens may include whole benign lymphoid follicles within the lamina propria. The villi overlying the follicles may be distorted, and the intervening mucosa may be normal or may show a variable abnormality in villous architecture. At one time it was thought that NLH was a separate entity from CVH. Today NLH is thought to be one of the manifestations of CVH, tending to develop as the patient grows older. NLH may represent "compensatory" hyperplasia of immature B cells that are genetically defective and unable to mature as plasma cells. Rarely, diffuse NLH is found in otherwise apparently normal individuals.

Electron microscopy of the intestinal mucosa in a patient with CVH and a flat small-bowel biopsy specimen has confirmed the absence of plasma cells.[41] The intestinal absorptive cells were shorter than normal and their microvilli were shortened and slightly decreased in number. There was a striking increase in intraepithelial lymphocytes. Absorptive cell organelles were normal except for modest dilatation of the endoplasmic reticulum and for a marked increase in multivesicular bodies. Strikingly, goblet cells contained no mucus of the usual variety but rather there was a membranous transformation of the portions of the cell that ordinarily contain mucus (Figure 5-19). Such a change has been described previously only in bacterial gastroenteritis of piglets.[42]

The most common adult immunodeficiency syndrome is that of selective IgA deficiency.[37] The incidence of this syndrome is approximately 1 in 700. Many individuals with selective IgA deficiency have no apparent clinical abnormalities, possibly because they have a "compensatory" increase in serum IgM and presumably of secretory IgM. Others with selective IgA deficiency, especially if deficient in the IgG subclasses IgG_2 and IgG_4, present with three types of illness – sinopulmonary disease, malabsorption syndrome, or autoimmune disease.[43] There is *no* predisposition to the intestinal parasite *Giardia*. Pathologically the intestinal mucosa is generally normal, although there is a decrease or absence of IgA-bearing cells in the gut. IgM- and IgG-staining plasma cells are present in normal numbers. A single intriguing patient has been reported with a severe villous abnormality of the small intestine associated with selective IgA deficiency.[44] The patient did not respond to a strict gluten-free diet. Indirect immunofluorescent studies of his serum revealed an IgG antibody that bound to epithelial cells in control intestine but not to cells in the patient's diseased mucosa. Following treatment with cyclophosphamide, the serum antibody titer slowly declined and the patient's intestinal mucosa returned to a normal appearance. These unique findings suggested an autoimmune etiology for the intestinal mucosal damage in this individual.[44] An earlier report may also represent this same syndrome,[45] although the presence of antibody to epithelial cells was not looked for. This report describes a 13-year-old with selective IgA deficiency, a flat jejunal mucosa, and severe fetal malabsorption. The child failed to respond to gluten restriction or to intravenous infusions of fresh frozen plasma, to antibiotics, or to hyperalimentation.[45] The incidence of celiac sprue may be increased in patients with selective IgA deficiency. These patients have a severe intestinal mucosal lesion and respond to a gluten-free diet.

Structural alterations of the intestinal mucosa may sometimes be found in the large variety of immunodeficiency syndromes found in the pediatric age group.[46–48] Certainly the increased incidence of gastrointestinal symptoms in children with severe combined immunodeficiency and in some children with

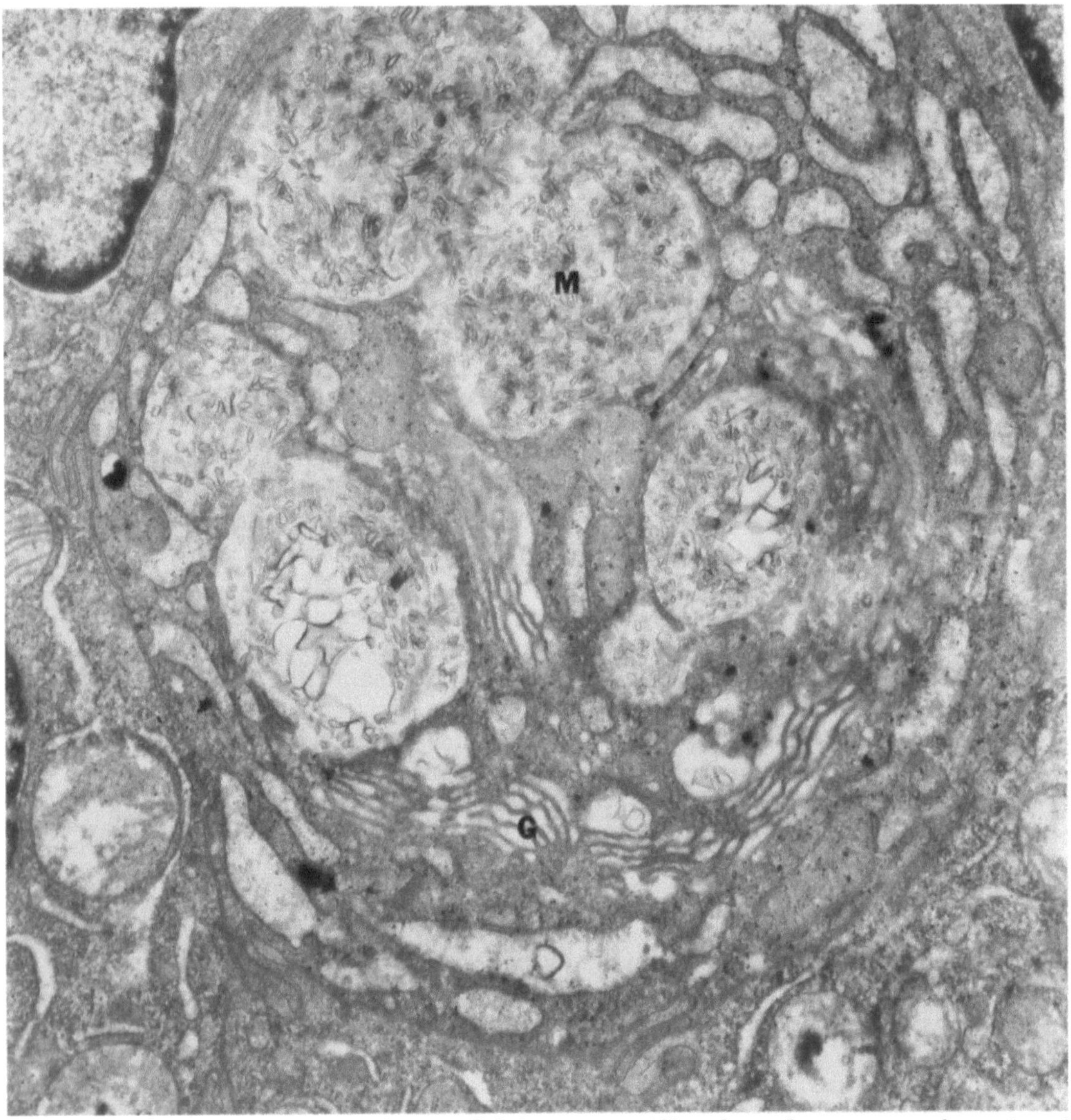

Figure 5-19. Electron micrograph showing Golgi apparatus (G) and membranous transformation of mucus droplets (M) in intestinal biopsy specimen of a patient with severe deficiency of serum immunoglobulins. × 25,000. (Reproduced from Dobbins,[138] by permission of John Wiley & Sons, Inc. Copyright © 1978, John Wiley & Sons, Inc.)

isolated T cell deficiency suggests that an intact cellular immune system is essential for the normal function of the intestinal tract.[46] There is usually no histological abnormality other than absence of plasma cells in the lamina propria in the majority of these patients.[47] Plasma cells are not present in biopsy specimens of patients with infantile X-linked agammaglobulinemia and in most children with CVH. Plasma cells are easily recognized in biopsy specimens of patients with ataxia-telangiectasia, selective IgA deficiency, immunodeficiency syndrome with normal serum gammaglobulins, and X-linked immunodeficiency

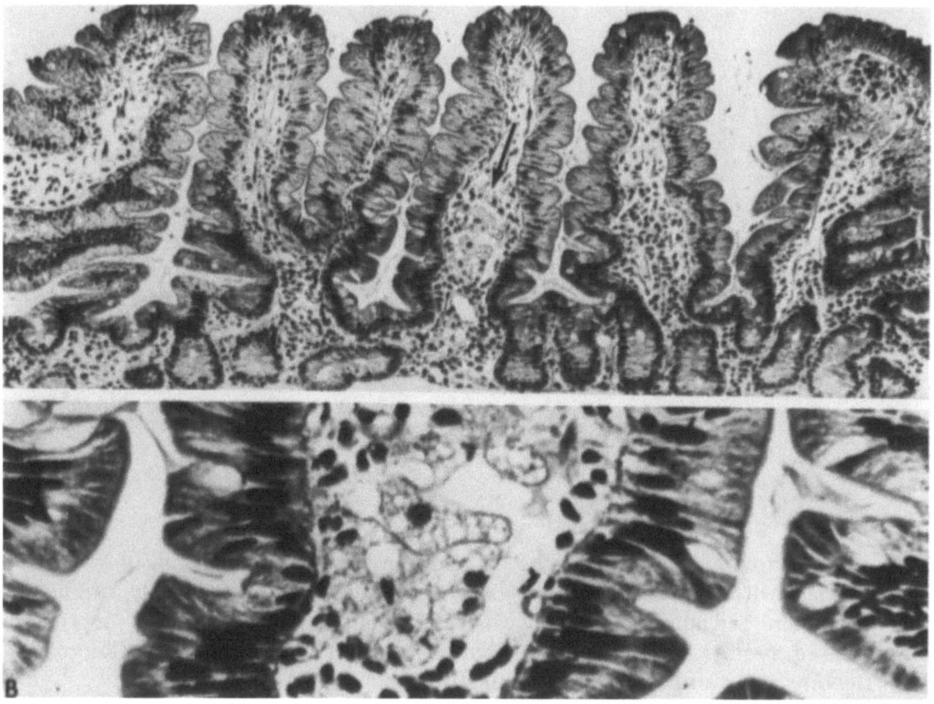

Figure 5-20. Section of an intestinal biopsy specimen obtained from a patient with chronic granulomatous disease. (A) Villous architecture appears normal. A cluster of macrophages is present in the lamina propria (arrow). × 184. (B) Higher magnification shows typical vacuolated appearance of the macrophages. H & E stain, × 736. (Reproduced from Dobbins,[16] by permission of Churchill Livingstone, Inc., New York. Original supplied by C.E. Rubin.)

with hyper-IgM. Multiple small-bowel biopsy specimens are necessary to assess the morphology in all symptomatic children with immunodeficiency because the villous architecture may vary from normal to severely abnormal in different biopsy specimens obtained at the same time from the same individual.[48]

Chronic granulomatous disease (CGD) of childhood is an inherited illness manifested by repeated bacterial infections.[46] The phagocytic cells of these patients can ingest but not kill certain microorganisms. Two of nine children with CGD had steatorrhea and six had mild to moderate vitamin B_{12} malabsorption. In seven or eight of the patients biopsied, there were clumps of characteristic vacuolated pigmented macrophages within the lamina propria adjacent to the crypts and occasionally within the villous core (Figure 5-20). Small-bowel morphology was otherwise completely normal. The macrophages were PAS positive and contained a yellowish-brown pigment. Unlike the case in untreated Whipple's disease, the villous architecture was normal and there were far fewer macrophages in the lamina propria.[46] Large foamy macrophages may be seen in the lamina propria of intestinal biopsy specimens of patients with macroglobulinemia. These macrophages are not pigmented and are not PAS positive (they may be faintly PAS positive) and should easily be distinguishable from the macrophages seen in CGD and in Whipple's disease.

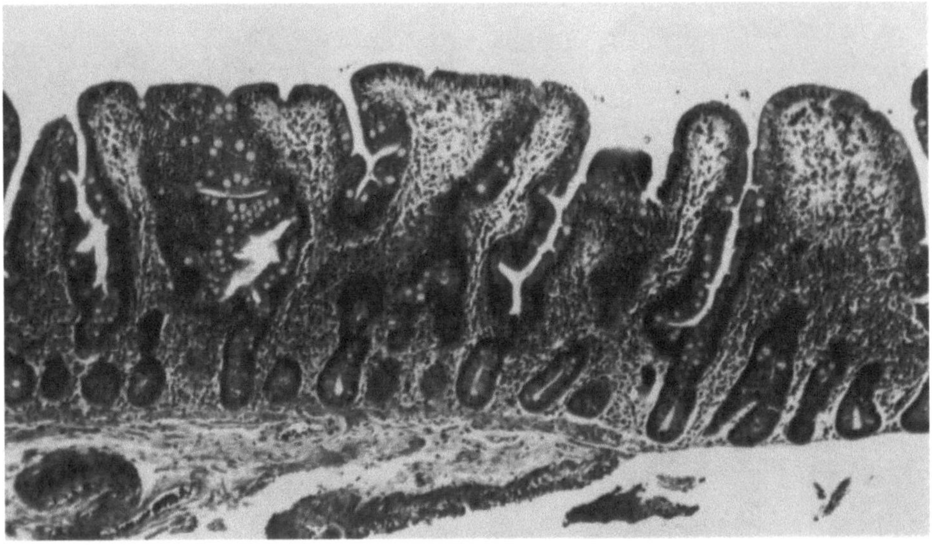

Figure 5-21. Proximal jejunal biopsy specimen showing a mild abnormality in a Puerto Rican patient with tropical sprue. The villous architecture is only slightly altered, but there is an increased number of lymphocytes and plasma cells in the lamina propria, and an increased number of intraepithelial lymphocytes. H & E stain, × 80. (Reproduced from Dobbins,[16] by permission of Churchill Livingstone, Inc., New York.)

Tropical Sprue Syndrome

Many asymptomatic people living in certain areas of the world, for example Ceylon, Puerto Rico, Haiti, India, and Hong Kong, have mild or moderate nonspecific abnormalities in the proximal jejunum. Some of these apparently clinically well people may even have a severe nonspecific abnormality that is indistinguishable from that seen in untreated celiac sprue.[49] Nonspecific mucosal abnormalities may be found in individuals who have lived in those areas, generally after a few years but sometimes after only a few weeks or months (Figure 5-21). The pathogenesis of this syndrome is not clear, but most evidence favors enterotoxin production by an infectious agent, in addition to intestinal injury caused by unknown influences in the environment.[50] When severe, the characteristic features of this syndrome include chronic diarrhea, steatorrhea, macrocytic anemia, glossitis, and emaciation, all due to intestinal malabsorption of fat, carbohydrates, vitamin B_{12}, and folic acid.

Jejunal biopsy in symptomatic patients, as well as in asymptomatic individuals residing in tropical areas, will show abnormalities of villous structure ranging from mild to severe (Figures 5-21, 5-22). A North American returning from a tropical area of the world with malabsorption who is found to have a severe nonspecific lesion may have either tropical sprue or celiac sprue. If the patient has tropical sprue, treatment with folic acid and a broad-spectrum antibiotic such as tetracycline generally results in a prompt remission of symptoms, often with improvement in intestinal structure. If the patient fails to respond to such treatment, a gluten-free diet should be tried to exclude celiac sprue.

It is important to repeat that biopsy specimens obtained from patients residing in the tropical areas of the world cannot be interpreted by the same criteria

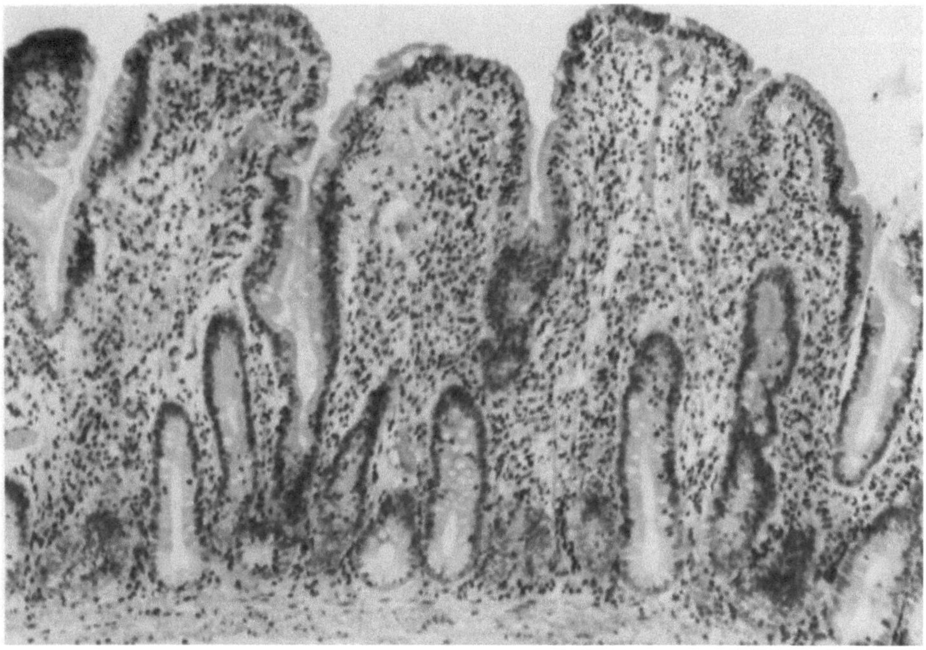

Figure 5-22. Proximal jejunal biopsy specimen showing a severe villous abnormality in tropical sprue. There is a marked increase in lamina propria lymphocytes and plasma cells. Note the intense infiltration of the epithelium by lymphocytes, a highly characteristic feature of both celiac sprue and tropical sprue (see Chapter 3). H & E stain, × 540. (Reproduced from Dobbins,[16] by permission of Churchill Livingstone, Inc., New York.)

as those used in North America.[16,49] The range of normal in small-bowel morphology of asymptomatic individuals residing in tropical areas must be established for each country. Intestinal biopsy specimens in normal asymptomatic individuals residing in tropical countries often show mild villous abnormalities compared with biopsy specimens of normal North Americans. Further, mucosa in both asymptomatic people and clinically ill individuals with malabsorption may show more evidence of injury in Haitians or Vietnamese than in Puerto Ricans.[49]

Electron microscopy of intestinal biopsy specimens of individuals with tropical sprue, just as in celiac sprue, show only nonspecific changes.[51] Changes described in epithelial cells include mild alterations in microvilli, a marked prominence of lysosomes, retention of lipid, and "degenerative" changes in endoplasmic reticulum and mitochondria similar to those seen in celiac sprue (Figure 5-23). There is also thickening of the basal lamina and retention of lipid within the basal lamina.[51] Putative etiological agents such as bacteria or viruses have not been identified in the specimens.

Childhood Kwashiorkor

Kwashiorkor is one of the most common illnesses of children in "undeveloped" areas of the world.[52] In growing children, protein deficiency per se has been

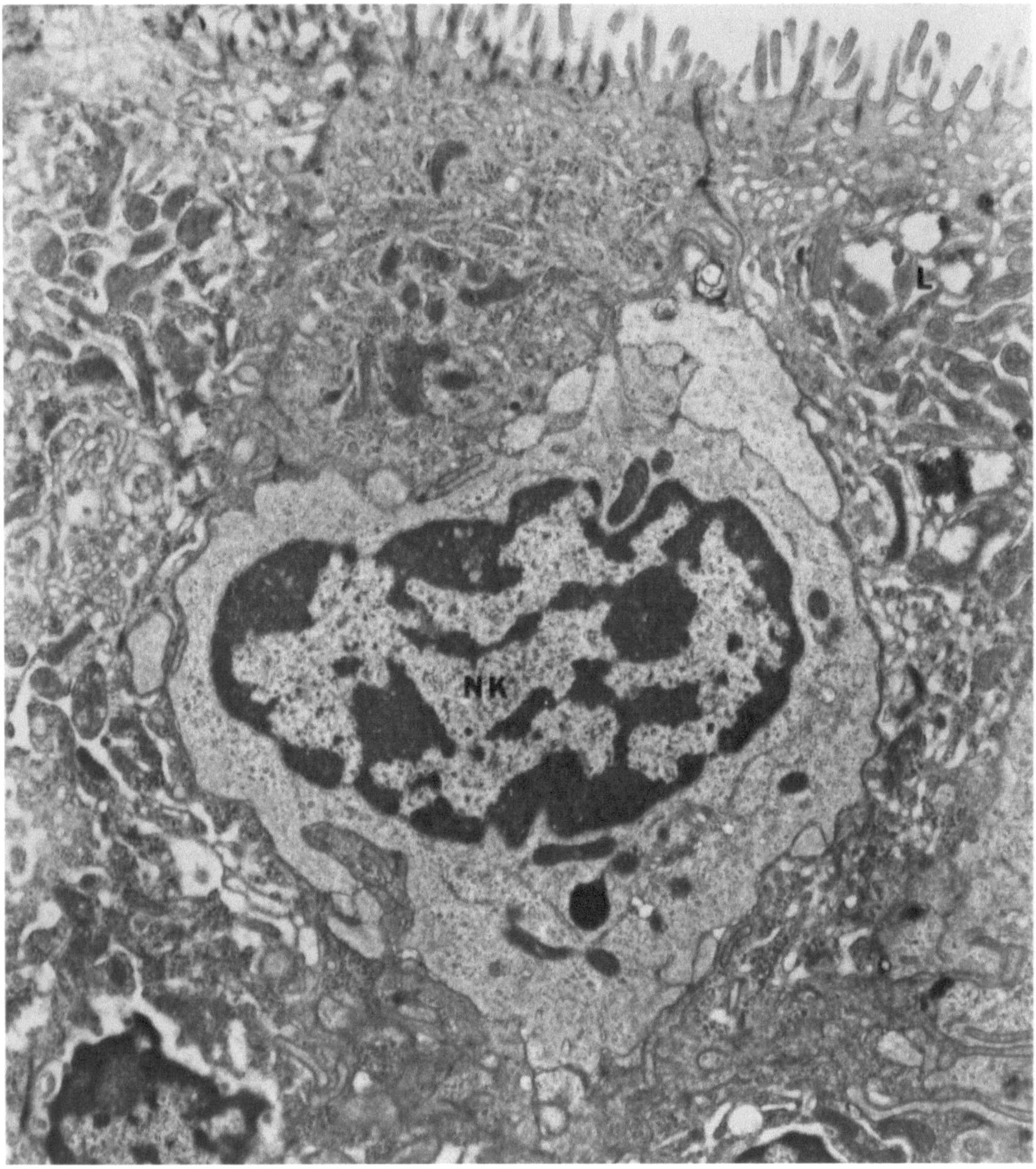

Figure 5-23. Electron micrograph showing apical portion of epithelial cells in intestinal biopsy specimen of a patient who acquired tropical sprue during a two-week visit to the Philippines. Note irregularity of microvilli, dilated endoplasmic reticulum, prominent lysosomes (L), and the intraepithelial lymphocyte that has the morphology of a large granular lymphocyte or of a natural killer (NK) cell. × 14,500

shown to cause severe nonspecific proximal jejunal abnormalities indistinguishable from those of untreated celiac sprue.[52] The same lesion can be produced experimentally in growing animals by depriving them of proteins. When diagnosed, these patients will respond to a normal diet containing adequate amounts of high-quality protein, with concomitant improvement in the appearance of the intestinal mucosa. The intestinal lesion in kwashiorkor may be patchy, mild to moderate in degree, and impossible to differentiate from abnormalities characteristically seen in tropical sprue. The diagnosis is established on the basis of response to protein feeding.

The structural changes consist of an abnormally thin intestinal mucosa with marked atrophy of villi, increased cellularity of the lamina propria, increase in number of intraepithelial lymphocytes, and decreased height of epithelial cells.[52] Fine structural changes observed include the presence of marked retention of lipid in epithelial cells in intestinal biopsy specimens of 9 of 16 patients with kwashiorkor.[52] There was absence of lipid droplets in Golgi vacuoles, and chylomicrons were diminished in number. These changes are very reminiscent of those seen in intestinal epithelial cells in congenital β-lipoprotein deficiency (Figure 5-57). This implies that in the severe lesion of kwashiorkor, as in β-lipoprotein deficiency, there is reduced synthesis of low-density lipoproteins. Following clinical recovery, the abnormal lipid accumulation in epithelial cells is no longer present. *Giardia lamblia* parasites were present in the intestinal lumen of 12 of the 16 patients with kwashiorkor.[52] The role, if any, of this parasite in the changes described is not clear. A fine structural study of the small-intestine mucosa in infantile marasmic malnutrition disclosed abnormalities of the brush border, large autophagosomes and residual bodies, the deposition of collagen filaments, and a dense finely granular material below the basal lamina.[53] These changes reversed following clinical recovery.[53]

Familial Enteropathy

This is a new syndrome described in five infants residing near Toronto, Canada.[54] It consists of protracted diarrhea from birth, failure to thrive, and intestinal mucosal changes comparable in severity to those seen in celiac sprue. No consistent hematological or immunological defects were found in the infants. Duodenal biopsy specimens showed a severe mucosal lesion, crypt hypoplasia without an increase in mitoses, absence of an inflammatory cell infiltrate in the lamina propria, and most remarkably, absence of the brush border of intestinal epithelial cells. Electron microscopy confirmed the absence of epithelial cell microvilli and demonstrated a striking increase in lysosomelike inclusions and the presence of autophagocytosis in intestinal epithelial cells. Celiac sprue was easily excluded because the infants either had not ingested gluten or failed to respond to total parenteral nutrition. There were some similarities to kwashiorkor. However, in kwashiorkor there is a marked increase in inflammatory cells in the lamina propria and patients respond to adequate nutritional intake, features lacking in familial enteropathy. Because other types of small-intestine cells (goblet cells, Paneth cells, and enteroendocrine cells) were normal, the abnormality may be restricted to the absorptive cells. The abnormality may be one of selective failure of normal maturation and differentiation of crypt cells.[54]

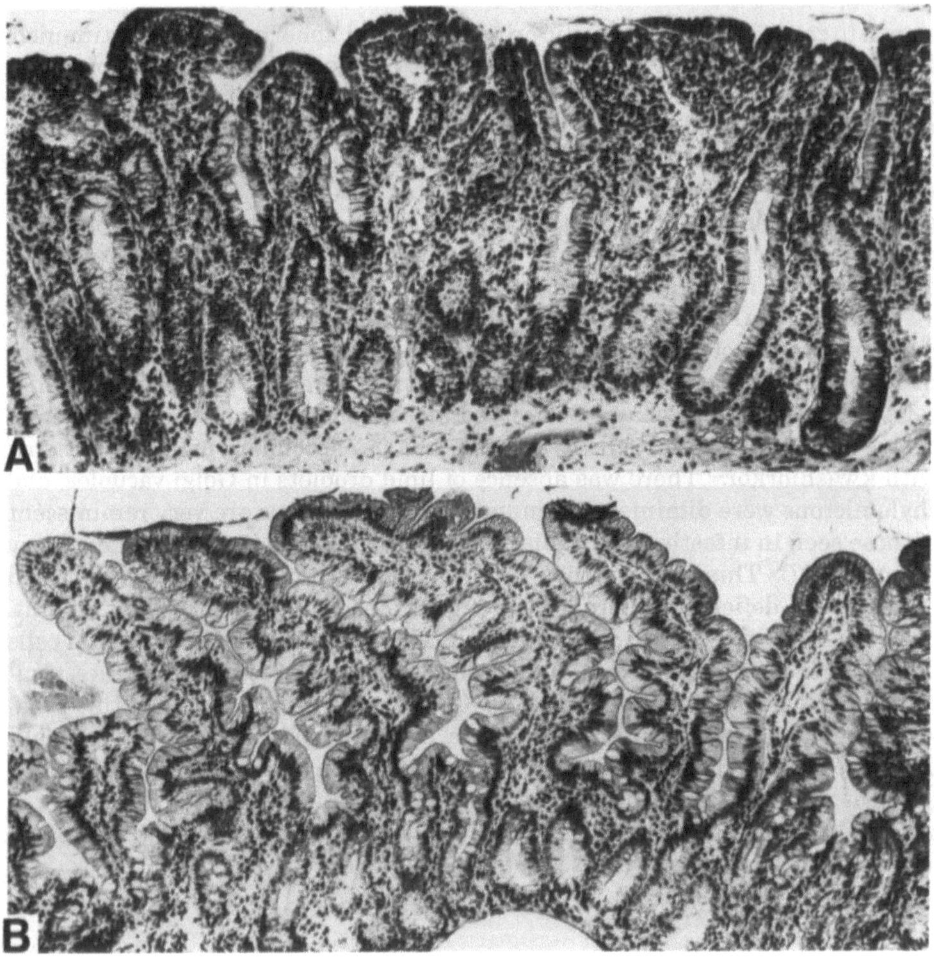

Figure 5-24. Proximal jejunal biopsy specimens from a 7-year-old child with acute gastroenteritis of unknown etiology. (A) Severe abnormality seen at time of diagnosis. H & E stain, × 184. (B) Normal specimen 6 weeks after the biopsy in (a) and 12 weeks after onset of symptoms. The patient was clinically well. H & E stain, × 184. (A and B reproduced from Dobbins,[16] by permission of Churchill Livingstone, Inc., New York. Original supplied by C.E. Rubin.)

Infectious Gastroenteritis

This entity is probably caused by a variety of viral and bacterial pathogens, most of which have not been clearly identified.[55,56] Jejunal lesions are generally mild to moderate in degree and are nonspecific, but may be severe with a flat mucosa (Figure 5-24). The clinical picture of acute gastroenteritis at the onset of symptoms and the usually temporary nature of symptoms suggests the diagnosis. Moderate to severe changes were documented in the intestinal mucosa of 31 infants and children with "nonbacterial" gastroenteritis.[28] Repeat biopsy specimens in some of the children when they became asymptomatic showed improvement or return to normality.[28] Histologically, there may be an acute inflammatory process with formation of early crypt abscesses. Perera et al.[1] have noted rare patients with infectious enteritis in whom malabsorption with acute inflammatory changes in the biopsy specimens continued for months before the

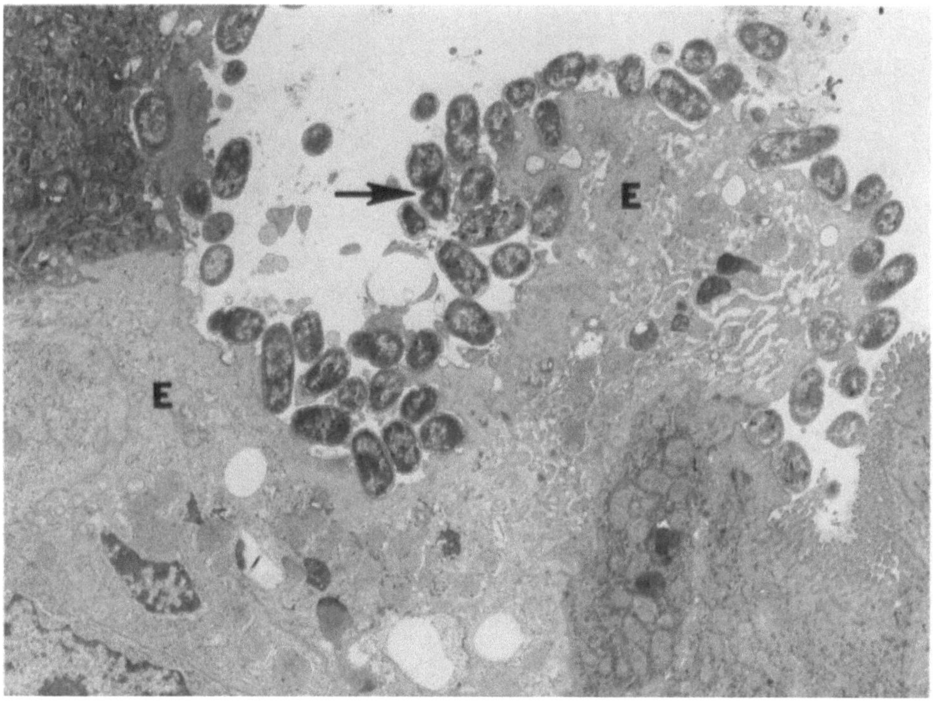

Figure 5-25. Electron micrograph showing adherence of *E. scherichia coli* 0119 (arrow) to intestinal epithelial cells (E) in an infant with severe diarrhea. Note complete absence of microvilli over the involved epithelial cells. See text for additional details. × 5,800. (Reproduced from Rothbaum et al.,[57] with permission of Hemisphere Publishing Corporation, New York.)

lesions and symptoms finally disappeared. Pathogenic *Escherichia coli* may produce diarrhea with a flat intestinal lesion and even death in infants.[55,57] Fluorescent antibody staining for the presence of *E. coli* may be required to establish the invasive nature of the organism.[1,55]

An ultrastructural study of enteropathogenic *E. coli* 0119 infection in 18 infants with protracted diarrhea focused on jejunal biopsy changes.[57] Presence of "villous atrophy" along with a chronic inflammatory cell infiltrate in the lamina propria was demonstrated. Further, there was absence of the glycocalyx and of microvilli at the site of bacterial adherence (Figure 5-25). Cellular damage was indicated by the presence of a dilated endoplasmic reticulum, mitochondrial changes, and cytoplasmic pallor. There was no evidence of bacterial invasion of the epithelium. The published illustration clearly shows the presence of a severe (flat small-bowel biopsy) mucosal lesion in these infants.[57]

The apparent attachment of bacteria to intestinal epithelial cells was first reported by Hampton and Rosario[58] in the mouse ileum. The area of attachment was restricted to the luminal side of the epithelial cell and was characterized by a thickening of the plasma membrane and a cup-shaped homogeneous density just beneath the membrane. There was no evidence of an inflammatory reaction in the underlying lamina propria, and the significance of the attachment was not apparent. The profuse attachment of bacteria to epithelial cells in *E. coli* 0119-induced diarrhea of infants clearly appears to be pathogenic. We have seen

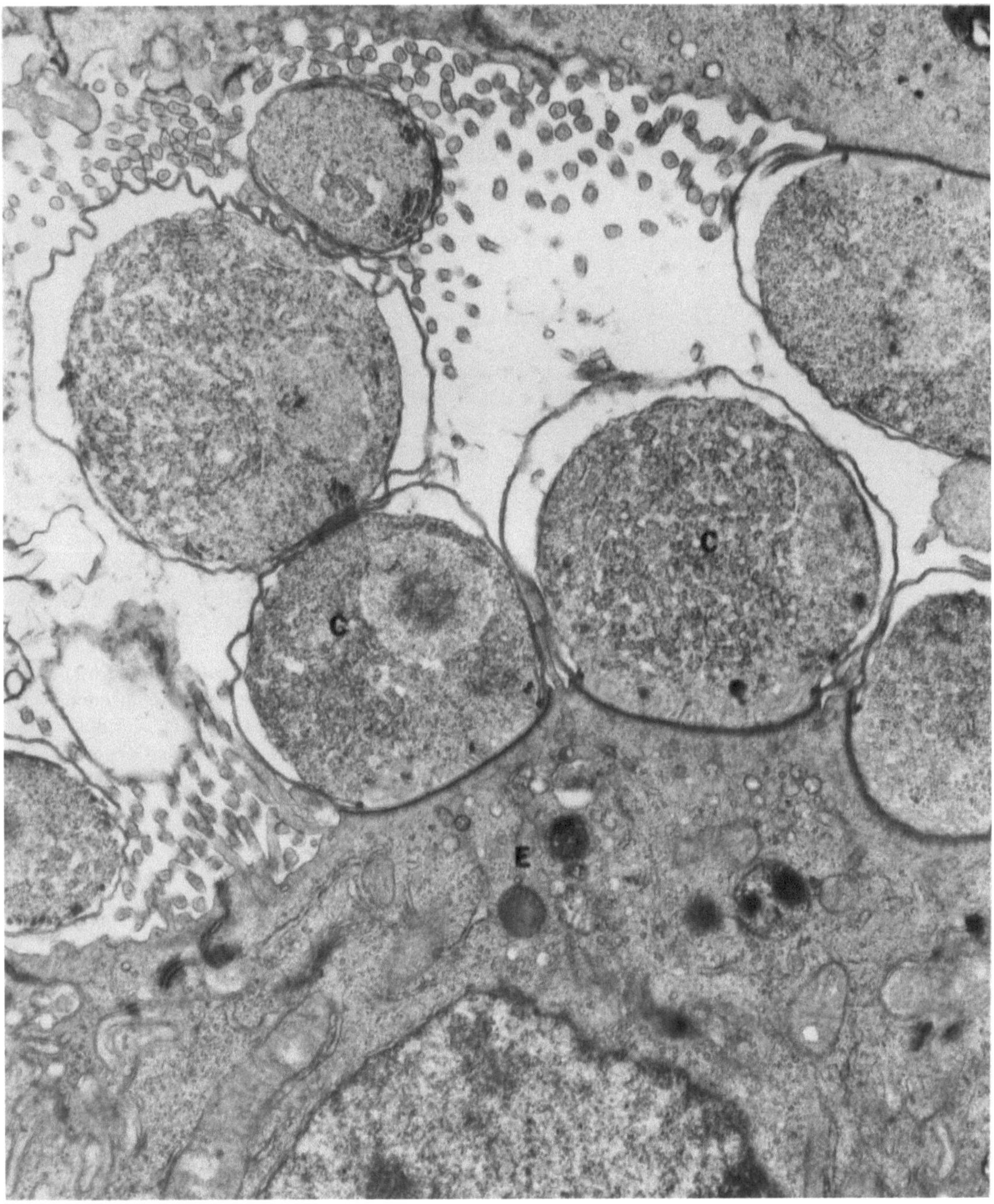

Figure 5-26. Electron micrograph showing adherence of cryptosporidia (C) to intestinal epithelial cells (E) in an adult with acquired immunodeficiency syndrome and chronic diarrhea. Note again the absence of microvilli over the involved cells. × 14,500

a similar profuse attachment of cryptosporidia to the intestinal mucosa of individuals with AIDS (Figure 5-26) and suggest that the resulting marked reduction in mucosal absorptive surface may contribute to the diarrhea in AIDS.[59]

Interactions of Microorganisms with Intestinal Epithelial Cells

Morphological alterations in the microvillus border of intestinal epithelial cells may be produced by a variety of microorganisms. Some of these alterations are clearly the morphological representation of adherence. However, these alterations may not represent clinically important pathological conditions. A number of organisms colonize the intestine. These organisms are particularly likely to adhere to blind invaginations of the epithelium or to enter the crypts of Lieberkuhn, only then establishing a form of contact relationship with the epithelial surface. Such organisms include protozoans, spiral-shaped microorganisms, yeast, fusiform bacteria, and lactobacilli.

Electron microscopy and freeze-fracture studies show that the attachment sites of spirochetes and of flagellated bacteria to the apical surface of absorptive cells differ dramatically.[60] The attached end of the spirochete lies embedded in the host cell terminal web, adhering to adjacent microvilli. The position of the spirochete outer membrane that participates in the attachment site is marked in freeze-fracture replicas by a series of circumferential P-face particle arrays.[60] In contrast, the bacterium lies in a shallow indentation of the intestinal cell membrane and does not adhere to the microvilli. Freeze-fracture replicas show that the narrow attachment site of the bacterium is characterized by a loose band of heterogeneous intramembranous particles.[60] This suggests that distinctive bacterial surface proteins are involved in the host cell interaction.

The best example of an organism that attaches directly to epithelial cells is that of a segmented filamentous bacterium commonly found in the rat ileum.[58,60] These bacterial filaments may cover the surface of ileal villi in a dense hairlike mat. The individual bacterial filament attaches to epithelial cells by a specialized attachment segment, or "hold fast." At the attachment site the hold fast displaces microvilli and invaginates, but does not penetrate, the epithelial cell membrane. The invagination takes place only to the level of the terminal web. The cell wall of the bacterium is closely applied to the outer leaflet of the trilaminar plasma membrane of the epithelial cell. The attachment site is further modified by the presence of two dense layers in the apical cytoplasm. The outer layer is represented by a dense homogeneous plaque, 30 to 50 nm thick, which is found just below the area of contact between bacterial cell walls and epithelial cell membranes. The dense plaque itself is surrounded by a filamentous layer, the filaments apparently being derived from those of the terminal web. The morphological changes in intestinal epithelial cells appear to be limited to the terminal web and microvillus area; they do not seem, at least as determined morphologically, to affect the epithelial cells adversely. When there is very dense colonization, alterations at the microvillus border are very severe, and it seems likely that the function of these cells may therefore be altered.

An association of bacteria and spirochetes to the luminal plasma membrane of colonic epithelial cells in humans has been noted[61] (Figures 5-27, 5-28). Even though bacteria and spirochetes are aligned parallel to and between microvilli of epithelial cells, there is no evidence of cell injury similar to that seen in *E. coli*

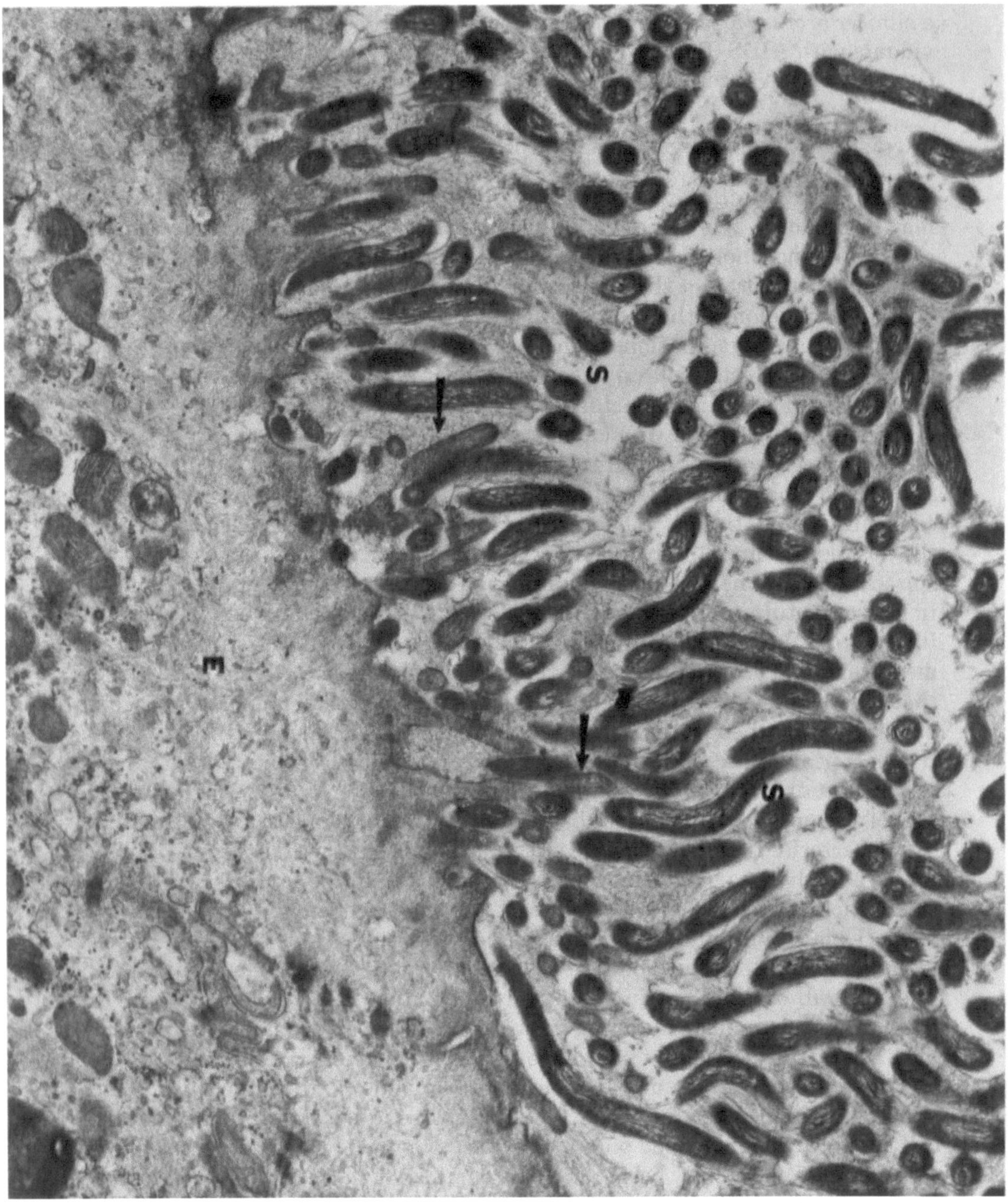

Figure 5-27. Electron micrograph showing close association of spirochetes (S) with colonic epithelial cells (E). The spirochetes are so profuse that the epithelial cell microvilli (arrows) are difficult to discern. × 21,700

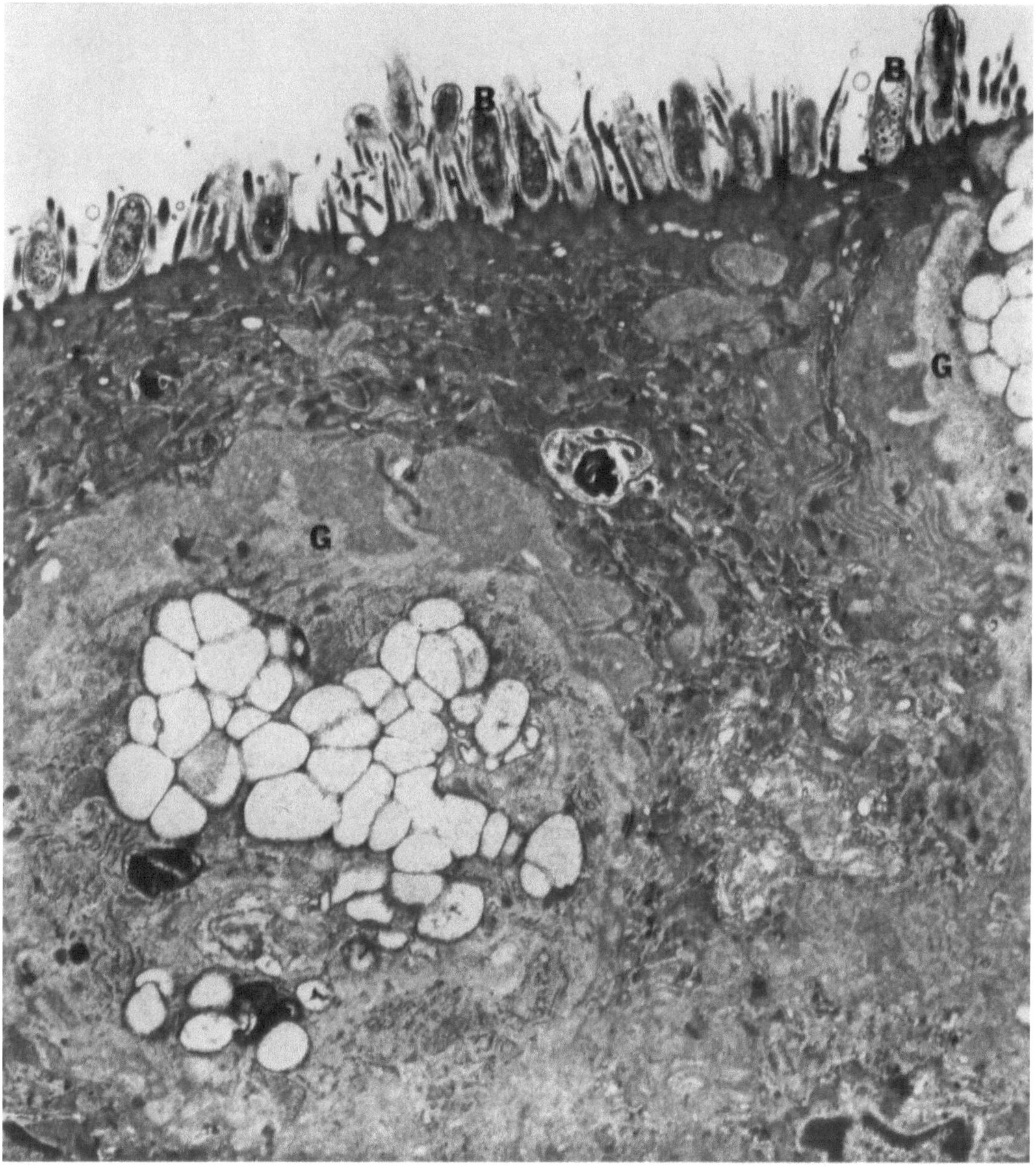

Figure 5-28. Electron micrograph of normal colonic epithelial and goblet (G) cells showing parallel alignment of a bacterial organism (B) with the epithelial cell microvilli. There is no evidence of cell injury related to the bacterial species. × 9,000. (From Dobbins,[138] reprinted by permission of John Wiley & Sons, Inc. Copyright © 1978, John Wiley & Sons, Inc.)

$0119B_{14}$-induced diarrhea of infants.[55] This close association of bacteria and spirochetes with microvilli of colonic epithelial cells in adults has not been related to any apparent pathological condition, although mucosal penetration by spirochetes has been observed on one occasion.[62] Further, rectal spirochetosis can be observed at light microscopy and has been reported to occur in as many as 5% to 10% of colonic appendiceal specimens. It is found in rectal biopsy

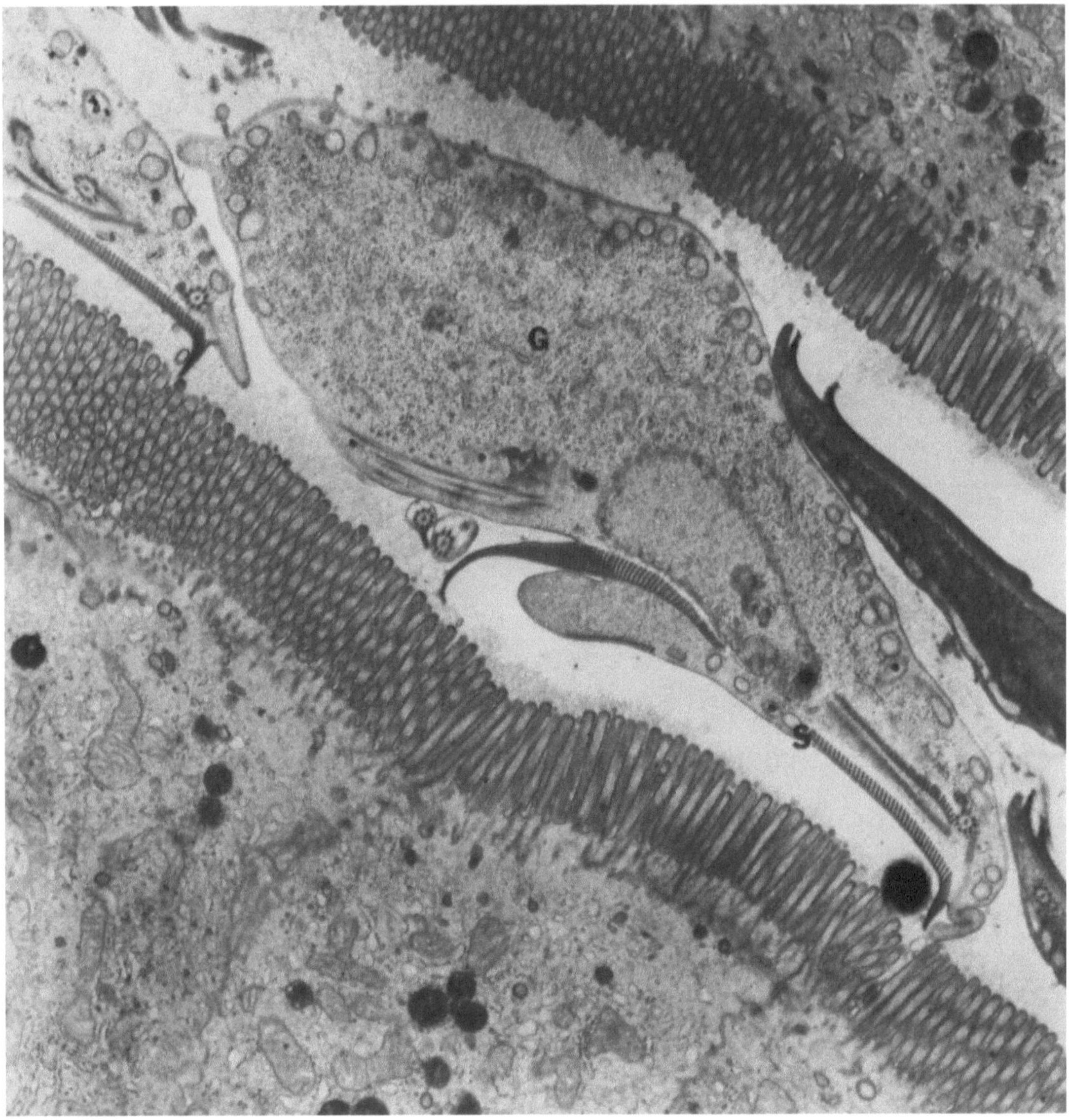

Figure 5-29. Electron micrograph showing luminal trophozoites of *Giardia lamblia* (G) adjacent to intestinal epithelial cells of a patient with chronic diarrhea. The sucker plate (S) of the organism appears to be penetrating the epithelial cell microvilli. × 16,000. (From Dobbins,[138] reprinted by permission of John Wiley & Sons, Inc. Copyright © 1978, John Wiley & Sons, Inc.)

specimens of one third of sexually active gay men.[63] Attachment of spirochetes and bacteria to the small intestine has *not* been observed in adults.

The human intestine is often parasitized by the protozoan *G. lamblia.* This may occasionally result in extensive injury to the intestinal mucosa, the crypt cells being particularly susceptible to injury. Figure 5-29 shows several trophozoites of *G. lamblia* adjacent to epithelial cells in a human intestinal biopsy specimen. The sucker plate of one of the organisms appears to be penetrating microvilli. Following penetration, these organisms may invade the epithelial cell and even be found within the lamina propria.[1]

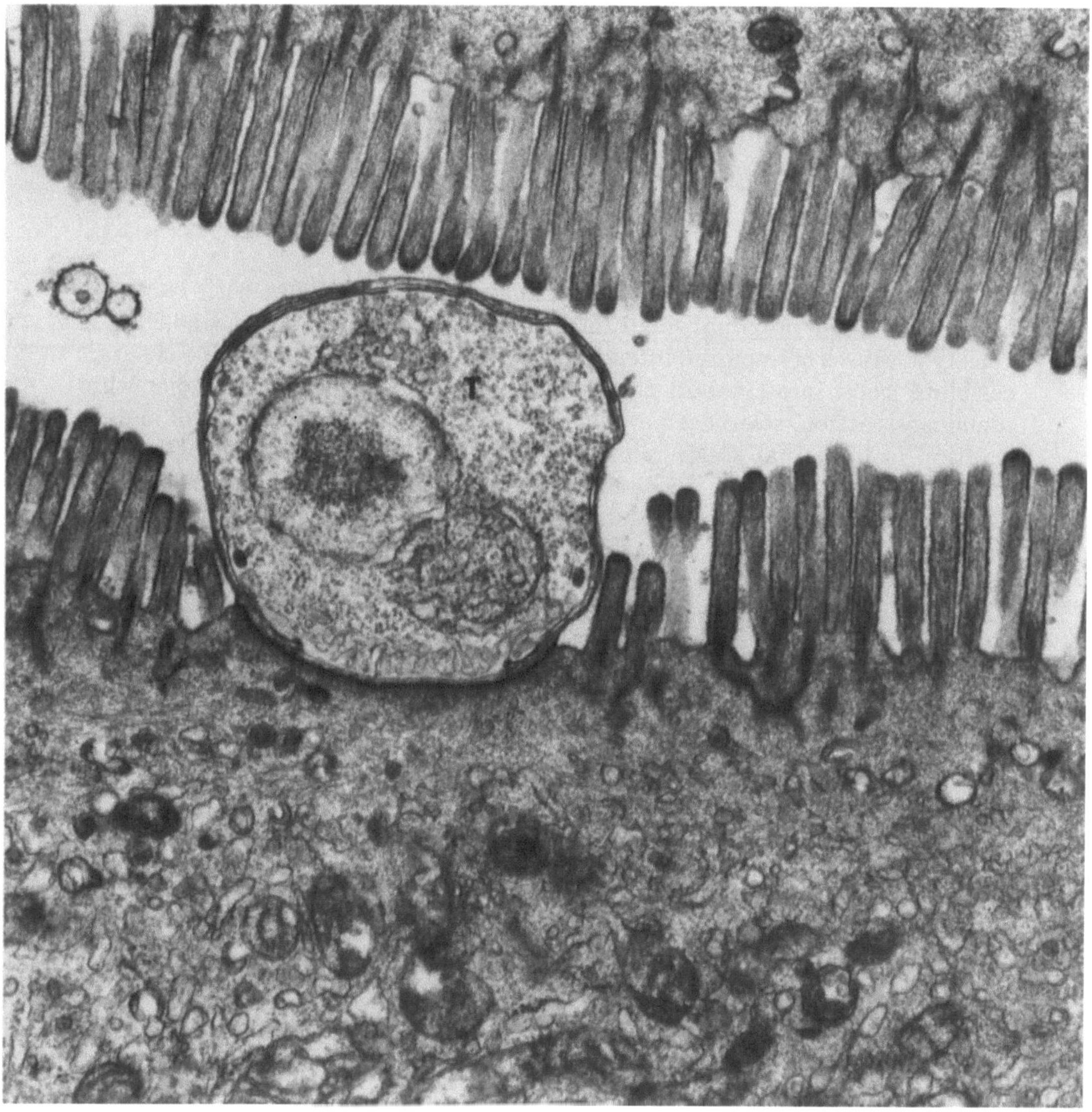

Figure 5-30. Electron micrograph showing a single cryptosporidial trophozoite (T) intimately attached to the plasma membrane at the apex of an intestinal epithelial cell in a patient with AIDS and diarrhea. Even though the microvilli are displaced at the site of attachment there is no evidence of epithelial cell injury. × 31,900

Cryptosporidia are protozoan parasites that may produce mucosal abnormalities of the small intestine in a variety of species, including humans.[59,64] The cryptosporidia appear to be intimately attached to the microvillus plasma membrane of intestinal epithelial cells, as illustrated at electron microscopy (Figures 5-26, 5-30). Cryptosporidiosis is now frequently found in patients with AIDS.[59,64,65] Its clinical manifestation is profuse, watery diarrhea. The diarrhea has been speculated to result from the unique attachment of cryptosporidia and alteration of the sodium pump (cyclic adenosine monophosphate?) of the epithelial cell. Marcial and Madara[66] have clearly shown, using freeze-fracture techniques, that the cryptosporidia in guinea pig intestine are encased within the plasma membrane of the epithelial cell and hence are truly intracellular in location. In contrast, *Isospora belli*, another member of the coccidia family that includes the cryptosporidium, invades deeply into the intestinal epithelial cell and is found in an intracytoplasmic position. Coccidia and other parasites that invade intestinal epithelial cells will be discussed later in this chapter.

Stasis Syndromes

Steatorrhea and/or macrocytic anemia may arise in patients with bacterial overgrowth in the proximal small intestine.[67] This overgrowth is secondary to a variety of causes of impaired intestinal transit, including the presence of surgical blind loops, intestinal strictures, jejunal diverticulosis, chronic idiopathic pseudoobstruction, scleroderma, and diabetes mellitus. The etiology of malabsorption in intestinal stasis is probably multifactorial. Ament et al.[67] found bacilli in the submucosa in 4 of 30 biopsy specimens obtained from their patients with intestinal stasis; thus direct injury to the mucosa may play a role. In some of these patients, intraluminal bacterial conversion of conjugated to unconjugated bile salts may be of importance. Fagundes-Neto et al.[68] studied the effects of unconjugated bile salts on rat jejunal absorption of the macromolecular tracer horseradish peroxidase, and demonstrated that relatively low levels of unconjugated bile salts (0.5 mM) resulted in increased permeability at the epithelial cell tight junction. Patchy intestinal epithelial cell abnormalities (Figure 5-31), whether due to toxicity of unconjugated bile salts or the presence of the bacteria themselves, are often present in varying severity and may contribute to the malabsorption.[67]

The diagnosis of stasis syndromes is based on three criteria: (1) presence of increased volumes of proximal jejunal contents, (2) demonstration of increased concentrations of bacteria in the proximal jejunum, and (3) clinical and laboratory responses to an appropriate antibiotic. Most normal individuals will have bacterial counts in the proximal jejunum of 10^3/ml or less, whereas most individuals with stasis syndromes will have counts greater than 10^7/ml. It is important to make a firm diagnosis of stasis syndrome because some patients may have surgically correctable lesions while others may require periodic administration of antibiotics.

Mild, moderate, and even severe mucosal lesions similar to those seen in celiac sprue are common in the pseudoobstruction syndrome chronic intestinal pseudoobstruction (CIP) and in scleroderma.[69–73] The lesions appear to result from chronic intestinal stasis with intraluminal bacterial overgrowth, but this relationship is not fully settled.[71] Many patients have a dramatic clinical response (cessation of diarrhea and weight gain) following therapy with oral antibiotics.[71] A clinical trial with a gluten-free diet may be necessary to exclude celiac sprue in some patients with severe (flat) intestinal mucosal lesions.

Intestinal pseudoobstruction is a chronic illness characterized by recurrent obstructive symptoms not explained by mechanically correctable lesions. It is due to ineffective intestinal peristalsis. Pathologically there is usually an abnormality of the smooth muscle or of the myenteric plexus.[72] The form of CIP called familial visceral myopathy (FVM), and progressive systemic sclerosis (PSS) or scleroderma, have many similarities but are usually histologically distinguishable. In familial visceral myopathy, degenerating muscle cells and fibrosis are found throughout the muscularis but are most prominent in the

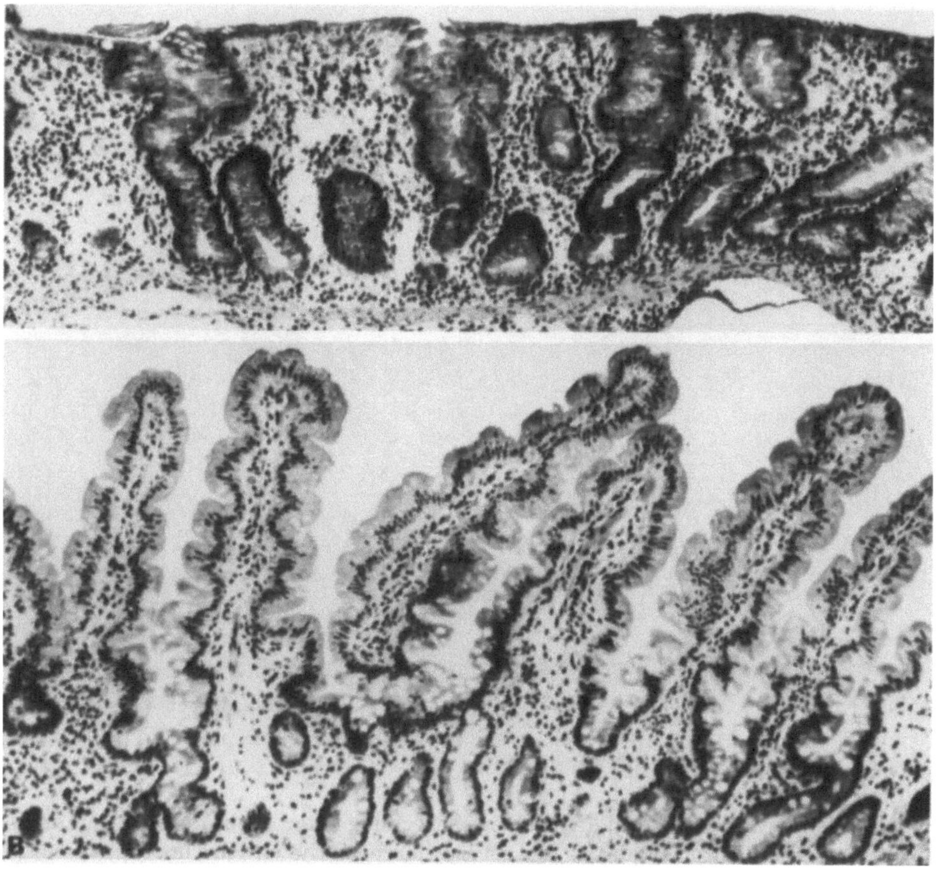

Figure 5-31. Biopsy specimens from a patient with intestinal stasis, steatorrhea, bacterial overgrowth, and intestinal pseudoobstruction. (A) Severe proximal jejunal lesion. H & E stain, × 184. (B) Normal specimen from site adjacent to specimen in (A), illustrating the patchiness of the lesion. H & E stain, × 184. (A and B reproduced from Dobbins,[16] by permission of Churchill Livingstone, Inc., New York. Original supplied by C.E. Rubin.)

muscularis externa. Electron and light microscopy show that the generating muscle cells have a pale cytoplasm with a fragmented and vacuolated appearance. Neural and vascular structures are normal. The lesion in PSS tends to be patchy in distribution, while it is diffuse in FVM. Fibrosis is more extensive in PSS, involving the circular muscle more than the externa. Vascular abnormalities may be present in PSS. Most important, the remaining muscle fibers in PSS are largely normal in appearance and have no vacuolar change.[72] In the neuropathic form of CIP, the muscularis propria is normal at both light and electron microscopy.[73]

Zollinger-Ellison Syndrome

One third of patients with ZES have diarrhea; 7% present with diarrhea as the major manifestation.[74] Steatorrhea is sometimes present in both groups of patients, and patchy nonspecific abnormalities may be seen in the proximal segment of the jejunum.[74,75] The diagnosis is based on a finding of elevated serum

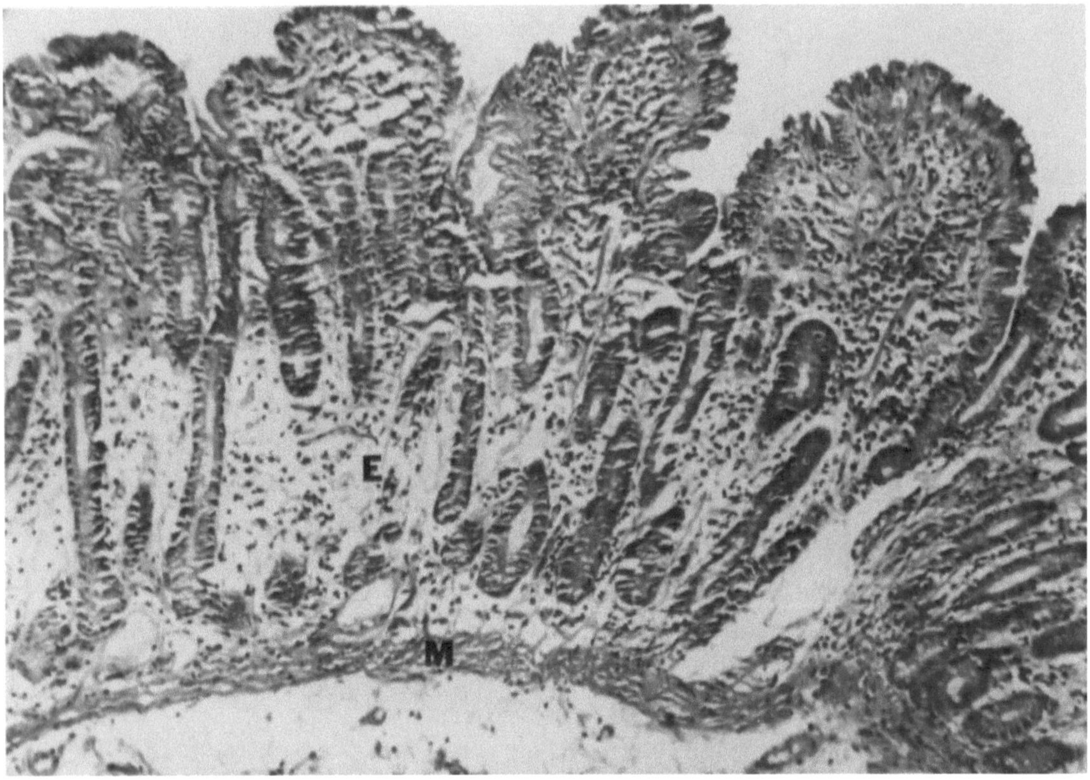

Figure 5-32. Moderate villous abnormality in biopsy specimen taken at the duodenojejunal junction of a patient with Zollinger-Ellison syndrome and severe diarrhea. The villi are broadened and there is a mild increase in number of plasma cells in the lamina propria. The surface epithelium is slightly flattened and irregular in appearance. There is moderate edema (E) of the lamina propria just above the muscularis mucosae (M). H & E stain, × 132. (Reproduced from Dobbins,[16] by permission of Churchill Livingstone, Inc., New York.)

gastrin levels in individuals with hypersecretion of gastric acid. Because patients may have evidence of severe mucosal abnormalities in the proximal small bowel on barium studies of the upper gastrointestinal tract, physicians may obtain small-intestine biopsies when they do not suspect the diagnosis of ZES clinically. The patchy lesions identified in the biopsy are generally mild in degree (Figure 5-32), and very uncommonly severe in degree. Three structural responses of the intestinal mucosa to excessive gastric acid production have been described. The most common is an inflammatory response with surface microulcerations, PMN leukocyte infiltration of the lamina propria, and edema of the lamina propria and submucosa.[75] Plasma cells may be increased in the lamina propria, but not to the degree seen in celiac sprue. The second most common is that of gastric surface mucous cell metaplasia of intestinal absorptive cells.[74] The third response is injury to intestinal absorptive cells, seen in electron microscopy but not in light microscopy.[74] Steatorrhea in these patients is largely related to acid inactivation of intraluminal enzymes, the structural abnormalities in the intestinal mucosa probably playing little role in this regard.

Graft-vs.-Host Disease

Graft-vs.-host disease (GVHD) is an immunological illness that results in skin, liver, and gastrointestinal mucosal damage. It occurs in humans as a pathogenic process following bone marrow transplantation.[76] Transplantation of marrow stem cells is now an appropriate treatment for aplastic anemia, refractory leukemia, and selected severe immunodeficiency syndromes.[76] Prior to transplantation patients must be given immunosuppressive therapy so that their body will not reject the marrow graft. This therapy generally consists of a combination of total body irradiation and chemotherapeutic drugs such as cyclophosphamide. Immunosuppressive therapy often results in damage to the liver and to the intestinal mucosa. Changes in the intestinal mucosa seen in this circumstance are discussed in the section of this chapter dealing with macrocytosis (impaired epithelial cell replacement). Changes secondary to immunosuppressive therapy are present for only up to 20 days after cessation of therapy.[77]

Acute GVHD starts three to seven weeks after transplantation and may be accompanied by crampy abdominal pain, profuse watery diarrhea, and jaundice.[76] Lesions range from necrosis of individual crypt cells to total loss of mucosa. The most severe involvement occurs in the ileum and right colon. There may be sheets of bacteria, fungi, and debris lining the few epithelial cells remaining. The earliest and mildest change seen in GVHD is that of necrosis of individual crypt cells, a change also seen following chemotherapy and irradiation injury. Changes of grade II severity include crypt abscess and crypt cell flattening, with or without crypt cell necrosis. In grade III severity there is, in addition to the above changes, dropout of one or more whole crypts in a biopsy specimen. In grade IV severity there is total denudation of the epithelium. The lamina propria does not appear to be altered histologically, but with appropriate immunofluorescent staining a marked depletion of IgA- and IgM-bearing plasma cells can be demonstrated. Intestinal infections due to bacteria, viruses, fungi, and parasites are the most troublesome complication for patients. Evidence of their presence may be obtained from appropriate examination of mucosal biopsy specimens. Clinically, rectal biopsy is more useful, and safer, than proximal small-intestine biopsy for detecting mucosal changes in GVHD.[76,77]

Small-intestine involvement is uncommon in patients with chronic GVHD. Chronic GVHD develops insidiously 3 to 12 months after transplantation in up to 40% of long-term survivors of allogeneic marrow transplants. Intestinal involvement in chronic GVHD is manifested by intractable diarrhea, malabsorption, abdominal pain, and severe malnutrition. Understandably, few peroral intestinal biopsies have been obtained in these patients, but ones obtained have been normal. Rubin has observed abnormal intestinal biopsy changes in one patient with chronic GVHD (Cerubin, Personal communication). Autopsy specimens may show focal fibrosis in the lamina propria and segmental fibrosis of the submucosal and serosal layers, extending from stomach to colon.[76]

Ultrastructural changes in the gut in GVHD have been reported only in rectal biopsy specimens.[77] All such specimens were obtained after 21 days, when the diffuse and severe chemoradiotherapy changes had cleared. Epithelial injury and lymphocytic infiltration predominated in the bases of crypts in mild GVHD and extended to the surface epithelium in severe GVHD. Lymphocytes formed the predominant cellular infiltrate and indented the cytoplasmic membranes

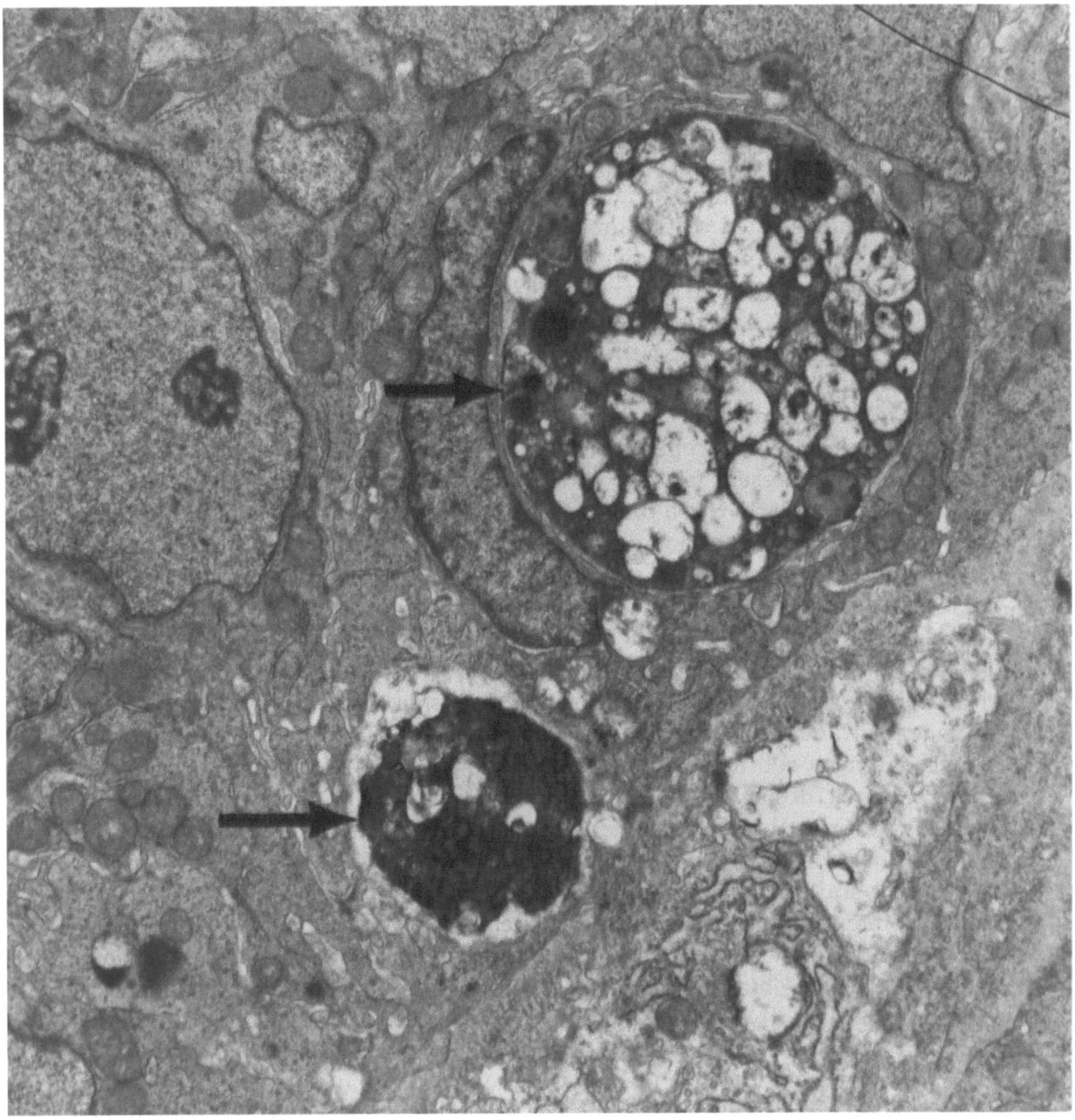

Figure 5-33. Electron micrograph showing presence of single-cell necrosis (arrows) of crypt epithelial cells in an intestinal biopsy specimen of a patient with acquired immunodeficiency syndrome, intestinal cryptosporidiosis, and cytomegalovirus infection. Note the membrane-bound collection (upper) of disrupted cellular constituents, including mitochondria and lysosomes, and the nonmembrane bound collection (lower), which appears to contain nuclear debris. There is no associated inflammatory reaction. × 14,500

of epithelial cells by point contact and by broad pseudopods, which projected deep into epithelial cell cytoplasm. The epithelial cell-lymphocyte point contacts were presumed to represent the recognition phase of lymphocyte-mediated cytolysis. Single-cell necrosis (apoptosis) of epithelial cells was prominant and presumably the result of lymphocyte-mediated injury.[77] Changes similar to those seen in GVHD have been described in intestinal mucosal biopsy specimens of individuals with AIDS and chronic diarrhea (Figures 5-33, 5-34). There

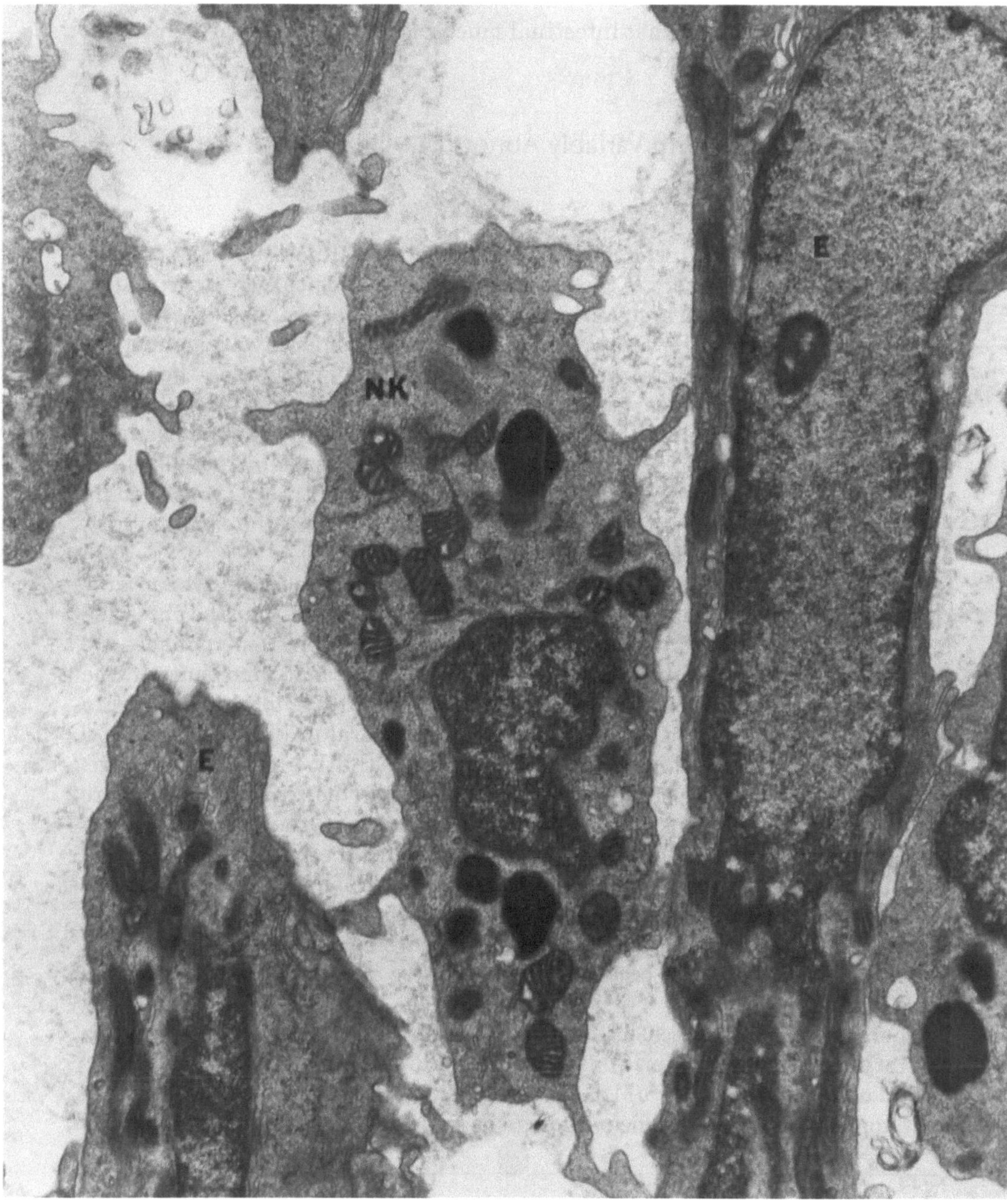

Figure 5-34. Electron micrograph of an intraepithelial lymphocyte with all the morphological features of a large granular lymphocyte or natural killer (NK) cell in the intestinal biopsy specimen of a patient with acquired immunodeficiency syndrome and chronic diarrhea. Note the presence of numerous mitochondria and the multiple large dense granules. In particular, note the cytoplasmic projections of the NK cell and their intimate relationship with adjacent epithelial (E) cells. × 19,600

is some reason to suggest that intestinal mucosal injury in AIDS is both pathogen and immune mediated.[78]

Diseases Associated with Variably Abnormal Villi

Whipple's Disease

Whipple's Disease is an uncommon, systemic, bacterial illness affecting primarily middle-aged white men. It is characterized morphologically by the infiltration of virtually all organ systems by macrophages that are intensely stained by the PAS stain. There is, however, a unique predisposition for involvement of the lamina propria of the small intestine, the mesenteric lymph nodes, the mitral and aortic valves of the heart, and the central nervous system.[79] Bacilli (most appropriately called Whipple bacilli) with all of the structural characteristics of bacteria have been observed within the intestinal mucosa, Kupfer cells of the liver, mesenteric and peripheral lymph nodes, heart, central nervous system, eye, and lung.[79,80] Patients may have impaired cellular-mediated immunity, although this has not been clearly established.[81]

Because the lamina propria of the proximal intestine almost always contains characteristic macrophages, biopsy examination of tissue from the intestine usually provides the definitive diagnosis. The typical Whipple macrophages are easily recognizable on routine H and E staining (Figure 5-35) and thus routine PAS staining is not necessary, although this is generally helpful to do (Figure 5-35B). It should be noted that macrophages in general are PAS positive, and therefore this feature alone does not distinguish the Whipple macrophage. However, the appearance of the PAS-stained intestinal biopsy specimen (Figure 5-35) is usually sufficient to establish the diagnosis in the vast majority of patients, and electron microscopy documentation of the presence of Whipple bacilli (Figure 5-35) is not essential. Occasional macrophages are found in the normal intestine. They may stain faintly (usually) or even strongly with the PAS reaction, but the inclusions are not sickle shaped. The experienced pathologist will not be misled by these occasional PAS-positive macrophages in the normal intestine.

There are three rare occasions when numerous PAS-positive macrophages in the intestinal lamina propria may be misleading: (1) in AIDS with *Mycobacterium avium-intracellulare* (MAI) infection, (2) in systemic histoplasmosis, and (3) in macroglobulinemia. The pathologist should easily distinguish the faintly

---

►

Figure 5-35. Untreated Whipple's Disease. (A) Severe proximal jejunal lesion showing marked infiltration of the lamina propria by macrophages that stain rather faintly (M). The surface epithelial cells are somewhat flattened. H & E stain, × 250. (B) Section adjacent to that shown in (A), showing intense staining of the macrophages. PAS and hematoxylin stain, × 250. (C) High-magnification light microscopy photograph of PAS-stained macrophages showing the rounded and sickle-shaped appearance of the individual macrophage inclusions. The appearance is virtually diagnostic of Whipple's disease, especially when found in the intestinal mucosa. PAS and hematoxylin stain, × 1000 (D) Electron micrograph of intestinal lamina propria showing free bacilli (arrows) and ingested bacilli within macrophages (M) and within a polymorphonuclear leukocyte (P). × 21,300

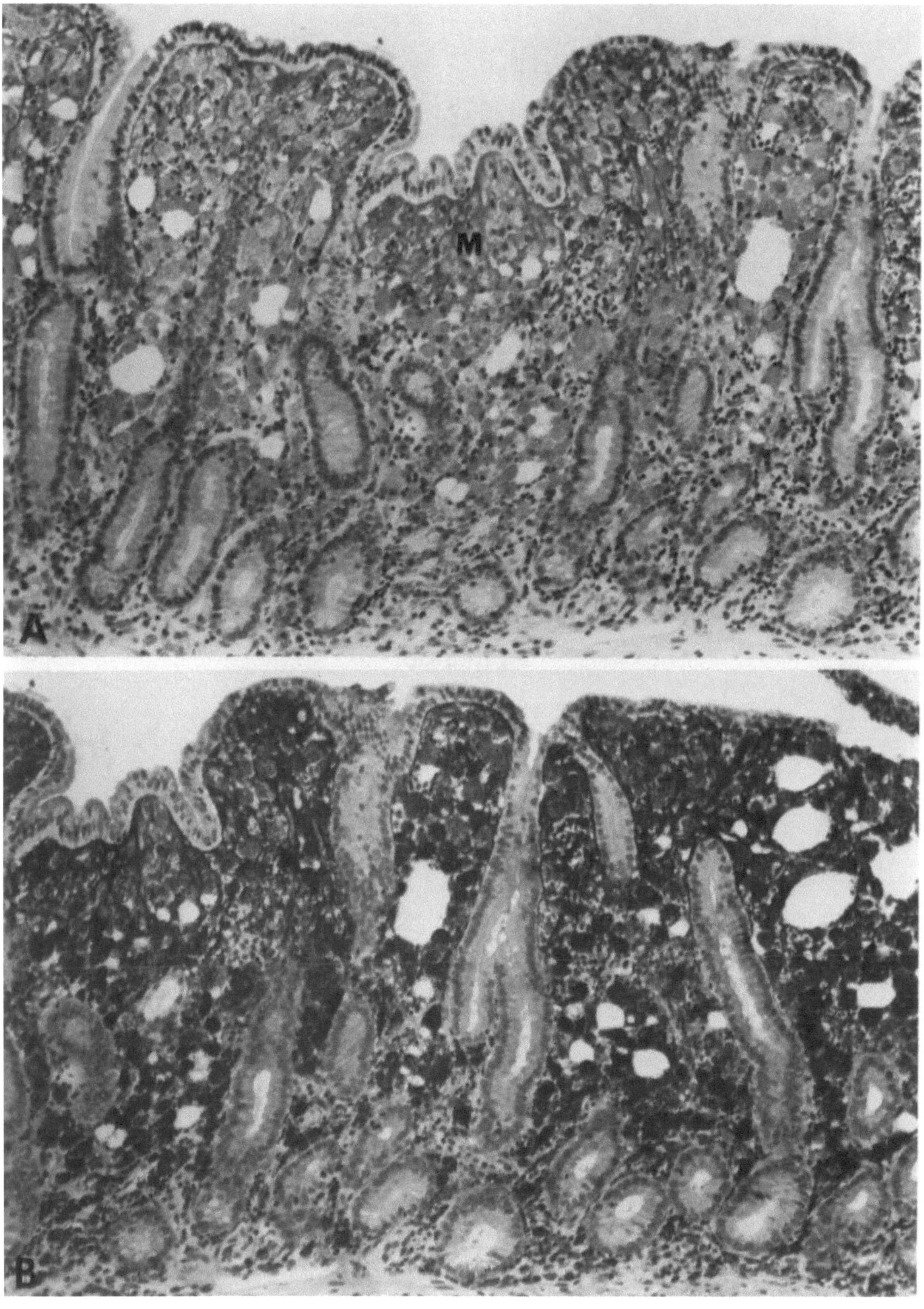
M
A
B

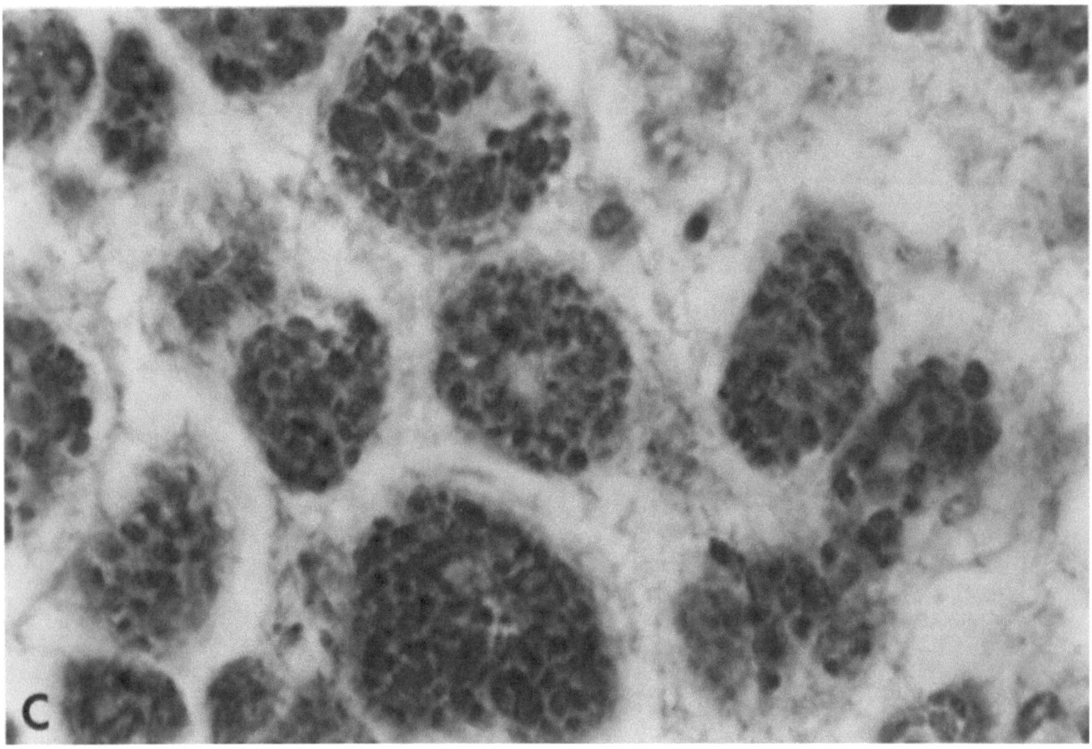

Figure 5-35C.

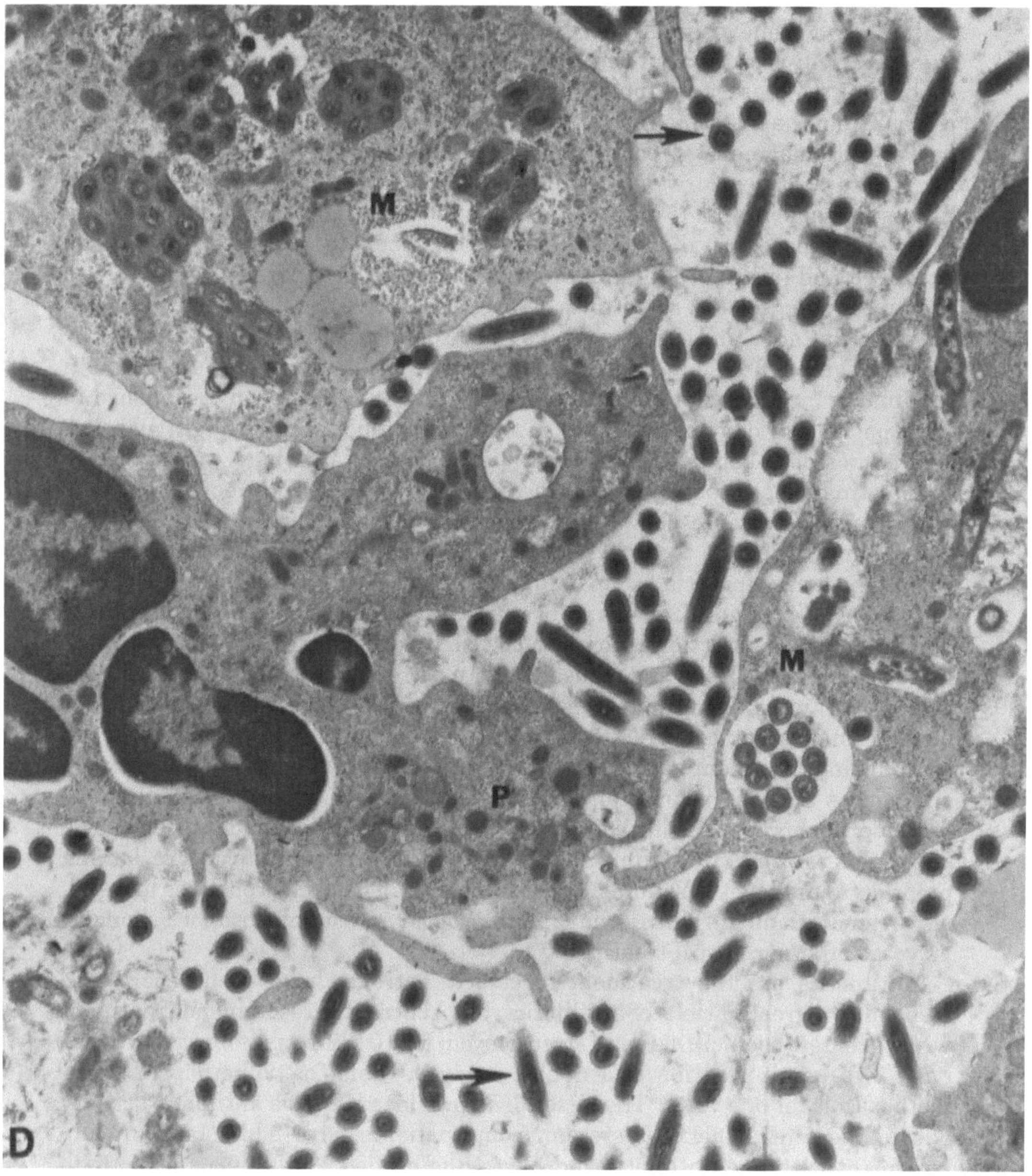

Figure 5-35D.

staining homogeneously PAS-positive macrophages of macroglobulinemia and the large, PAS-positive, rounded, encapsulated histoplasma organisms in macrophages from those of Whipple's disease.[82] It is harder to distinguish Whipple's disease from the intestinal mucosa associated with AIDS and a lamina propria packed with macrophages containing MAI (Figure 5-36).[83] In the latter case the bacilli are acid fast, are easily cultured, and have a characteristic electron microscopy appearance (Figure 5-37) different from that of Whipple bacilli. In Whipple's disease the villous architecture is distorted to varying degrees by the expanded and infiltrated lamina propria, the changes generally being mild to moderate in degree but sometimes severe. Absorptive cells at the apex of villi are often flattened and can be shown at electron microscopy to be infiltrated by bacilli (Figure 5-38). There are several recent reports of Whipple's disease with minimal or *no* intestinal involvement.[79] The diagnosis has been established in these patients by demonstration of presence of Whipple bacilli by light and electron microscopy examinations of lymph nodes or of brain biopsy specimens.

Intestinal biopsy specimens obtained during antibiotic treatment of patients with Whipple's disease show improvement in the appearance of the surface epithelium within one to two weeks. Free bacilli have been reported to persist within the lamina propria for up to nine weeks. These bacilli can be observed at light microscopy using Brown and Brenn stains of paraffin-embedded sections, and in PAS-stained sections. However, the bacilli are more easily detected when biopsy samples are fixed and embedded for electron microscopy but "thick" sectioned at 1 μm for light microscopy examination.[80] The macrophages, which at electron microscopy are filled with bacilli in various stages of degeneration, clear during treatment at a much slower rate than do free bacilli, so that even six months and one year after treatment the lamina propria often contains prominent macrophages, although free bacilli are no longer identifiable. The macrophages remaining after treatment tend to predominate in the lamina propria around crypt bases and in the submucosa. These macrophages may persist in small numbers within the lamina propria and submucosa for as long as 11 years after successful treatment.[79]

Antibiotics are the treatment of choice. Patients have been reported to respond to a great variety of antibiotics, but I currently recommend treatment with parenteral penicillin G and streptomycin for 10 to 14 days followed by twice daily oral administration of a double-strength tablet of trimethoprim-sulfamethoxazole for one year.[79] The latter drug provides effective antibiotic activity within the central nervous system. Some patients have had relapse during treatment with one antibiotic and have thus needed a different antibiotic to complete successful treatment. These problem patients are probably best followed by periodic examination of 1-μm plastic sections of intestinal biopsy specimens for the presence of bacilli free within the lamina propria.[79]

▶

Figure 5-36. Light microscopy photographs of intestinal mucosa at necropsy of a patient with AIDS and *Mycobacterium avium-intracellulare* involvement of the gut. (A) The lamina propria is packed with macrophages that, with H & E staining, clearly resemble those seen in Whipple's disease. H & E stain, × 500. (B) PAS and hematoxylin stain of the same specimen as in (a) again reveals similarity to Whipple's disease. × 500. (C) Acid-fast stain shows the macrophages to be packed with large bacillary structures. Whipple bacilli are not stained with this method. H & E stain, × 500

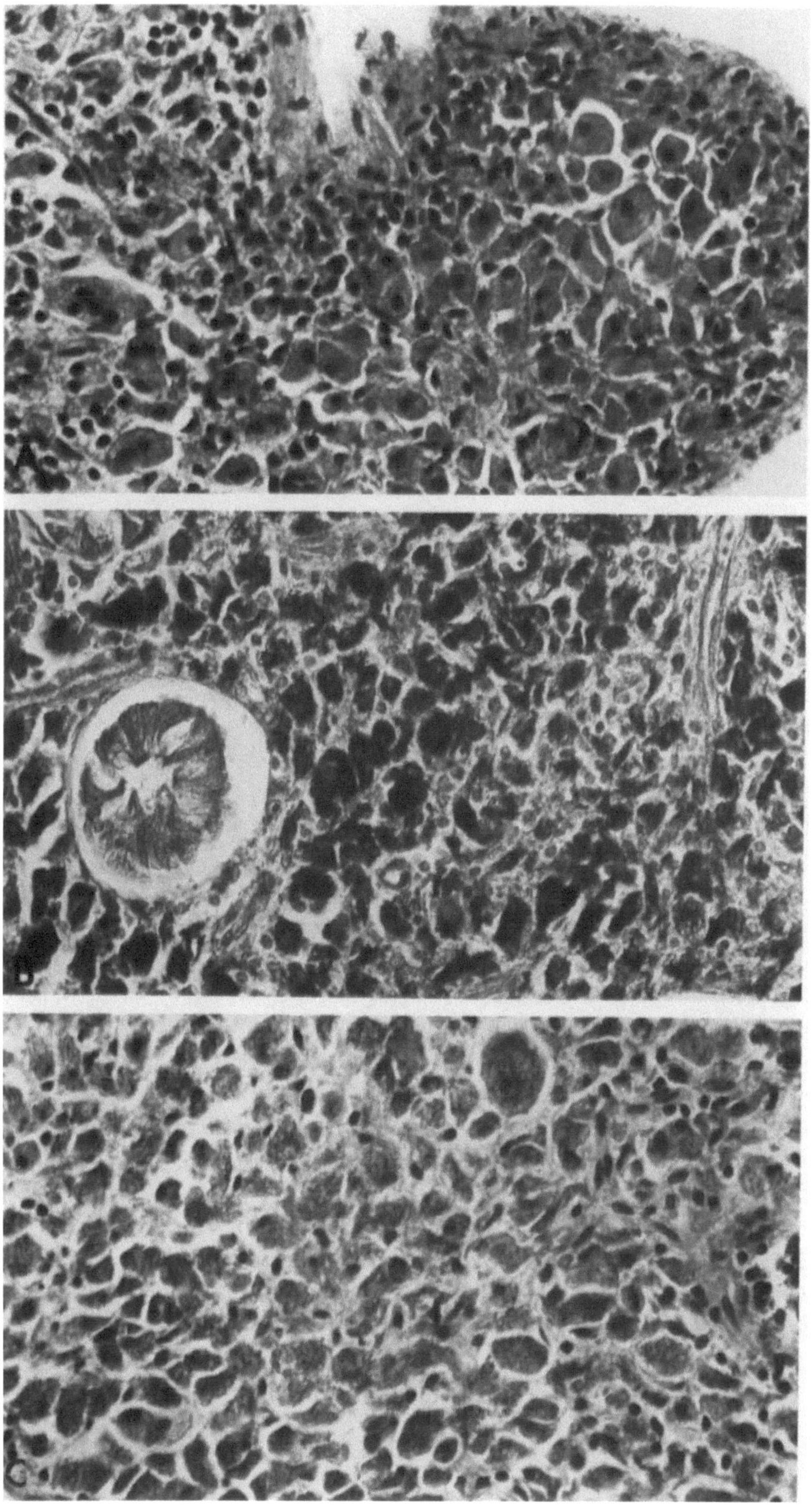

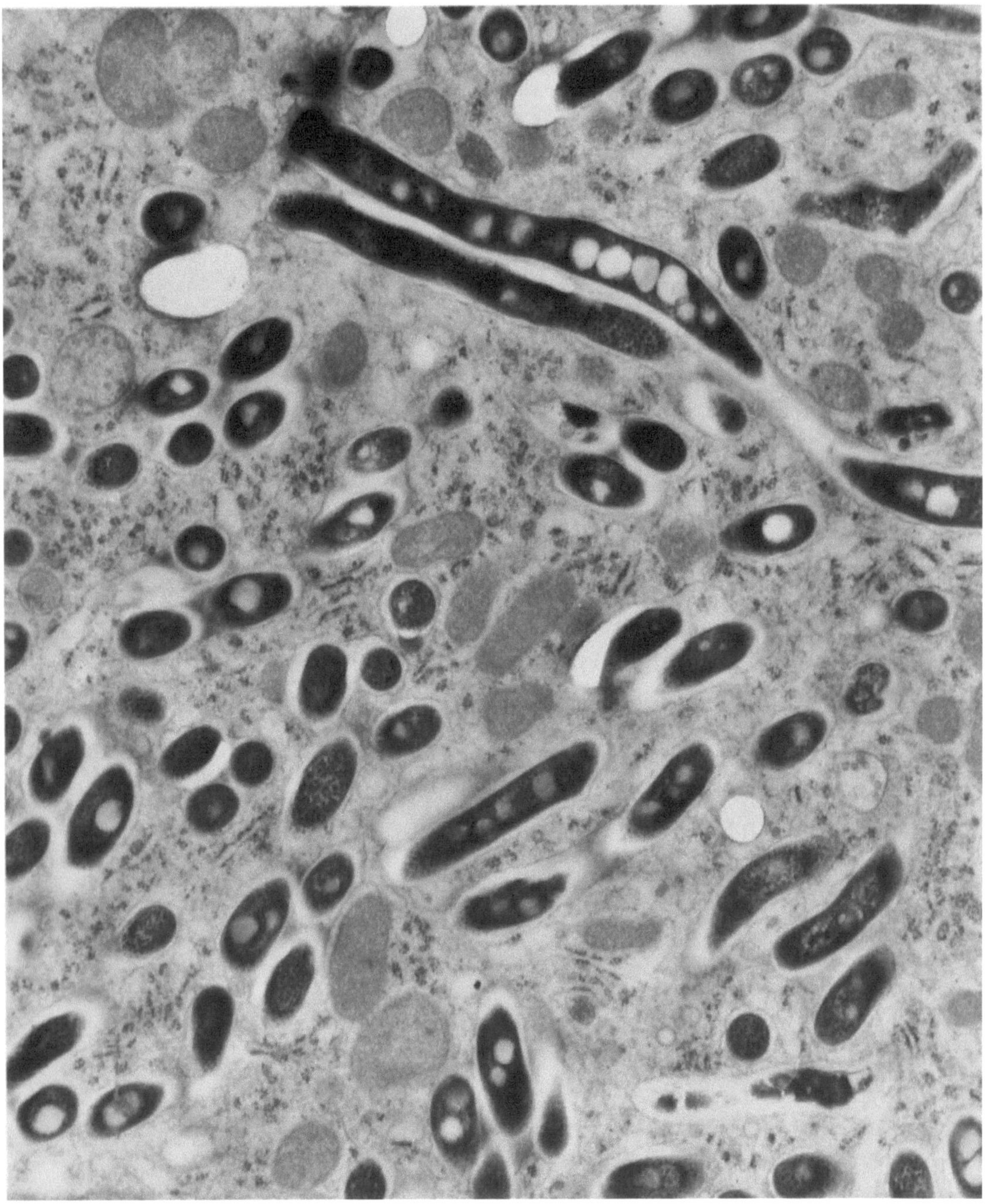

Figure 5-37. Electron micrograph showing typical appearance of *Mycobacterium avium-intracellulare* within an intestinal mucosal macrophage. Note that these organisms are obligate intracellular (macrophage) organisms, whereas viable Whipple bacilli are largely extracellular. × 25,000

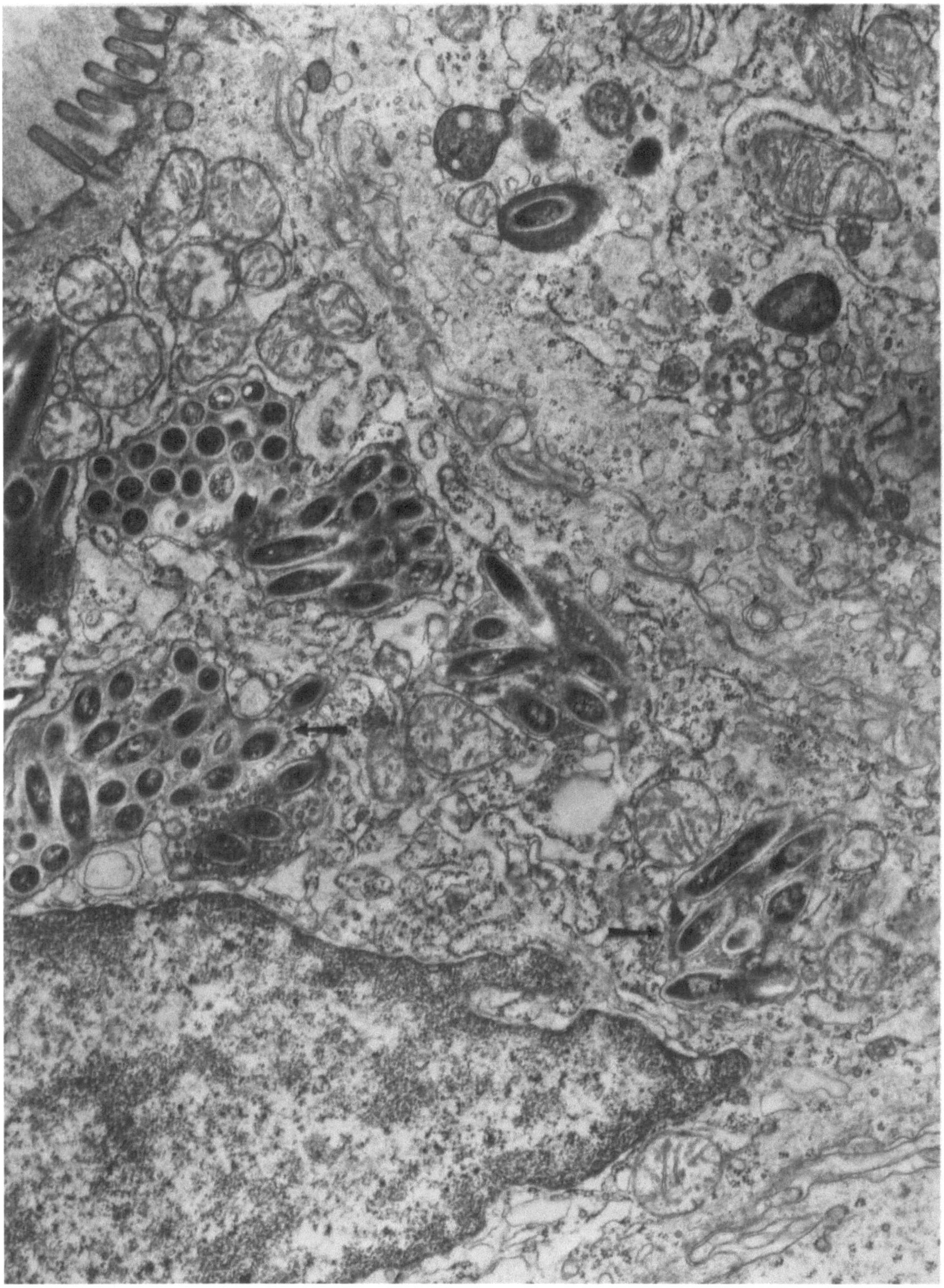

Figure 5-38. Electron micrograph showing intense invasion of intestinal epithelial cells by Whipple bacilli (arrow) in an untreated patient. × 21,700

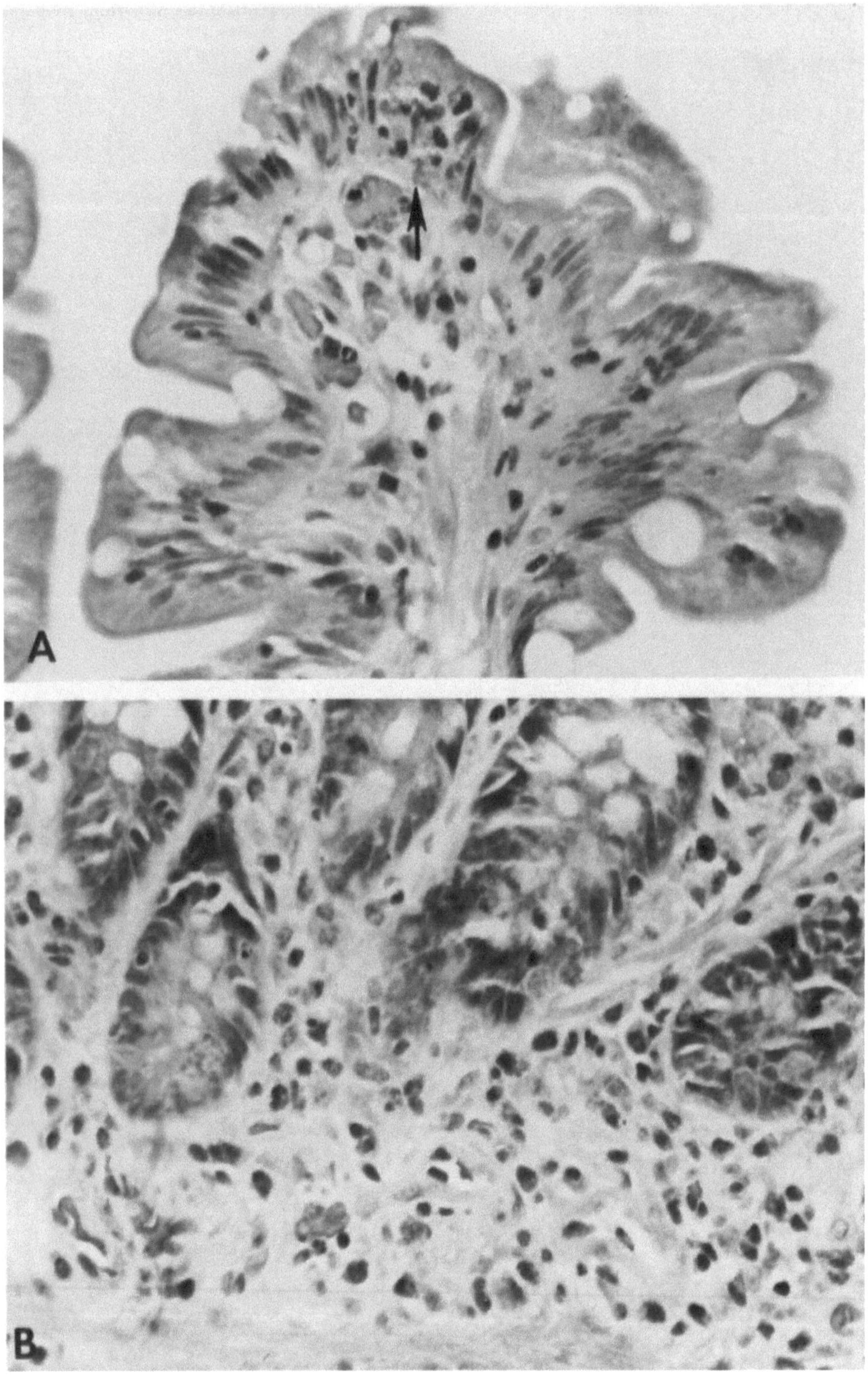

Figure 5-39. Biopsy specimen from patient with eosinophilic gastroenteritis and protein-losing enteropathy. (A) Apex of villus shows eosinophilic infiltration of the lamina propria and a cluster of eosinophils in the epithelium (arrow). H & E stain, × 500. (Reproduced from Dobbins,[16] by permission of Churchill Livingstone, Inc., New York.) (B) Base of crypts in biopsy section illustrated in (A) showing marked eosinophilic infiltrate, most of the cells in the lamina propria being eosinophils. H & E stain, × 500

Eosinophilic Gastroenteritis

Patients with eosinophilic gastroenteritis generally have substantial eosinophilic leukocytosis in the peripheral blood and gastrointestinal symptoms characterized by nausea, vomiting, diarrhea, abdominal pain, and sometimes steatorrhea and protein-losing enteropathy.[84] Eosinophilic infiltration of the intestine may be extramucosal (in the muscularis externa or subserosa), or the infiltration may be found in more distal portions of the small bowel, where peroral biopsy is difficult. When the eosinophilic infiltration involves the mucosa of the proximal jejunum, its presence can be established by peroral biopsy. The mucosal lesion tends to be very patchy, and multiple biopsy specimens may be needed to detect the eosinophilic infiltration. Leinbach and Rubin[85] have recommended that a minimum of eight specimens be examined before one concludes that there is no involvement of the proximal jejunum. Eosinophilic leukocytes are, of course, normally present in the lamina propria, but in this disease they are abnormally numerous and may occur in clumps within the lamina propria, submucosa, and/or epithelium (Figure 5-39). There may be massive infiltration by eosinophilic leukocytes, with complete loss of villi.[85,86]

The pathogenesis of eosinophilic gastroenteritis, a condition often attributed to allergy, is unclear. Only about half of the cases occur in atopic patients, and in this group food allergy is uncommon. Patients with the allergic form of eosinophilic gastroenteritis have an atopic history and food sensitivity, and may have very high circulating levels of serum IgE. Fluorescent staining of biopsy specimens may show a marked increase in IgE-staining plasma cells.[87] This suggests that this form of eosinophilic gastroenteritis is associated with IgE-mediated reactions arising in the gut. This does not establish a role for food allergy, however. Some patients appear to respond to removal of certain foods from their diet, particularly milk and meat products, while many require long-term treatment with low doses of corticosteroids.[84,85] The gastric mucosa may be more consistently involved than the intestinal mucosa, and hence gastric mucosal biopsy may establish the diagnosis when intestinal mucosal biopsy specimens are equivocal or negative for eosinophilic infiltration.[88] Rarely there may be esophageal involvement.[89]

Primary Intestinal Lymphoma

This disease usually presents with malabsorption and abdominal pain in the second or third decade of life.[90,91] It is not generalized lymphoma, in which small-intestine involvement may or may not be found. It is confined to the bowel and abdominal lymph nodes, and is more likely to be found in "undeveloped" areas of the world. It is often associated with an abnormal production of alpha heavy chain. Hence it is sometimes called heavy chain disease.[91]

The Mediterranean form of primary intestinal lymphoma may present initially as a diffuse and intense lymphoplasmacytic infiltration of the intestinal mucosa. Thus it is frequently termed immunoproliferative small intestine disease (IPSID).[92-94] This proliferation may be secondary to intense bacterial antigenic stimulation, and antibiotic treatment may inhibit evolution of an apparently benign disorder into a malignant stage.[92,93] The most constant histological finding in IPSID is a dense infiltrate of the lamina propria by mature-appearing plasma cells. Often these cells contain alpha heavy chain without

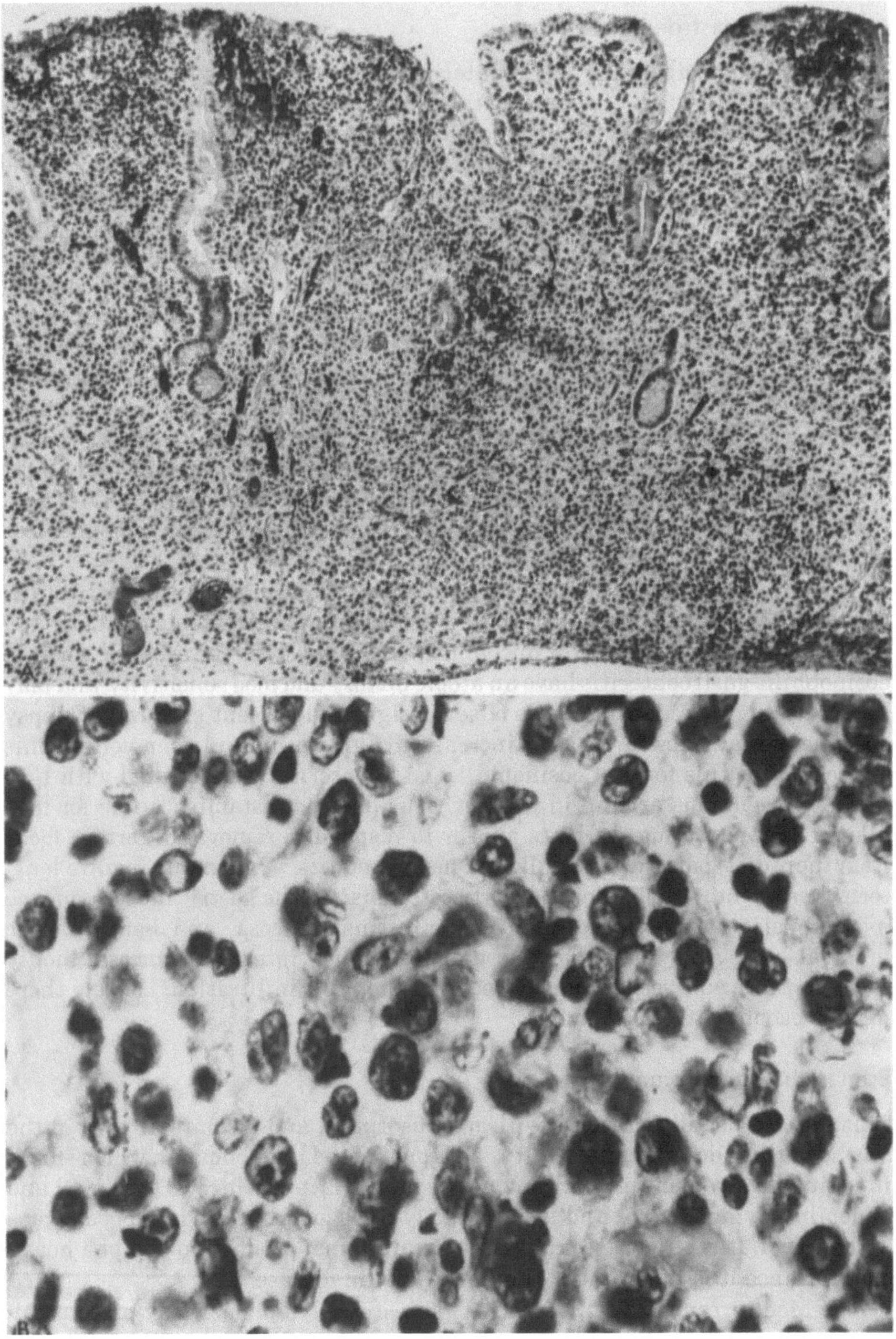

Figure 5-40. Primary intestinal lymphoma. (A) Absence of villi, loss of crypts, and striking round cell infiltration of the lamina propria. H & E stain, × 184. (B) Higher magnification shows that the infiltrate consists of malignant lymphoid cells. H & E stain, × 1,840. (b reproduced from Dobbins,[16] by permission of Churchill Livingstone, Inc., New York. Original supplied by C.E. Rubin.)

light chain. If malignant lymphoma develops, the superficial lamina propria often retains its dense plasma cell infiltrate, while the cytologically malignant cell is seen in the deeper portions of the mucosa, with invasion of crypts and involvement of submucosa and muscularis. The malignant lymphoma seen in IPSID is variously designated as lymphoplasmacytoid, follicular, diffuse follicular-center cell, or B-immunoblastic lymphoma. In the majority of these neoplasms the malignant cells do *not* contain light or heavy chains. Lesser degrees of lymphoplasmacytic infiltration of the intestinal mucosa are common in individuals from Southeast Asia, India, and Africa (tropical sprue?) and should not be confused with IPSID.

If there is extensive disease with involvement of the proximal jejunum and/or duodenum, peroral biopsy often allows the diagnosis (Figure 5-40). The malignant lymphoid infiltrate is found throughout the lamina propria, crypts are sparse, and the villi may no longer be present. Patchy involvement of the proximal intestine by the lymphoma can be detected, but this often requires scanning of many biopsy specimens to find the few malignant cells that make the diagnosis. The changes may be indistinguishable from those seen in celiac sprue, with no malignant cells being found in some sections and sparse malignant lymphoid cells being found in other sections. Eidelman, Parkins, and Rubin[90] found only scattered malignant lymphoid cells in a few serial sections from eight biopsy specimens of the proximal jejunum of a patient with primary intestinal lymphoma.

Pseudolymphoma (focal lymphoid hyperplasia) involving the small intestine alone is distinctively uncommon.[95] There are two reports of jejunal pseudolymphoma and approximately 20 reports of ileal pseudolymphoma.[95] The key histological feature is a polymorphous cell population of infiltrating lymphocytes, usually with a pronounced, reactive-appearing, follicular component throughout the lesion. Local resection, when required for obstructive symptoms, appears to be curative.

Parasitic Diseases

The four parasitic diseases that may cause proximal intestinal mucosal abnormalities in North Americans include giardiasis, coccidiosis, microsporidiosis, and strongyloidiasis. Outside North America, schistosomiasis, leishmaniasis, and capillariasis must be considered. Infection with the flagellated protozoan *G. lamblia* is now recognized as a common cause of diarrhea in both adults and children. Many patients with this infestation, particularly adults, are asymptomatic, but others have watery diarrhea, frank steatorrhea, and chronic abdominal pain. The symptoms may easily be confused with those of irritable bowel syndrome. Giardial cysts may be found in the stool, and stool examination is the first diagnostic procedure to be performed. However, if giardiasis is strongly suspected and cysts are not found in the stool, the diagnostic procedure is duodenal biopsy with examination of duodenal contents and of mucus adherent to the biopsy for the presence of the protozoan by Giemsa stain of a dried smear.[67] The parasite may with difficulty be found in H & E-stained sections (Figure 5-41). The use of PAS and hematoxylin staining brings out the parasite very nicely. Sometimes the parasite can be found only in mucus adhering to the intestinal specimen, and it is advisable to prepare smears of intestinal juice when giardiasis is suspected. The smears can then either be stained routinely with the Giemsa method or stained at a later date if the parasite is not found on the routine

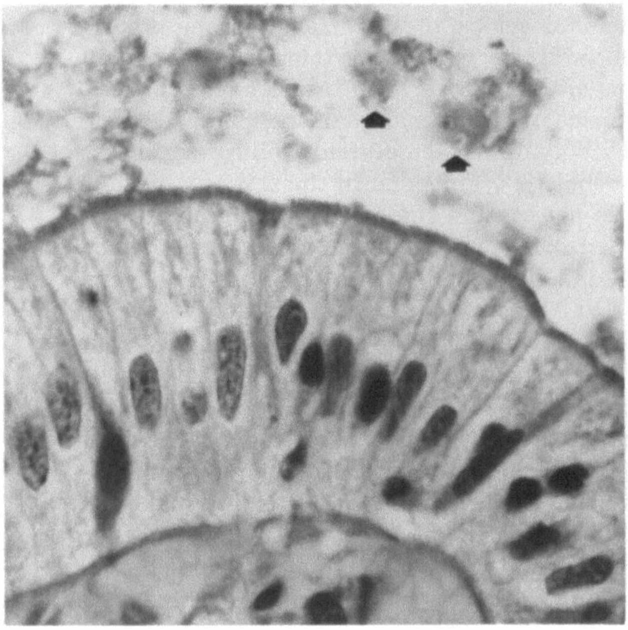

Figure 5-41. High-magnification photomicrograph of a villous tip from a biopsy specimen of a patient with giardiasis. The organisms are free in the lumen (arrows). H & E stain, × 1250. (Reproduced from Dobbins,[16] by permission of Churchill Livingstone, Inc., New York.)

sections. Mucus and intestinal juice obtained at the time of biopsy should be spread thinly on a glass slide and air dried for at least two hours. The smears can then be fixed in absolute methanol for 20 minutes and stained for 20 minutes in dilute Giemsa solution. This technique should be performed in all patients in whom small-intestine biopsy is being performed because of a suspected diagnosis of giardiasis.

Villous architecture in patients with giardial infestation is usually normal. Sometimes patchy lesions of mild to moderate severity may be seen, and exceedingly rarely a severe mucosal lesion may be detected.[96] The organism may rarely invade the intestinal mucosa, an observation first noted by Brandborg at light microscopy and later confirmed by other investigators at electron microscopy.[1,97] Giardiasis is the most common cause of diarrhea and steatorrhea in patients with late-onset immunodeficiency (CVH).[37] It also accounts for most of the morphological abnormalities of the intestinal mucosa in CVH (see discussion earlier in chapter). Giardiasis is not increased in frequency in patients with selective IgA deficiency.[37] All patients with giardiasis should be treated with quinacrine hydrochloride (Atabrine), or with metronidazole if quinacrine fails to eradicate the infestation.

Within the subphylum Sporozoa and the subclass *Coccidia* are two protozoan parasites (*Cryptosporidium* and *I. belli*) that may produce mucosal abnormalities of the small intestine.[59,64,98,99] Patients with coccidial infestations present with unrelenting chronic malabsorption or with chronic low-grade to overwhelming watery diarrhea. Intestinal biopsy is probably the best method for diagnosing this disease, even though it is sometimes difficult to find the causative organisms microscopically (Figure 5-42). *Cryptosporidium* is located intracellularly just below the plasma membrane and appears to be intimately attached to the microvillous plasma membrane of intestinal epithelial cells when examined at electron microscopy (Figures 5-26, 5-30) (see earlier section in this chapter dealing with this topic). Jejunal biopsies show mild to severe injury and attachment

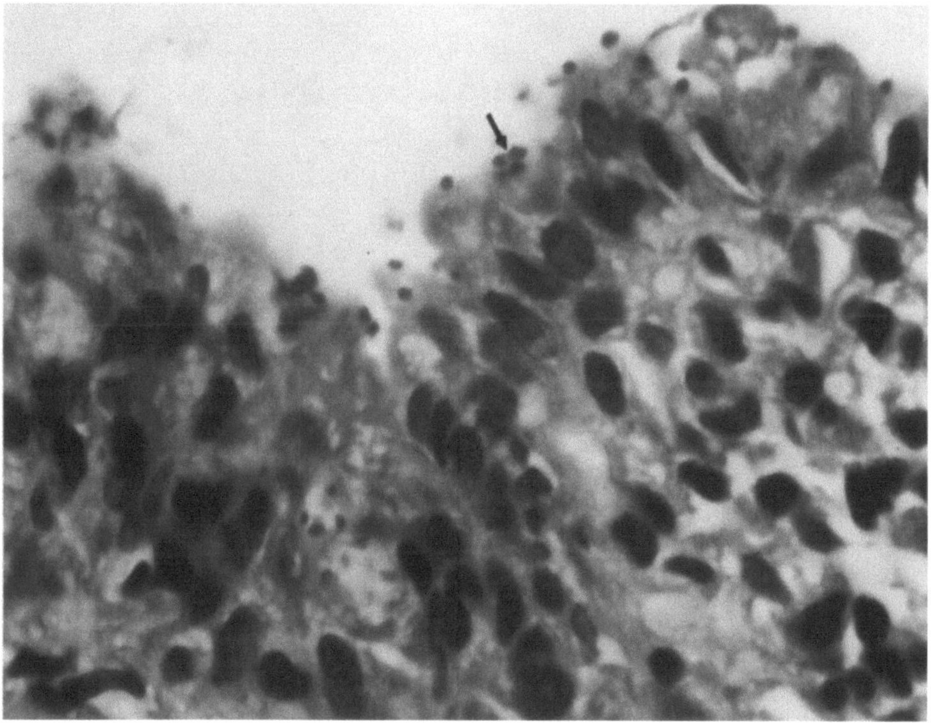

Figure 5-42. Light microscopy appearance of cryptosporidial organisms (arrow) attached to the brush border of epithelial cells of a jejunal biopsy specimen from a patient with profuse watery diarrhea. H & E stain, × 1,250. (Reproduced from Dobbins,[16] by permission of Churchill Livingstone, Inc., New York.)

of numerous tiny 2- to 4-μm organisms on the epithelial surface (Figure 5-42). Treatment is generally ineffective, though in one individual the cryptosporidiosis cleared after discontinuation of an immunosuppressive agent.[99]

In contrast, *I. belli*, another member of the coccidia family, invades the intestinal epithelial cell and is found in an intracytoplasmic position. The villous architecture varies from normal to severely abnormal, and the lesion is patchy. Various developmental forms of the coccidia may be seen within the epithelium or immediately beneath it (Figure 5-43).[98] The various stages of schizogony are easiest to see, and a colophonium-Giemsa stain may highlight the schizonts. Multiple sections must be examined at oil immersion microscopy to detect these organisms. The organism can be more clearly defined at electron microscopy.[100] Diagnosis is important because improvement may follow treatment with a combination of pyrimethamine and sulfadiazine with cotrimoxazole.[100]

Microsporidia occasionally cause diarrhea and malabsorption in patients with AIDS.[59] Microsporidia, like coccidia, are intracellular parasites that frequently infect invertebrates, have been reported to infect all classes of vertebrates, but, until the advent of AIDS, were an exceedingly rare cause of disease in humans.[101] The organisms invade intestinal epithelial cells (Figure 5-44), possibly with the aid of unique coiled polar filament (Figure 5-45) contained within microsporidial spores. The intracellular organisms are too small and too poorly stained to be seen easily at light microscopy and must generally be identified at electron microscopy.[59,101] The spores are too small to be identified in feces.

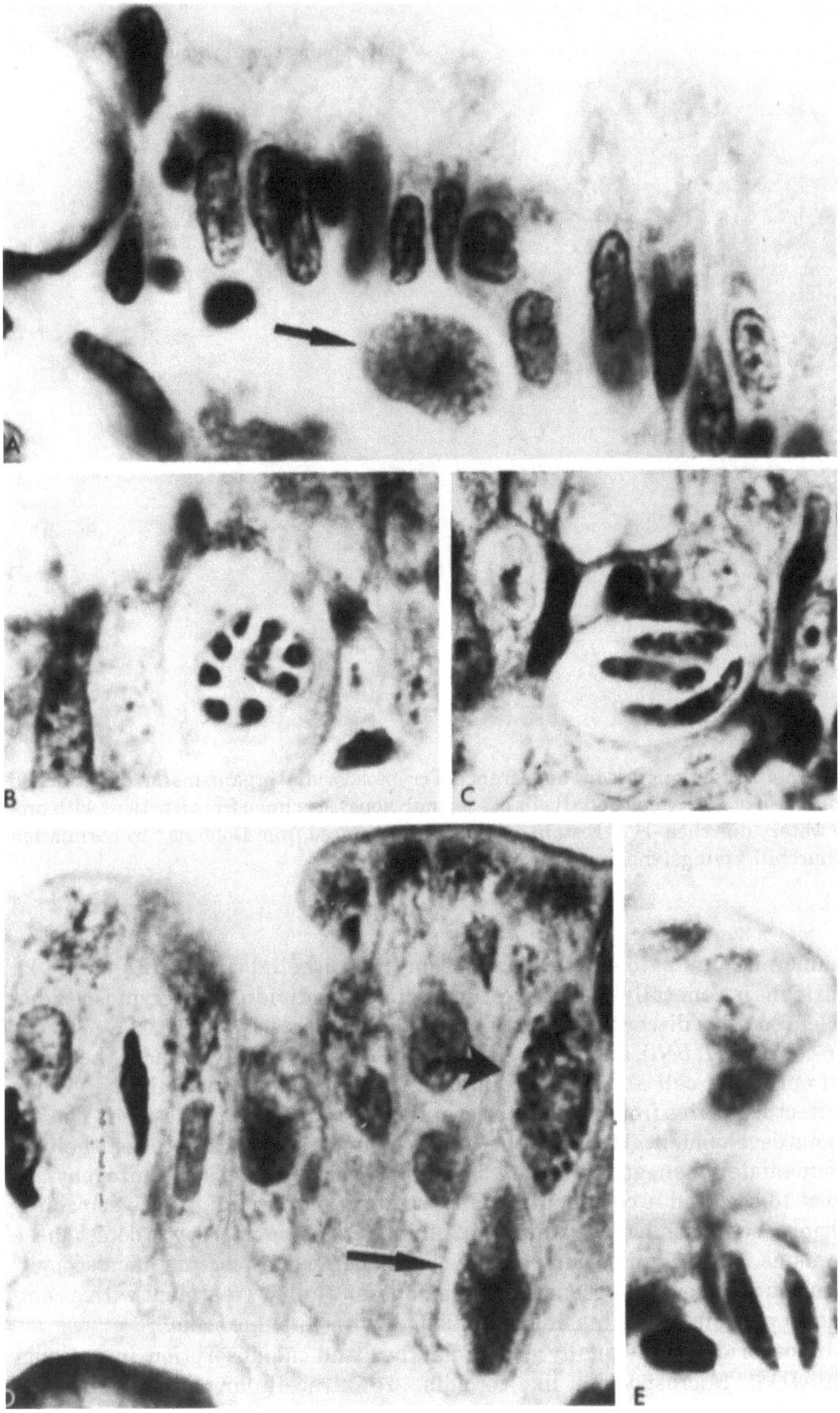

Figure 5-43. Various stages of *Isospora belli* in the intestinal epithelium. (A) Immature schizont below epithelial nuclei (arrow). H & E stain, × 1500. (B,C) Developing merozoites within schizonts. Colophonium-Giemsa stain, × 1500. (D) Two merozoites on the left in epithelial cells (darkly stained, elongated structures). A macrogametocyte (thick arrow) and an immature schizont (thin arrow) can be seen on the right. Colophonium-Giemsa stain, × 1500. (E) Two merozoites within a schizont Colophonium-Giemsa stain, × 1500). (All reproduced from Dobbins,[16] by permission of Churchill Livingstone, Inc., New York. Original supplied by C.E. Rubin.)

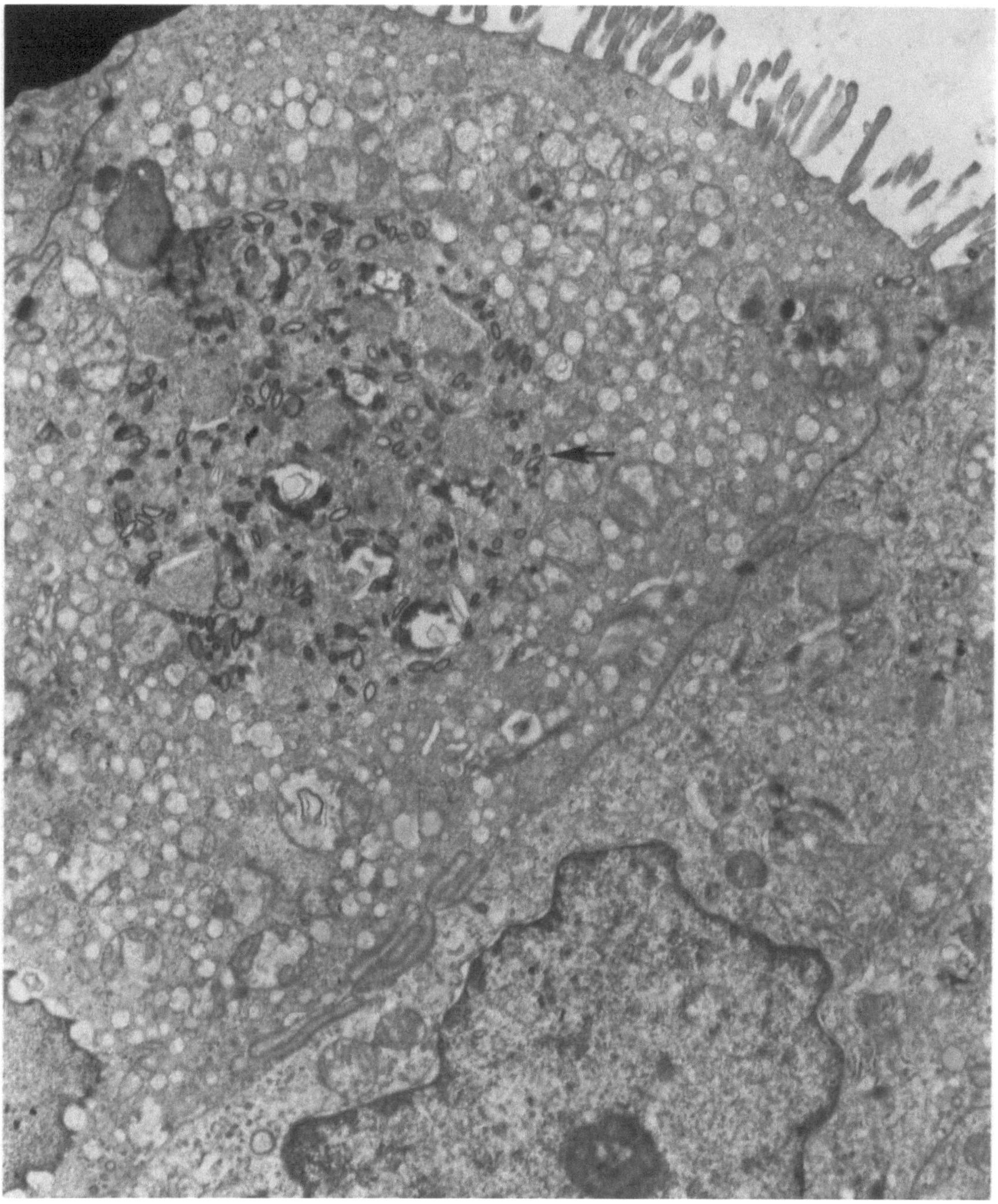

Figure 5-44. Electron micrograph of intestinal epithelial cells in a patient with AIDS and chronic diarrhea. One cell is invaded by a microsporidial meront (arrow). Note disrupted mitochondria and dilated endoplasmic reticulum of the epithelial cell. The organism has a single plasma membrane and numerous disk-shaped densities, is multinucleated, and has no mitochondria. × 14,500

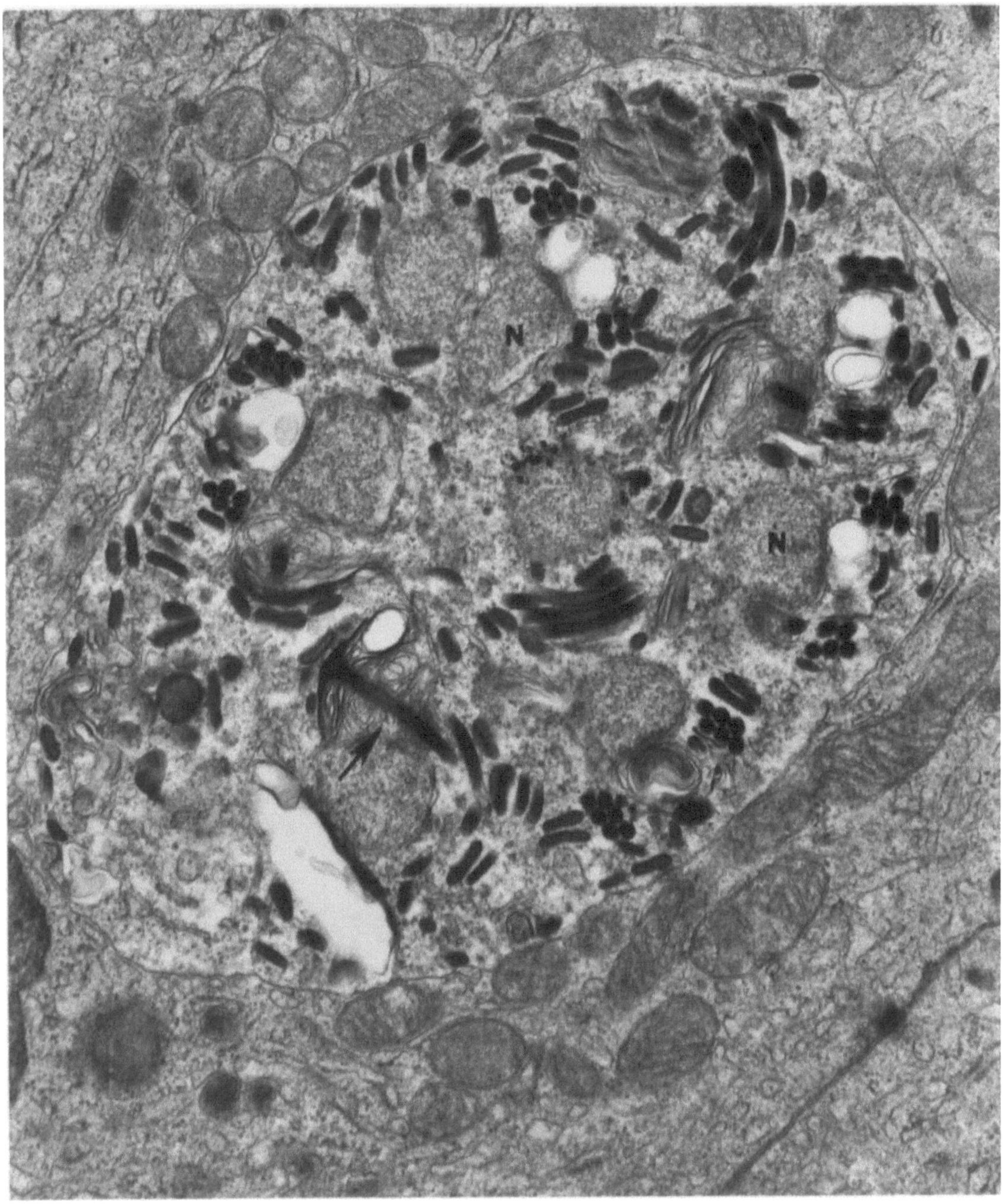

Figure 5-45. Electron micrograph at higher magnification than in Figure 5-44 showing microsporidial meront within an epithelial cell. Note the numerous disk-shaped densities, the multiple nuclei (N), and the polar filament (arrow). × 31,900

Strongyloidiasis, resulting from infestation by the nematode *Strongyloides stercoralis*, most commonly causes mild gastrointestinal symptoms with little change in the villous architecture. The infestation may result in severe injury to the intestinal mucosa with a flat lesion and complete loss of villi, and thereby produce severe malabsorption.[102] The illness on occasion can be quite serious. The diagnosis is made by confirming the presence of adult worms and larvae within intestinal crypts.[1,103]

The nematode *Capillaria philippinensis* inhabits the intestinal lumen and can be seen adhering to the tissue in biopsy sections.[104] There are no changes in the villous architecture that are not seen in uninfected control subjects living in the Philippines.

Schistosome ova may be seen in small-bowel biopsy specimens of patients with schistosomiasis, although villous structure is not altered from that seen in normal individuals living in endemic areas of schistosomiasis.[1] Schistosomiasis is regarded as a disease predominantly involving the colon, although autopsy studies show that there may be intestinal involvement in up to 40% of patients.[105] Duodenoscopy may show presence of white, papular lesions and a friable mucosa. Endoscopically obtained biopsy specimens may show the presence of organisms characteristic of *Schistosoma mansoni* and associated with chronic inflammation of the lamina propria.[105] The yield on duodenal biopsy may be increased by preparation of a wet mount and crushing of the biopsy specimen between two glass slides.

Visceral leishmaniasis (kala azar) is generally characterized by fever, anemia, and splenomegaly. A minority of patients have diarrhea and even steatorrhea. Weight loss and hypoalbuminemia may be extreme. The causative agent is a genus of parasitic protozoa of the family *Trypanosomatidae*; it is found predominantly in macrophages of the skin and viscera. Leishmania parasites were found in 5 of 10 pretreatment duodenal biopsy specimens and in none of posttreatment specimens in a report from Kenya.[106] Parasites were found in macrophages mainly in villous tips and rarely in the submucosa, and were free in the lamina propria on only one occasion. The epithelium was normal and there was a mild to moderate lymphocytic infiltrate. Plasma cells were sparse. The mucosal lesion ranged from mild to moderate in degree and was patchy in distribution.

Fungal Diseases

Gastrointestinal involvement occurs in up to 75% of patients with disseminated histoplasmosis but is clinically obvious in only about 25% of patients.[82,107] The definitive diagnosis depends on culture of the organism from involved tissues, but presumptive diagnosis may be made by biopsy evidence of yeast in the diseased tissues. The fungus was reported to involve the small bowel and to cause malabsorption in one patient in whom the intestinal lamina propria was greatly distended with macrophages that contained encapsulated fungi.[82] Treatment with amphotericin was successful, and a repeat intestinal biopsy specimen showed reversion of the mucosa to normal appearance.

The most frequently encountered fungal infection in immunosuppressed patients is due to *Candida* species. Mucosal lesions in the gastrointestinal tract are generally found in the esophagus; gastric and duodenal candidiasis are most unusual.[108] There is one report[108] of an immunosuppressed patient with thickened duodenal and jejunal mucosal folds in whom duodenal candidiasis was diagnosed when small white patches were observed endoscopically in the entire

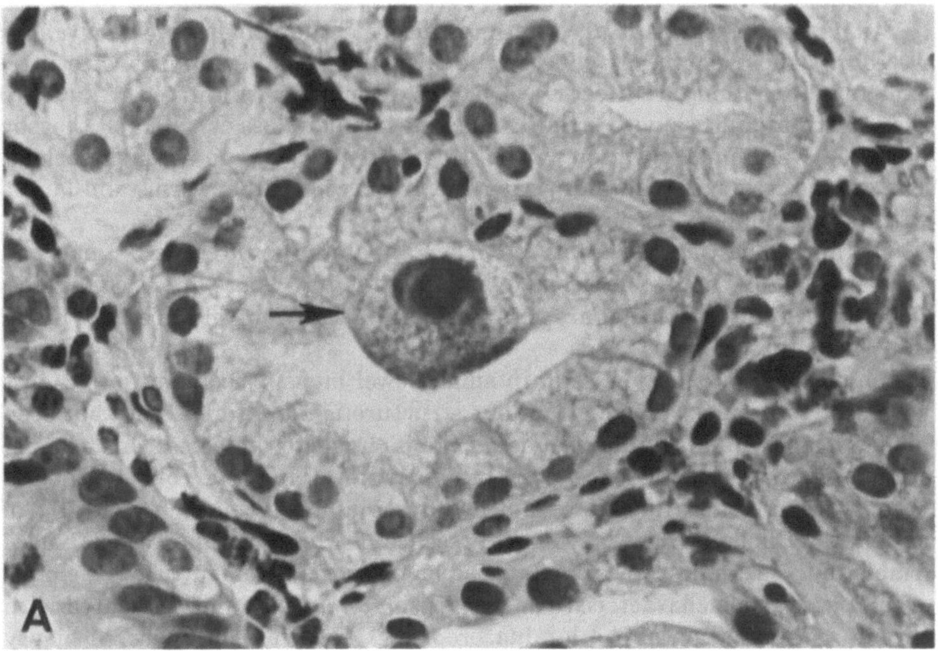

Figure 5-46A. Light microscopy appearance of cytomegalovirus (CMV) involvement (arrow) of Brunner glands in a proximal duodenal biopsy specimen. Note the enlarged cell with the large oval intranuclear inclusion and prominent surrounding halo. H & E stain, × 1,250.

duodenum. Biopsy specimens showed no alteration in villous structure other than acute inflammatory changes in the lamina propria.[108] Direct smear of brushings from the duodenal plaques demonstrated yeast and pseudohyphae of *Candida* species. Treatment with amphotericin B resulted in a return to a normal mucosal appearance. Duodenal candidiasis has been an occasional finding in patients with AIDS.

Viral Diseases

Two viral diseases–cytomegalovirus (CMV) and herpes simplex virus–may produce characteristic and diagnostic alterations in the intestinal mucosa. Patients who are debilitated or immunosuppressed are susceptible to gastrointestinal involvement by cytomegalovirus.[109] Its characteristic appearance is difficult to miss if sought but may be easily overlooked if not sought (Figures 5-46, 5-47). Cells invaded by cytomegalovirus are enlarged, with a large oval intranuclear inclusion and a prominent surrounding halo. There is a single report of a patient with late-onset immunodeficiency who had giardiasis and CMV infestation simultaneously.[109] Intestinal biopsy specimens were variable in the degree of villous abnormality with moderate and severe lesions. On reevaluation two months after eradication of the *Giardia*, small-intestine biopsy specimens still showed mucosal lesions. It seems likely that the CMV infection contributed to the mucosal injury, although this could not be established unequivocally.[109] CMV infection occurs in up to 90% of patients who have

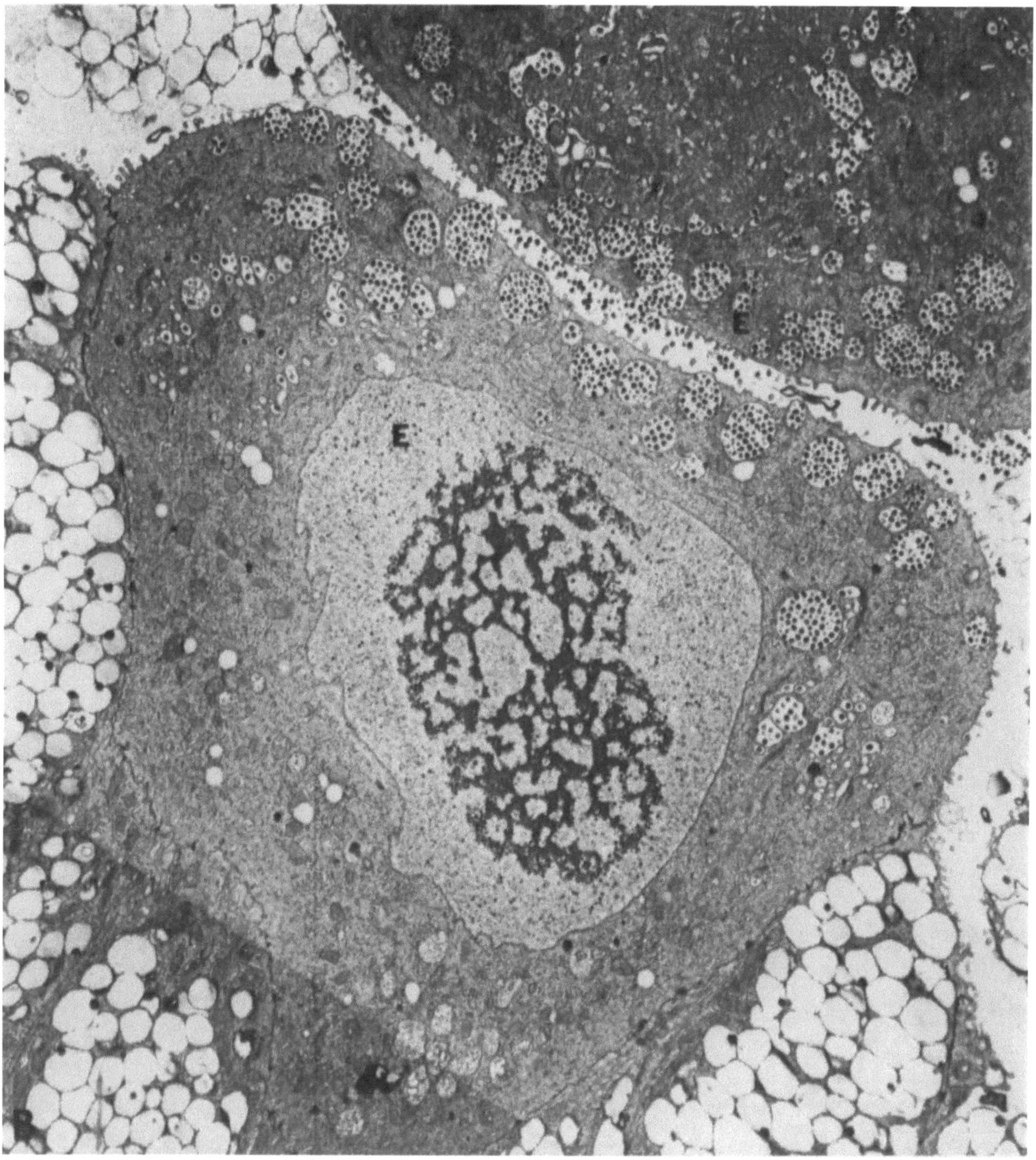

Figure 5-46B. Electron micrograph showing two adjacent epithelial cells (E) in Brunner glands invaded by CMV. Note that the involved cells no longer secrete mucus but are secreting fully formed CMV into the duodenal lumen. × 4,500

had renal transplantation and are maintained on immunosuppressive therapy.[110] Duodenal Brunner glands seem to be the elective site for gut involvement. There is no correlation between CMV presence with symptoms or with other morphological changes. The CMV is probably not a pathogen in this setting. CMV does appear to be pathogenic to the gut in patients with AIDS, in

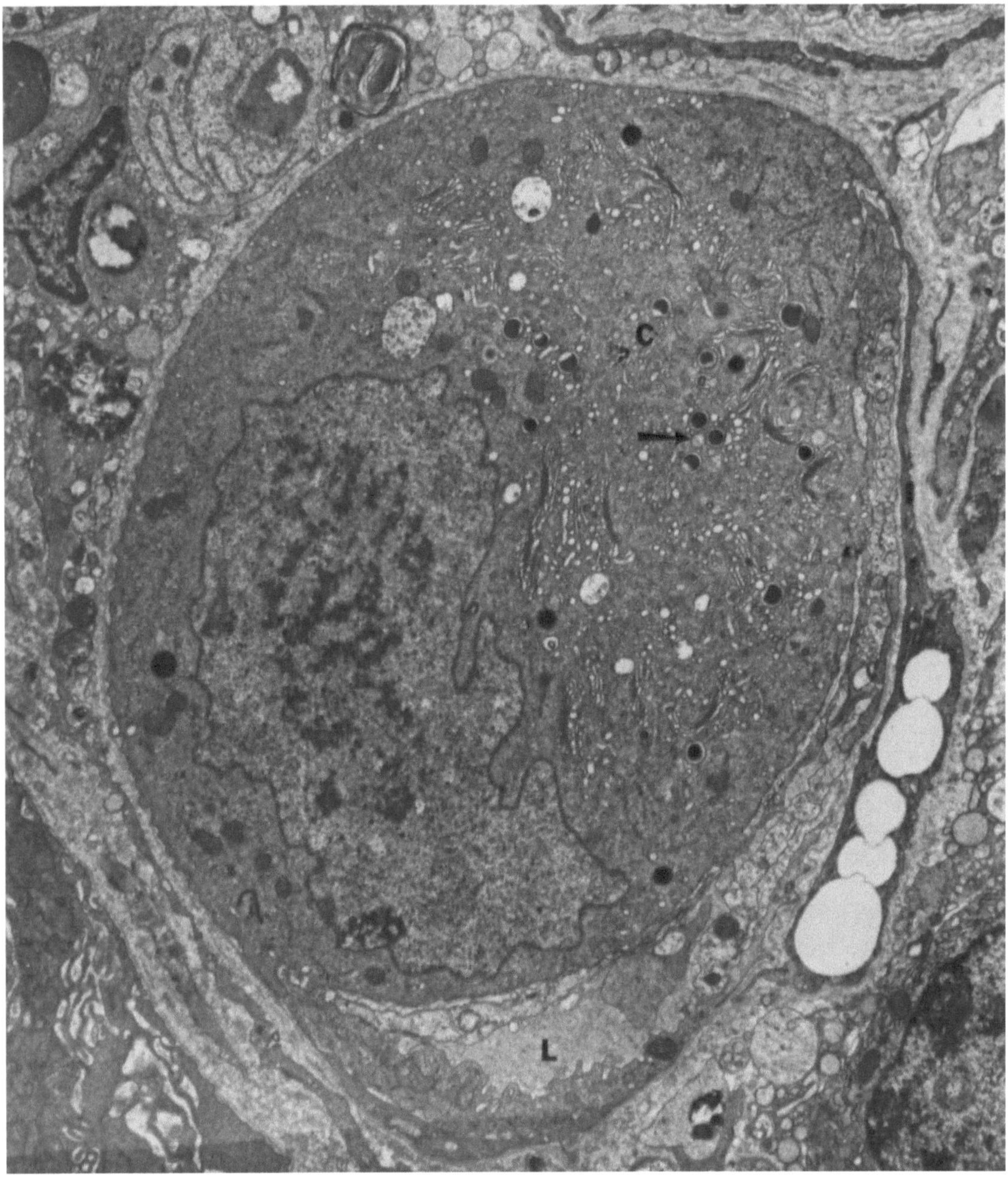

Figure 5-47. Electron micrograph showing effect of cytomegalovirus (CMV) on intestinal capillary endothelial cell (C) in a patient with acquired immunodeficiency syndrome. The endothelial cell is so enlarged that the lumen (L) is virtually occluded. Intranuclear CMV particles are present, while the cytoplasm contains large dense inclusion (arrow) characteristically seen in CMV infections. × 8,400

whom there may be diffuse gastroenterocolitis with extensive presence of CMV in the stomach, duodenum, and colon.[111]

Norwalk agent often results in clinical gastroenteritis and mildly abnormal intestinal histology characterized by mucosal inflammation, absorptive cell abnormalities, shortening of villi, crypt hypertrophy, and increased numbers of epithelial-cell mitoses.[56,112] The virus has not been detected within involved mucosal cells by electron microscopy, possibly because of the small size of the virus and its patchy distribution. The mucosal lesion disappears within two weeks.[56]

Duodenal biopsy specimens obtained from infants and young children with rotavirus diarrhea also reveal mucosal abnormalities similar to those described with Norwalk virus. Unlike Norwalk virus, rotaviral particles can be visualized by electron microscopy of the intestinal epithelial cells in biopsy specimens.[113]

Macroglobulinemia

This syndrome is considered a malignant neoplasm of the lymphoreticular system, similar to lymphoma and multiple myeloma. It is distinguished from multiple myeloma by its longer more benign course and by its lack of bone lesions, renal complications, and amyloidosis. Two cases have been reported in which involvement appeared to be limited to the small bowel and bone marrow.[1,16] The patients presented with malabsorption. In another report a patient presenting with diarrhea was found to have macroglobulinemia with intestinal involvement.[114] The diagnostic feature in intestinal biopsy specimens is the presence of hyaline, amorphous masses in an extracellular location within the lamina propria (Figure 5-48). This same hyaline material is often found between the epithelial cells as well. Its accumulation may give the villi a clubbed shape. The hyaline material is moderately to intensely eosinophilic and is faintly PAS positive. It has been shown by specific immunofluorescent stains to be composed of a monoclonal IgM macroglobulin. Lymphangiectasia secondary to macroglobulinemia with "increased viscosity" of lymph consequent to its high IgM content has been reported in two patients.[115]

Lymphangiectasia

Primary lymphangiectasia is a rare congenital obstructive defect of the lymphatics that was first described in 1961 by Waldmann et al.[116] They reported a group of patients with "idiopathic hypoproteinemia" who had protein-losing enteropathy and steatorrhea and whose intestinal biopsy specimens demonstrated the presence of marked lymphatic dilatation. The lymphatic abnormality may be generalized so that patients also have chylothorax, chyluria, chylous ascites, or a lymphedematous limb, either singly or in combination. When there is intestinal involvement, numerous dilated lymphatics may be seen throughout the mucosa and submucosa. Some give the appearance of rupturing into the lumen. In the area of lymphangiectasia overlying villi may be distorted, while in areas devoid of dilated lymphatics the villous architecture is usually completely normal (Figure 5-49). Patients may respond to a low-fat diet, particularly when the diet is supplemented with medium-chain triglycerides. Occasional patients have responded to surgical lymphovenous anastomoses.

Secondary lymphangiectasia may be seen in a variety of diseases including constrictive pericarditis, primary myocardial disease, Behçet's disease, intesti-

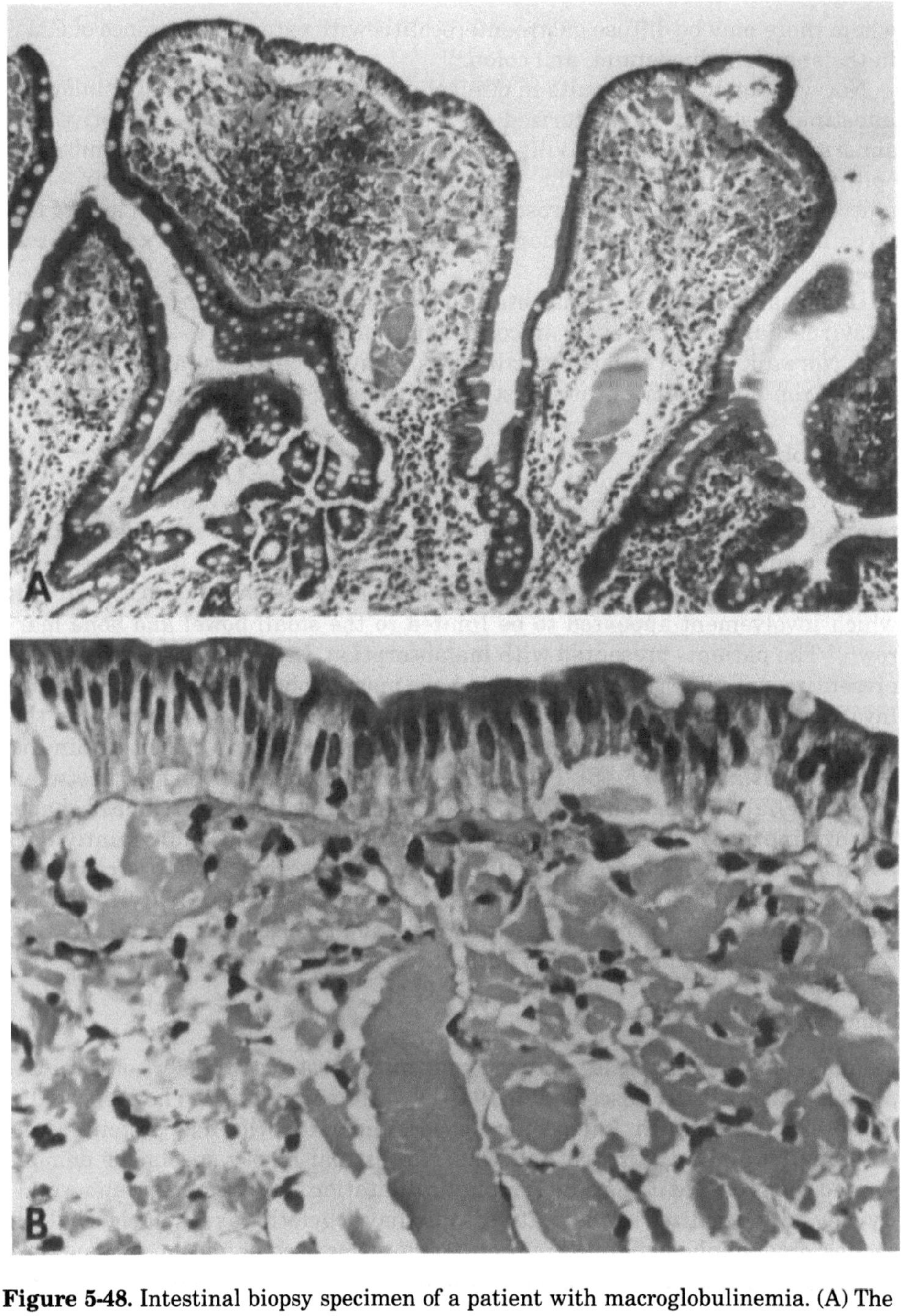

Figure 5-48. Intestinal biopsy specimen of a patient with macroglobulinemia. (A) The lamina propria is filled with amorphous hyaline masses. H & E stain, × 150. (B) Higher magnification shows the extracellular masses of macroglobulin. H & E stain, × 540

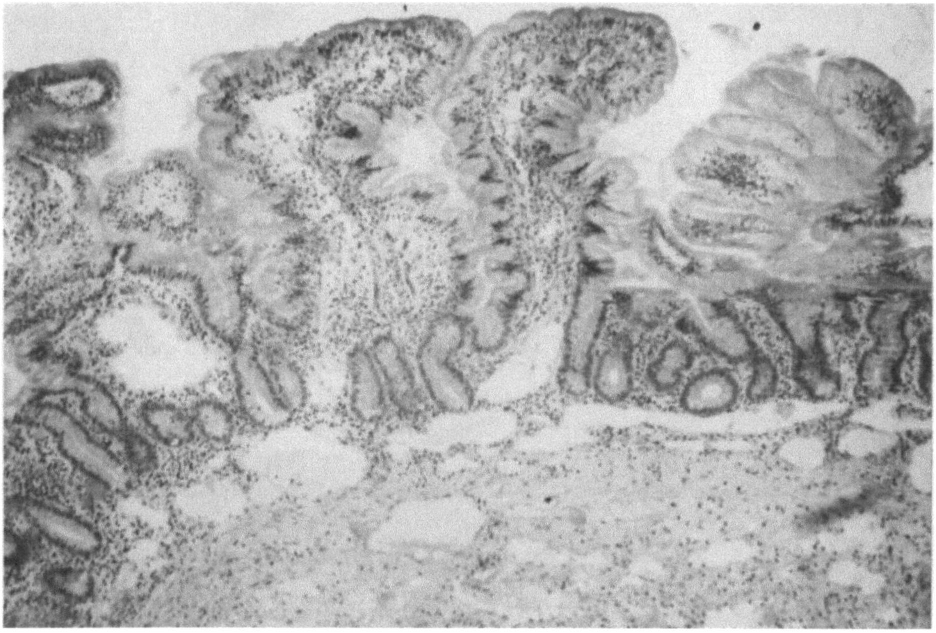

Figure 5-49. Proximal jejunal biopsy specimen from a patient with congenital lymphangiectasia manifested by protein-losing enteropathy and chylous ascites. Dilated lacteals distort villous structure in a patchy fashion, while dilated lymphatics are present diffusely throughout the submucosa. H & E stain, × 132. (Reproduced from Dobbins,[16] by permission of Churchill Livingstone, Inc., New York.)

nal lymphoma and carcinoma, retractile mesenteritis, and even sarcoidosis.[117] Symptoms in patients with constrictive pericarditis may mimic those of primary intestinal lymphangiectasia in every respect, and thus it is important to show the presence of normal central venous pressure in all patients thought to have primary intestinal lymphangiectasia. Localized lymphangiectatic cysts of the small bowel may be found in random locations at postmortem examination. Most are submucosal and thus would be missed by suction biopsy.

Electron microscopy shows that intestinal lacteals in patients with lymphangiectasia differ somewhat from the normal. Lymphatic endothelial cells in lymphangiectasia contain unusually prominent intracellular microfilaments and a more prominent basal lamina than do normal lacteals (Figure 5-50). Further, collagen and reticulin fibers, fibrocytes, and smooth muscle are more prominent about the lacteal basal lamina in lymphangiectasia than in the normal condition. The increased number of endothelial microfilaments may be related to "work hypertrophy" caused by increased hydrostatic pressure within lymphatic lumina. The prominent collagen fibers and cells closely apposed to the basal lamina may perform a "supporting" function (Figure 5-50). As expected, there is marked retention of chylomicrons throughout the lamina propria and in the intercellular spaces of intestinal epithelial cells (Figure 5-51). Interestingly, lymphatic junctional complexes are closed in lymphangiectasia, whereas junctional complexes in normal lacteals are occasionally open.[118] In contrast to

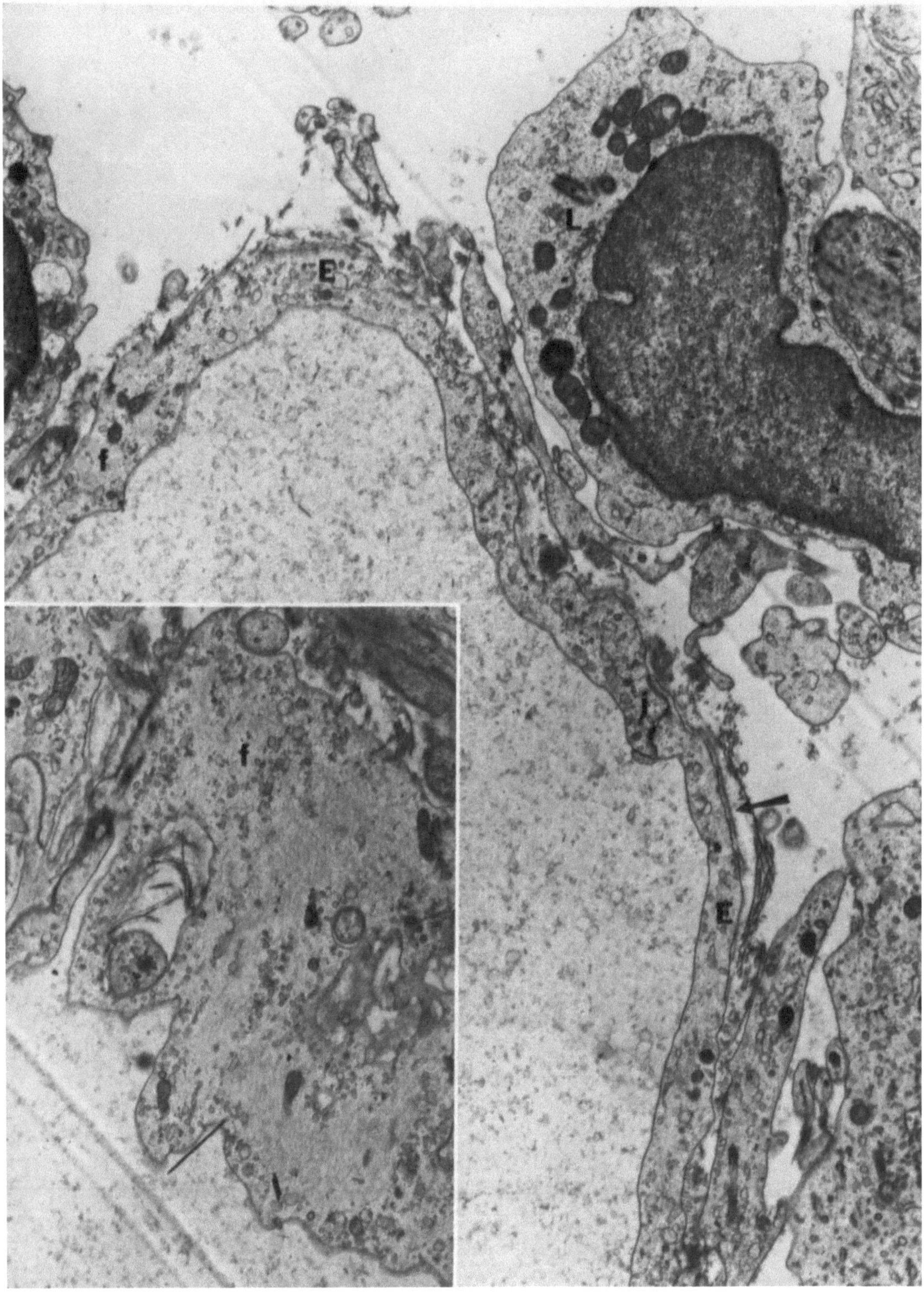

Figure 5-50. Electron micrograph of lacteal endothelium (E) in intestinal lymphangiectasia. Cytoplasmic microfilaments (f) are prominent; the basal lamina (arrow), though discontinuous, is better developed than in normal subjects. The inset shows clearly the increased cytoplasmic filaments (f). Note the well-developed junctional complex (j) containing desmosomes between adjacent endothelial processes. Lymphocyte (L). × 10,500.

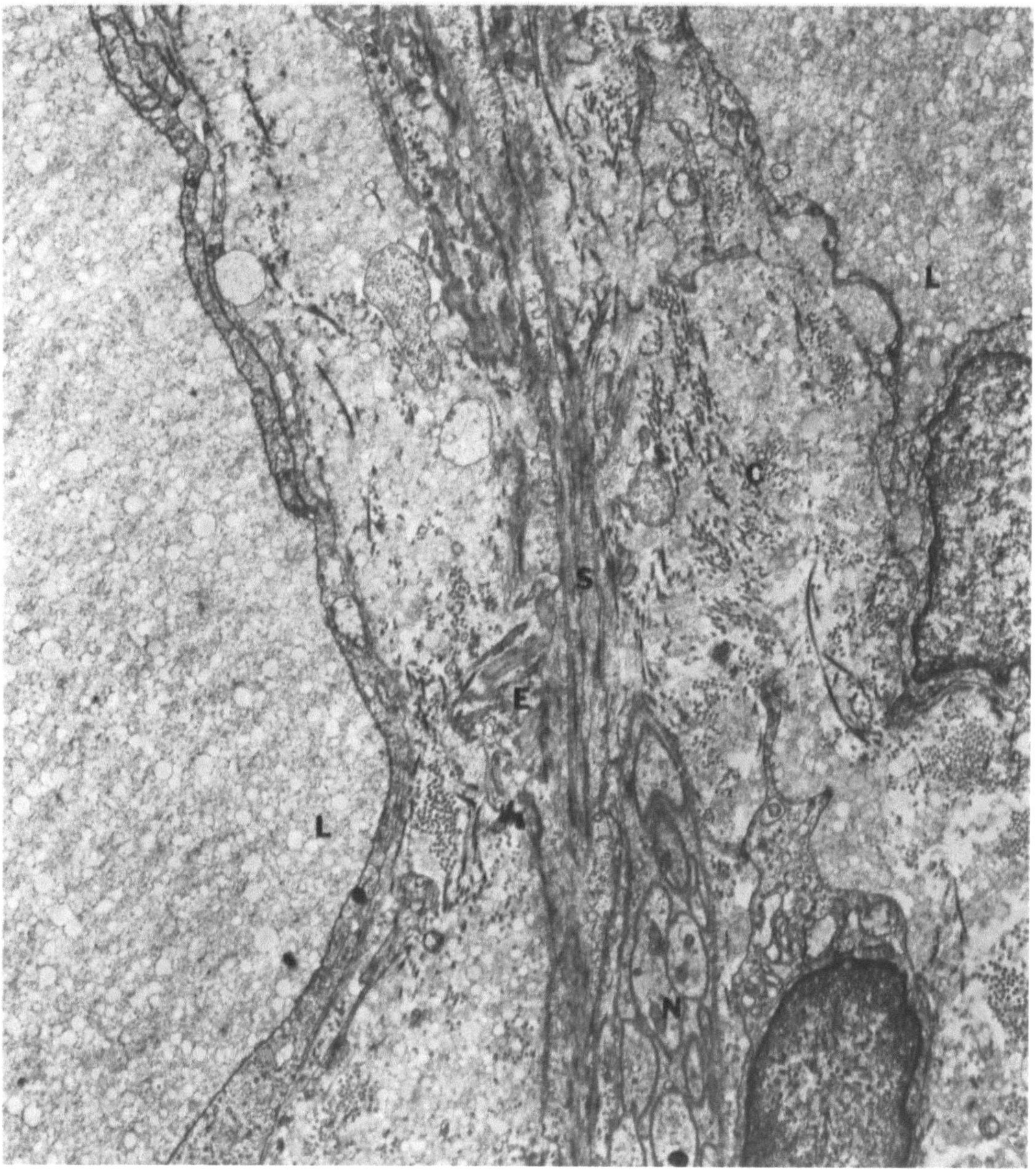

Figure 5-51. Electron micrograph in lymphangiectasia showing retention of chylomicrons and proteinaceous material within lymphatic lumina (L) and in extracellular spaces of the lamina propria. Smooth muscle (S), nerve fibers (N), collagen (C), and elastic fibers (E) are also present. × 10,500

normal, lipid droplets are prominent within the lymphatic endothelium, and macrophages packed with lipid droplets (lipophages) are frequently noted within the lymphatic lumen.

One patient with primary hypo-β-lipoproteinemia was found to have intestinal lymphangiectasia that appeared to be secondary to obstruction of lymphatic drainage caused by excess lipid storage within abdominal lymph nodes.[119] Lacteal junctions in this patient were also noted to be closed, and little lipid was found within the lymphatic endothelium. Dilatation of intestinal and abdominal lymphatics is a prominent finding in Whipple's disease. This dilatation is presumably related to obstruction to lymphatic flow caused by infiltration of abdominal lymph nodes and infiltration of the lamina propria of the intestinal mucosa by PAS-positive macrophages. Further, actual bacterial invasion of the endothelium of lacteals in Whipple's disease has been demonstrated.[120] Moreover, numerous open cell junctions in lacteals were found through which polymorphonuclear leukocytes and other cells appeared to enter the lacteals.

In experimentally induced intestinal lymphangiectasia,[121] lymphatic junctions are closed rather than open. Electron microscopy of mucosal and submucosal lymphatics in Crohn's disease in which lymphatic obstruction is prominent, shows changes similar to those seen in primary lymphangiectasia.[122] That is, the lymphatics in Crohn's disease have only closed junctions associated with abluminal accumulation of protein-rich fluid. Both features suggesting decreased permeability.[122]

Drug and Irradiation Injury

Severe vitamin B_{12} and folic acid deficiency, acute radiation reaction, and chemotherapeutic drugs used for treatment of malignancy all result in inhibition of DNA synthesis and impairment of epithelial cell replacement.[123-126] Crypt mitotic activity is reduced, epithelial cells are enlarged, and villous abnormalities vary from mild changes to complete loss of villi (Figure 5-52). Macrocytosis may be readily apparent or quite inapparent. When present, it is often irregularly distributed from villus to villus or crypt to crypt.

Even when a severe mucosal lesion is present, it is generally easy to differentiate it from that of celiac sprue because of the presence of macrocytosis, a decreased number of mitoses, and a less dense cellular content of the lamina propria. Following treatment with folic acid or vitamin B_{12}, or cessation of drug therapy, the changes in the intestinal mucosa revert to normal. Radiation-induced changes also revert to normal, unless the radiation has been excessive.

A variety of drugs have been implicated in production of intestinal malabsorption.[127-134] Administration of these drugs may result in malabsorption of specific substances such as vitamin B_{12} or folic acid by causing inhibition of the specific transport mechanisms in the intestine. Their administration does not ordinarily result in light microscopy evidence of morphological injury to the intestinal mucosa, but injury may be identifiable by electron microscopy. The most severe gut injury reported was produced by MER-29 (triparanol), a drug once used in treatment of hypercholesterolemia. This drug resulted in ectodermal changes including cataracts, ichthyosis, and alopecia.[127] An intestinal lesion indistinguishable from that of severe celiac sprue, with resulting malabsorption, was found in two subjects. The intestinal lesion was reversed by withdrawing the drug while patients were maintained on a normal diet. A lesion of similar severity, reversible on withdrawal of the drug, has been induced by the nonsteroidal antiinflammatory agent sulindac (Clinoril).[129] Sulindac also may induce marrow aplasia, Stevens Johnson toxic epidermal neurolysis syndrome, pneumonitis, pancreatitis, hepatitis, and life-threatening hypersensitivity.[129] Antimetabolites, colchicine,[130] and antibiotics, especially neomycin, have produced morphological injury to the intestinal mucosa easily identifiable at light microscopy. The effect of neomycin on the intestine has been thoroughly

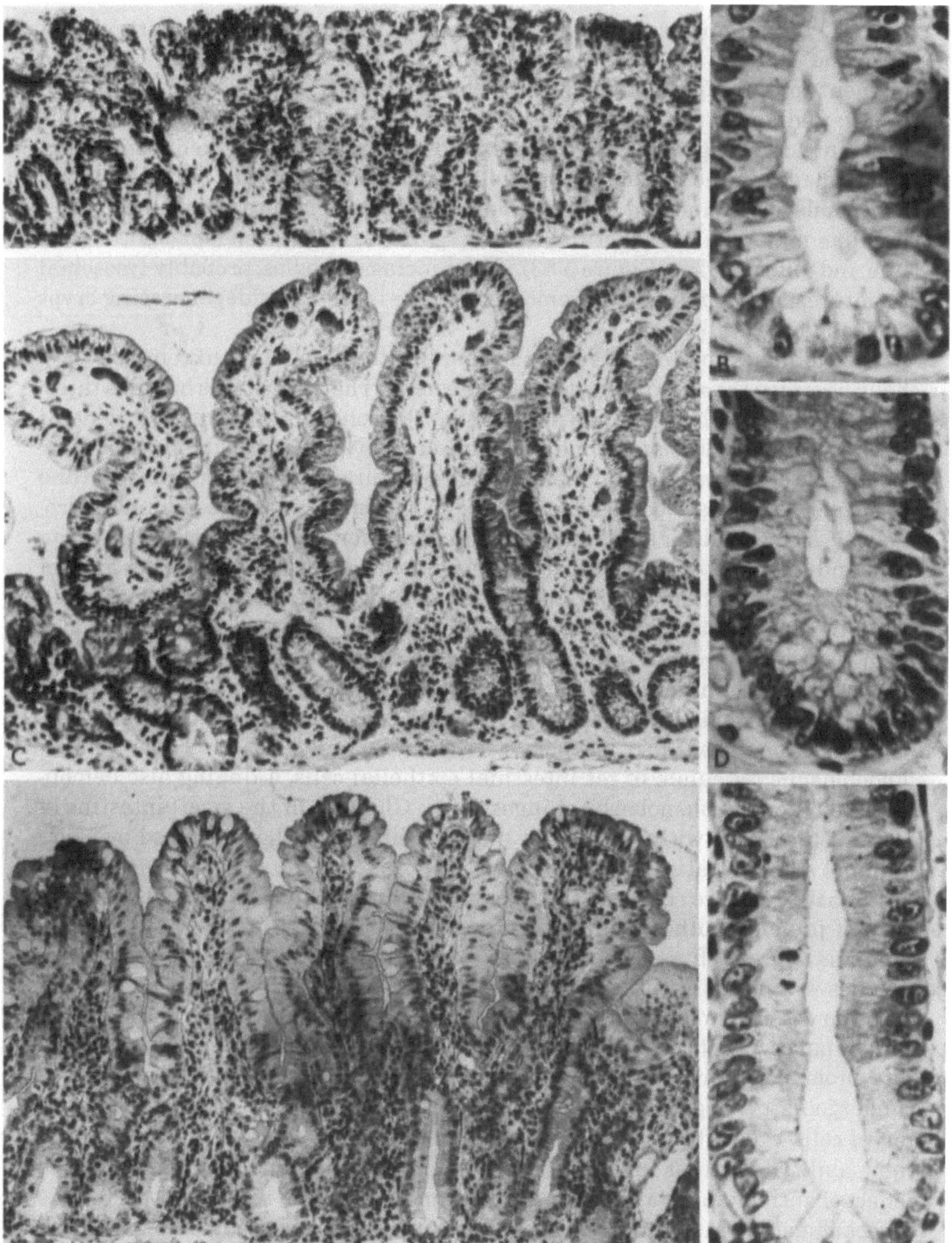

Figure 5-52. Small-intestine changes due to drug or irradiation injury. (A) Severe proximal jejunal lesion in a patient treated with busulfan for myelocytic leukemia. H & E stain, × 184. (B) Epithelial cells in a crypt from the biopsy specimen in (A) are macrocytic. H & E stain, × 736. (C) Normal jejunal biopsy specimen from the same patient eight weeks after discontinuation of treatment. H & E stain, × 184. (D) Epithelial cells in a crypt from the biopsy specimen in (C). H & E stain, × 736. (E) Proximal jejunal biopsy specimen in a patient with folic acid deficiency showing minimal villous abnormalities. H & E stain, × 184. (F) Epithelial cells in a crypt from the specimen shown in (E) are macrocytic. H & E stain, × 736. (A through F reproduced from Dobbins,[16] by permission of Churchill Livingstone, Inc., New York. Original supplied by C.E. Rubin.)

studied.[131] Light microscopy changes induced by neomycin include shortening of villi and infiltration of the lamina propria with inflammatory cells and pigment-containing macrophages. Neomycin induces considerable crypt cell injury: undifferentiated crypt cells show marked dilatation of the endoplasmic reticulum, swelling of mitochondria, and loss of microvilli. Pigment-containing macrophages in the lamina propria following neomycin injury to the intestine are membrane bound and contain finely granular dense areas, myelin figures, lipidlike material, and identifiable cell organelles such as endoplasmic reticulum and mitochondria (Figure 5-53). These inclusion bodies, probably lysosomal in nature, presumably result from macrophage ingestion of degenerating crypt cells and other cells.

Alcohol ingestion in concentrations similar to those found in common alcoholic beverages can produce not only acute hemorrhagic gastritis but also intestinal damage in both humans and rats.[132] Alcohol may produce a general malabsorption syndrome including malabsorption of amino acids, glucose, fatty acids, folic acid, and other nutrients. Ultrastructural changes thought to be due to alcohol injury have been demonstrated in the small intestine of human alcoholics.[10] Mitochondrial abnormalities and dilatation of the endoplasmic reticulum and cisternae of the Golgi apparatus are easily observed following alcohol administration to human volunteers. These changes are identifiable in epithelial cells of crypts and villi. Evidence of focal cytoplasmic degradation is also demonstrable in human intestinal epithelial cells following alcohol ingestion. The alterations are similar to those produced by ethanol in the liver and suggest that ethanol is injurious to the small intestine in the absence of nutritional deficiency. Alcohol has also been shown to cause proliferation of the smooth endoplasmic reticulum of rat intestinal epithelial cells, a finding also demonstrated following phenobarbital ingestion.[133] Changes in the small intestine of chronic alcoholics with cirrhosis of the liver similar to those induced by acute ethanol administration are reported.[9]

A vast literature concerns the effects of antimetabolic agents and of inhibitors of protein synthesis on structure and function of the small intestine in animals. Inasmuch as this review is confined to the human intestine, these studies will not be cited here except for one particularly important paper[7] concerning the effect of protein synthesis inhibitors, such as cycloheximide, on the intestinal crypts of rats. This study showed that intestinal crypt cells can tolerate profound alterations in their metabolic and functional activity without undergoing cytological or degenerative changes. It indicated that intestinal crypt cells, even after structural alterations have been induced, have a remarkable ability to recover function. Trier,[134] in an electron microscopy study of the effect of antimetabolic agents on the intestinal mucosa in humans, reported the effect of methotrexate in patients being treated for psoriasis and multiple basal cell carcinomas of the skin. No light microscopy changes were noted, but marked fine structural changes were revealed by electron microscopy. The changes were apparent within six hours and lasted for at least 48 hours. The cells involved included absorptive cells and crypt epithelial cells. The changes were spotty, with many cells being spared. Damaged cells showed swollen, fragmented mitochondria and marked dilatation of the Golgi apparatus and of the endoplasmic reticulum. Some cells showed clumping of nuclear material and contained intracytoplasmic inclusions thought to be epithelial cellular remnants (submicroscopic apoptosis?). That undamaged cells frequently were seen

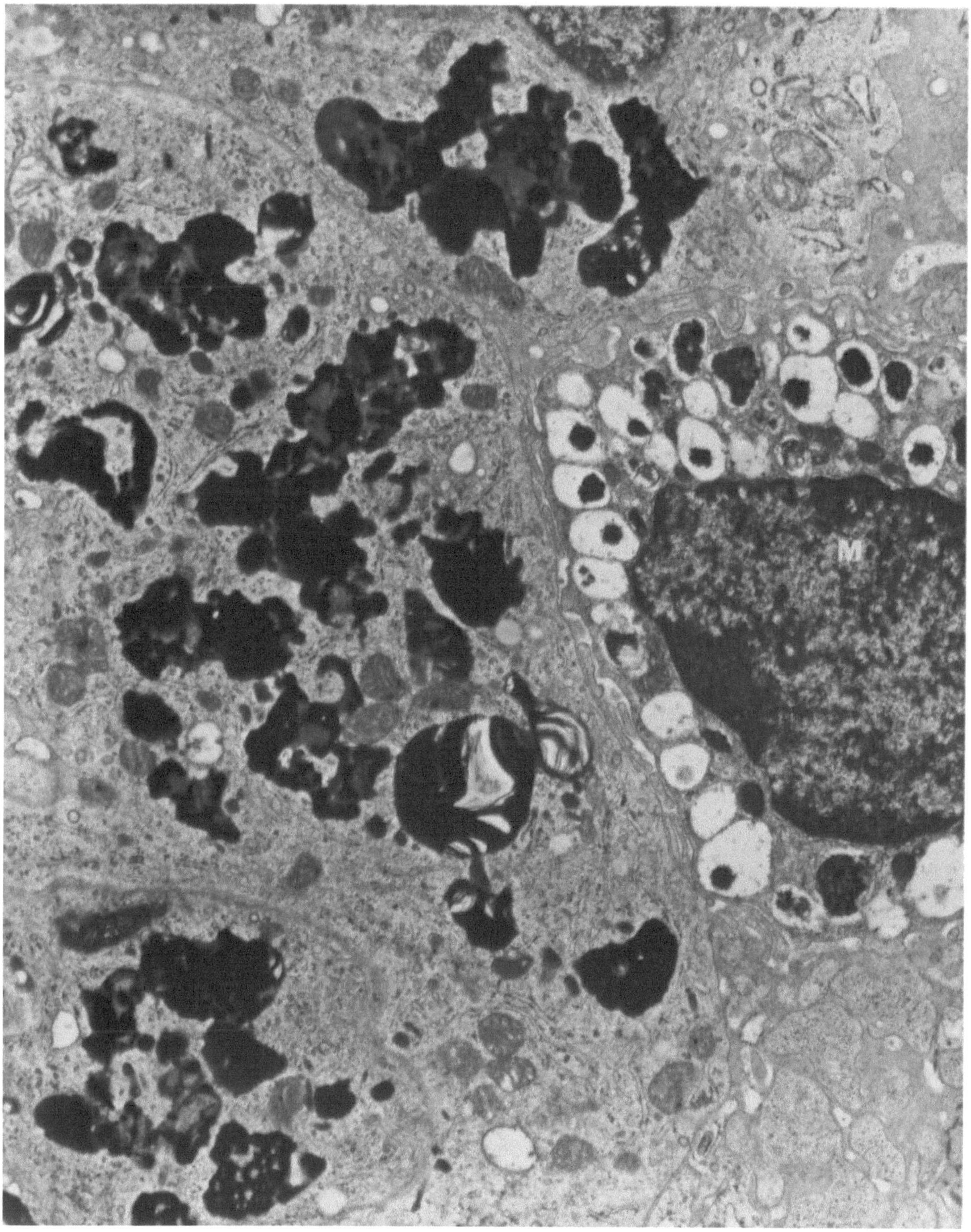

Figure 5-53. Electron micrograph showing characteristic macrophage inclusions in neomycin-induced malabsorption. A mast cell (M) is also present. See text for details. × 15,000

adjacent to severely injured cells suggested a marked variation in sensitivity of the cells to methotrexate. The changes were not specific and were similar to those seen following cell injury due to a variety of agents.

A number of reports document the effects of radiation on the intestine of laboratory animals; only a single electron microscopy study of the acute effects of radiation on the human intestine has been done.[5] In this study, peroral biopsies were obtained from patients before, during, and after abdominal x-ray therapy. Light microscopy changes observed included a reduction in number of mitoses in the crypts to 25% to 51% of control levels, marked shortening of villi with a resultant 39% to 48% reduction of epithelial surface area, megaloblastosis and megalocytosis of epithelial cells, infiltration of the lamina propria with plasma cells and polymorphonuclear leukocytes, and occasional crypt abscess formation. Electron microscopy revealed cytological epithelial cell changes including swelling and dilatation of mitochondria and of the endoplasmic reticulum, irregular distribution and shortening of microvilli, a scalloped nuclear border, and the presence of huge nucleoli. The lesion cleared rapidly and completely. Mitoses were normal within three days and villi were normal within two weeks after cessation of irradiation. The study indicated that extensive morphological damage to the small-intestine mucosa in humans is regularly induced by therapeutic doses of abdominal x-ray therapy. It was emphasized that these changes were associated with surprisingly few symptoms and were rapidly reversible. Even the severe mucosal lesion induced by radiation injury is rapidly reversible.[135] Chronic radiation injury occurs months to years after exposure and results in variable villous abnormalities and marked fibrosis, usually in the ileum.

Diseases Associated with Normal Villi—Diagnostic

a-β-Lipoproteinemia

a-β-lipoproteinemia, first described by Bassen and Kornzweig in 1950,[135a] is characterized by steatorrhea, retinitis pigmentosa, spiny-shaped red cells in the peripheral blood (acanthocytosis), and neurological changes similar to those seen in Friedreich's ataxia.[136] Patients are unable to synthesize apoprotein B, which is required for the synthesis and transport of very-low-density lipoproteins (β-lipoproteins) and chylomicrons. There is normal uptake of fatty acids by absorptive cells, but the fatty acids cannot be synthesized into chylomicrons. Absorptive cells thus are stuffed with large intracytoplasmic fat particles that are unable to exit into the lamina propria. Failure to absorb essential fatty acids may explain the defective lipid membrane of the red cell and account for its acanthocytic or thorny appearance. The progressive neurological disease may also be related to a deficiency of essential fatty acids.

Intestinal biopsy specimens show normal villous architecture, but the intestinal absorptive cells are packed with large droplets of triglyceride (Figure 5-54).[137] The diagnostic feature in biopsy specimens processed for routine histological examination is the extremely vacuolated cytoplasm of the absorptive cells covering the upper third of villi (Figure 5-54).[137] These vacuoles can be shown to contain fat in frozen sections stained with oil red O. Light microscopy sections of osmium-fixed epoxy-embedded specimens stained with toluidine blue better demonstrate the striking accumulation of lipid in varying sized

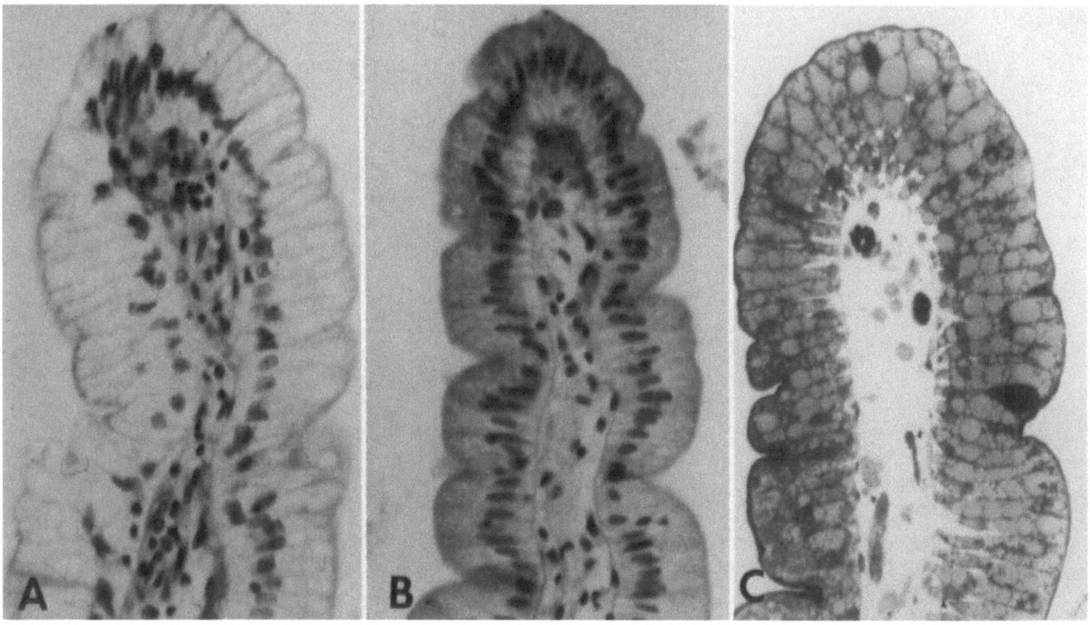

Figure 5-54. (A) Normal villus in a patient with a-β-lipoproteinemia. All of the absorptive cells covering the upper half of the villus are vacuolated, an appearance resulting from extraction of the excess lipid in these cells during processing. H & E stain, × 500. (B) Appearance of a villus from a normal individual for comparison. H & E stain, × 500. (C) Thick section of osmium-fixed epoxy-embedded villus from patient showing striking accumulation of lipid droplets in the epithelial cells. Toluidine blue stain, × 750. (A through C reproduced from Dobbins,[16] by permission of Churchill Livingstone, Inc., New York.)

droplets in intestinal absorptive cells (Figure 5-54). The intestinal lamina propria appears to be normal except for the presence of an unusual number of macrophages containing bizarre inclusions.[137,138] Electron microscopy changes found in a-β-lipoproteinemia are discussed in the section on lipid storage disorders.

Crohn's Disease

The distal ileum and/or colon is the usual site of involvement in this disease. The proximal intestine and duodenum may be involved rarely, permitting diagnosis by peroral small-bowel biopsy.[139,140] The proximal involvement may not be detected by x-ray examination or by endoscopy, but noncaseating epithelioid granulomas with giant cells may be found on biopsy (Figure 5-55). The presence of such epithelioid granulomas is very supportive of the diagnosis of Crohn's disease, especially if the clinical and radiological findings are compatible. The presence of granulomas is not specific for Crohn's disease—they may be found rarely in other conditions, such as sarcoidosis.[117,141] Perera et al.[1] have observed a solitary granuloma in a biopsy specimen obtained from a patient with unclassified sprue and in a biopsy specimen obtained from one patient with celiac sprue and dermatitis herpetiformis. Bjorneklett et al.[142] observed numerous epithelioid and giant cell granulomas in jejunal biopsy specimens of a patient with malabsorption and a moderate intestinal lesion. Because the patient had

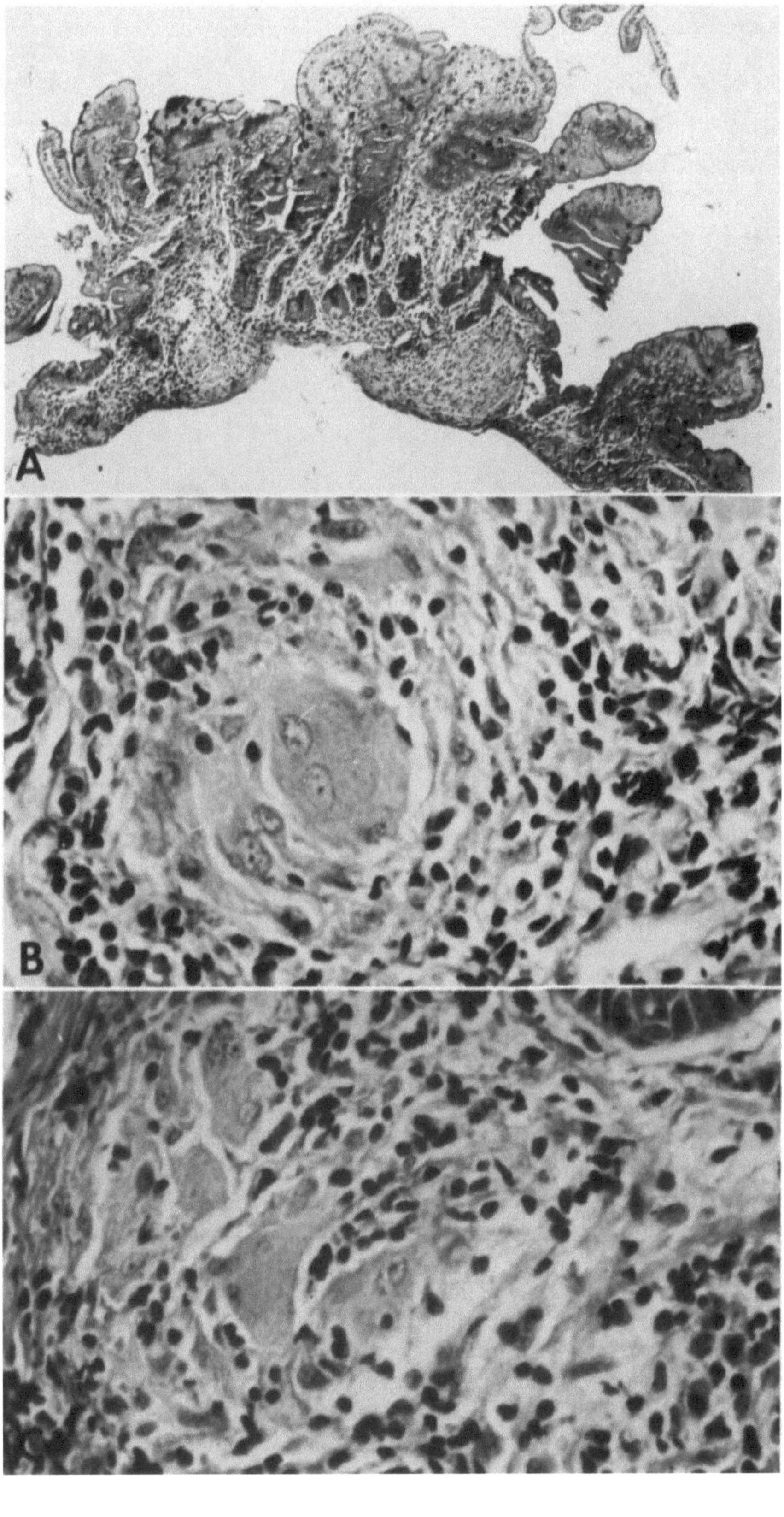
A
B

a complete clinical and histological remission on a gluten-free diet, the granulomatous lesion was considered to be a manifestation of celiac sprue. I have observed a solitary granuloma in an otherwise normal intestinal biopsy specimen obtained from a patient with postgastrectomy diarrhea. Finally, Douglas et al.[143] have reported that five patients had simultaneous celiac sprue and sarcoidosis—the gastrointestinal lesion of celiac sprue and pulmonary manifestations of sarcoidosis.

Schuffler and Chaffee[140] found a spectrum of abnormalities ranging from mild to severe, in the absence of granulomas, in peroral mucosal biopsy specimens obtained from patients with Crohn's disease involving the duodenum. The abnormalities were patchy and included flattened mucosa, abnormal surface epithelium infiltrated by large numbers of PMN leukocytes, increased numbers of plasma cells and PMN leukocytes in the lamina propria, crypt abscesses, erosions, and pyloric gland metaplasia.

Electron microscopy of the intestine in Crohn's disease, including scanning EM, has provided much information in regard to fine structure in this disease but has provided no new insights into pathogenesis of the disease.[122,144–151] The study by Richert and Carter[145] correlating light microscopy with scanning electron microscopy findings in the early lesion of Crohn's disease further documented that small aphthae (ulcers) are the earliest pathological change in Crohn's disease. In a series of studies, Dvorak et al.[144,146–148] observed nonulcerated portions of involved mucosa and found no consistent changes in goblet cells and epithelial cells except for a marked increase in cytoplasmic dense vesicles (Figure 5-56), the latter feature being explored by Marin et al.[150] in a freeze-fracture study of the lysosomal inclusions. The freeze-fracture analysis showed presence of intramembranous particles in the lysosomal membranes and a marked similarity of epithelial cell lysosomes to those of the lamina propria macrophages. EM showed degranulation of mast cells (which are increased in number) and eosinophils, endothelial cell injury and hypertrophy, and reduplication of the endothelial cell basal lamina.[147] There was smooth-muscle proliferation, myofibroblast transformation, and some necrosis, while nerve fibers showed proliferation of axons containing dense core granules (presumably adrenergic nerves) and some axonal necrosis.[148] Bishop, Polak, and Bryant[149] found proliferation of VIP (peptidergic)-containing nerves.

Lipid and Aminoacid Storage Diseases

Electron microscopy, histochemical, and biochemical studies of biopsy specimens may help make the diagnosis in lipid storage diseases.[138] These diseases may be suspected when the ganglion cells of Meissner's plexus are vacuolated. Vacuolated ganglion cells have also been reported in Fabry's disease,[152,153] and Perera et al.[1] have observed them in Niemann-Pick disease. In cholesterol ester

◄

Figure 5-55. Endoscopically obtained duodenal biopsy specimen from a patient with Crohn's disease involving the terminal ileum but without apparent disease of the upper intestine. (A) Normal villous architecture. Note the presence of granulomas below the crypts. PAS and hematoxylin stain, × 132. (B,C) High magnification showing that the granulomas consist of multiple epithelioid cells surrounded by a thin rim of round cells. H & E stain, × 270

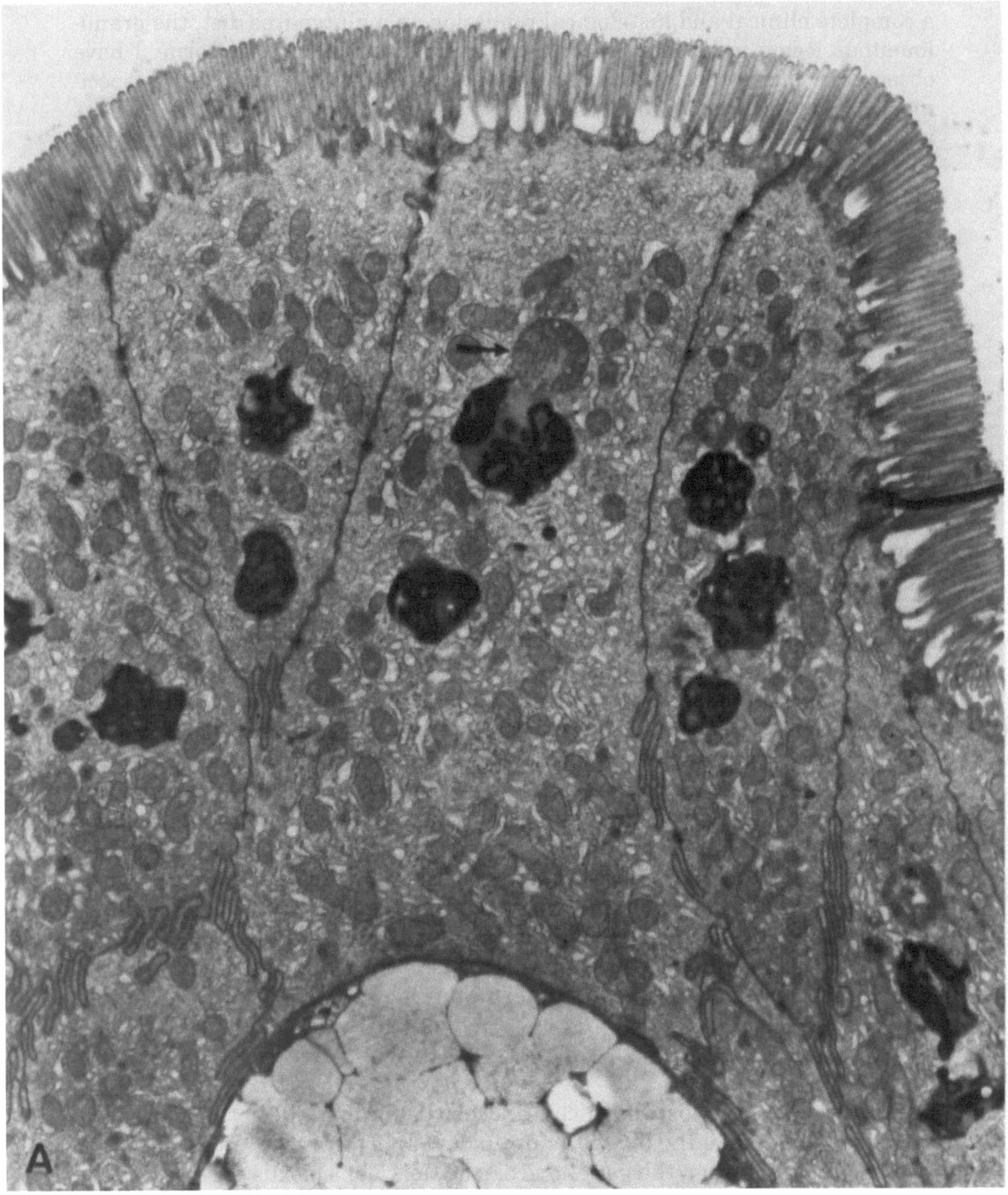

Figure 5-56. Electron micrograph of involved ileum in Crohn's disease. (A) Striking increase in number of "dense bodies" (lysosomes) in epithelial cells. These dense bodies are thought by some to be characteristic of Crohn's disease. The lysosomes here are mostly heterogeneous dense bodies. Also present is a marked increase in number of multivesicular bodies (arrow), some of which are quite large in size. × 10,000. (B) Similar dense bodies in the cytoplasm of a goblet cell (G). The epithelial cells (E) to the right and left of the goblet cell show "degenerative changes." × 25,000

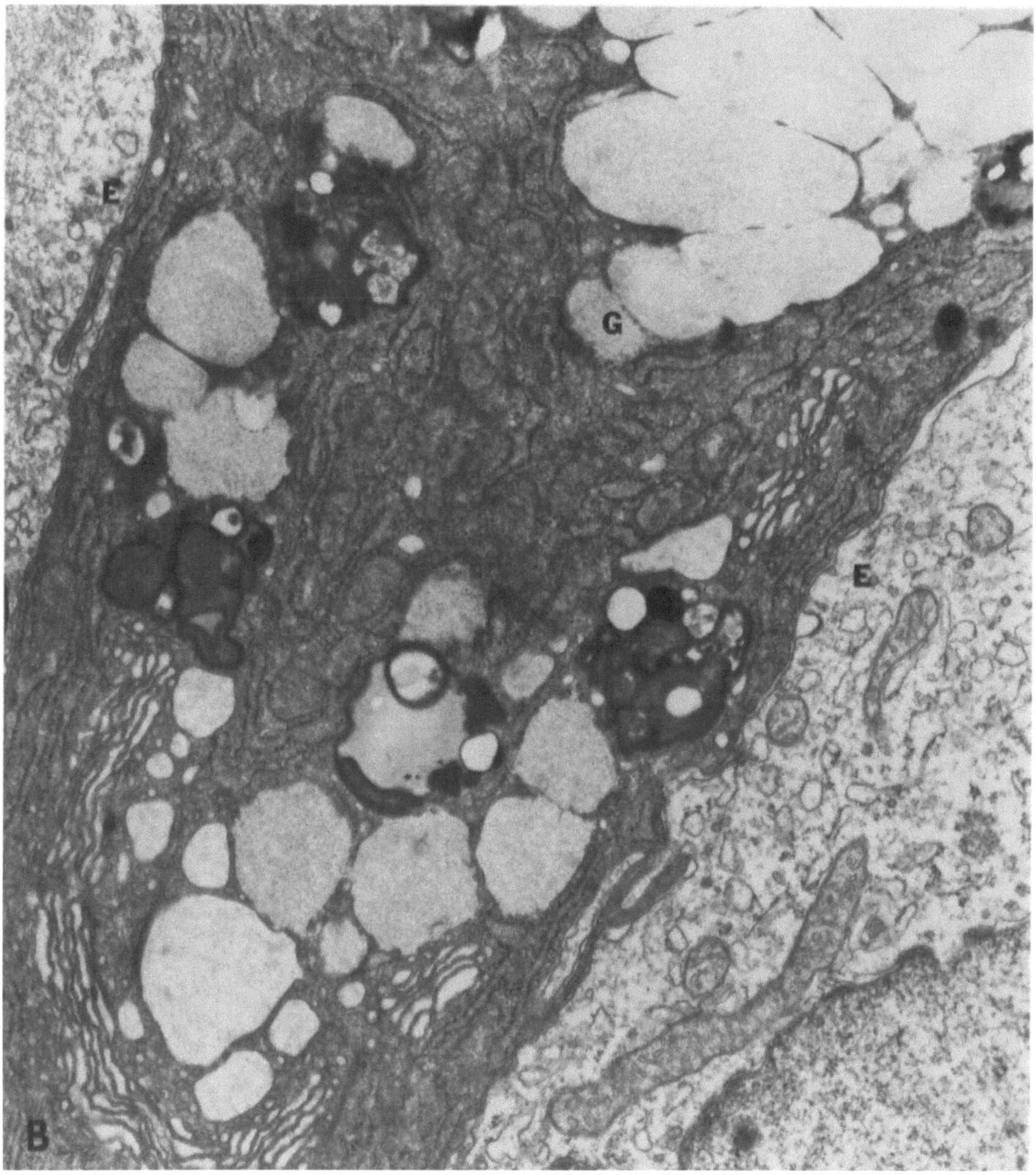

Figure 5.56B.

storage disease, the lamina propria may be greatly distended by numerous foamy-appearing macrophages, all filled with cholesterol esters and carotenes.[154] In Tangier disease, numerous foamy macrophages filled with cholesterol esters are also found in the muscularis mucosae, in the submucosa, and in the myenteric plexus.[138,155]

Lipid storage disorders, such as Tangier disease,[155] a-β-lipoproteinemia,[137] hypo-β-lipoproteinemia,[119] cholesterol storage disease,[154] and Fabry's disease,[153] commonly involve the intestinal mucosa. Intestinal biopsy specimens obtained

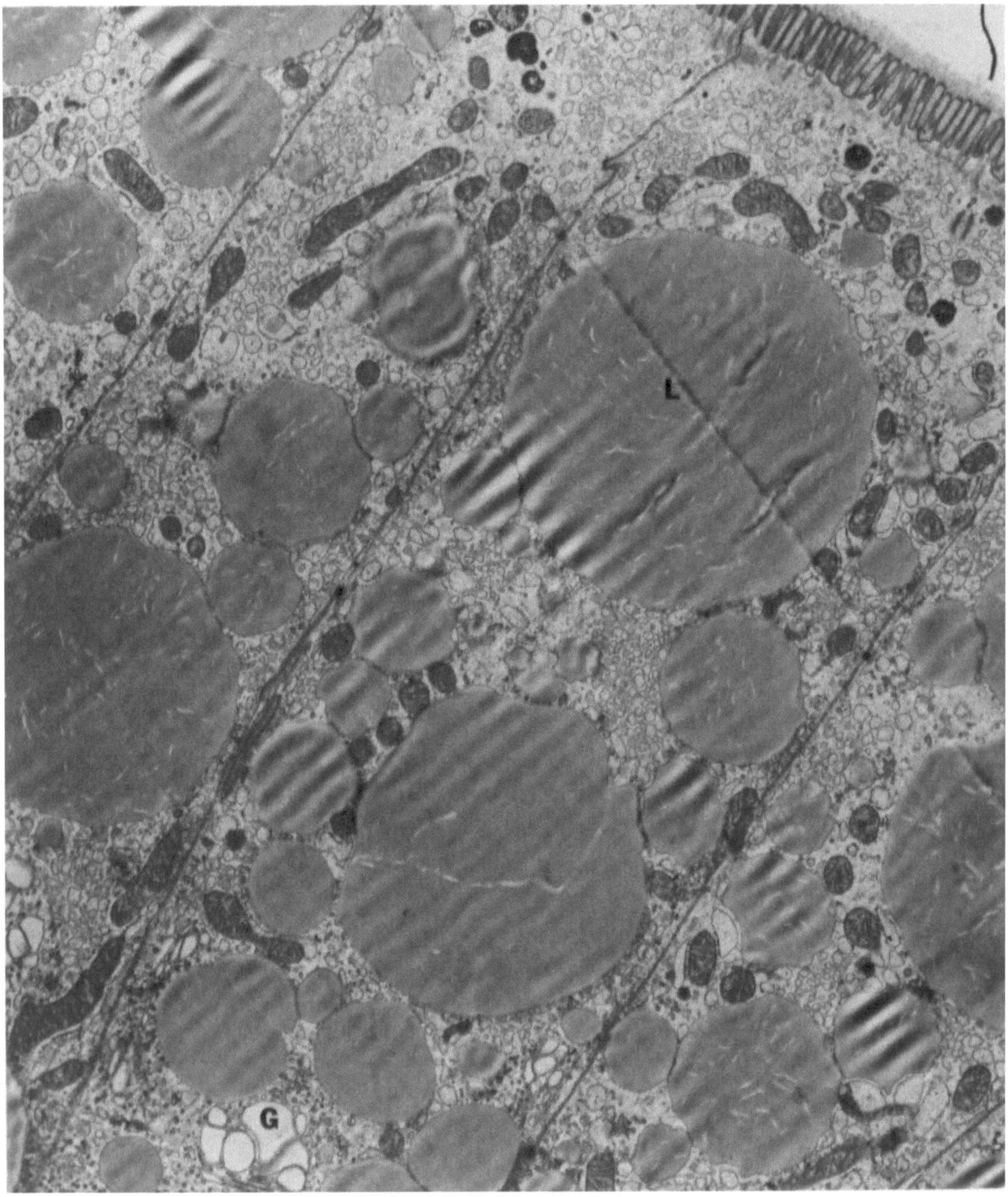

Figure 5-57. Electron micrograph of apex of intestinal absorptive cells of a patient with a-β-lipoproteinemia. The cells are massively packed with lipid (L) but otherwise are normal in appearance. Note that the Golgi (G) is devoid of lipoproteins. × 10,400

from patients with a-β-lipoproteinemia show normal villous architecture but absorptive cells packed with large droplets of triglyceride (Figure 5-57). These large droplets of lipid appear sometimes to be free within the cytoplasm of the absorptive cell and on other occasions to be enclosed within membranes of the endoplasmic reticulum. Intraluminal hydrolysis of triglycerides and mucosal esterification of free fatty acids and monoglycerides to triglyceride are normal in these patients. Following a fat meal, lipid droplets appear within the endoplasmic reticulum, but there is a striking absence of accumulation of lipid within Golgi vacuoles, and chylomicrons are not produced (Figure 5-58). The basic defect is inability of the intestinal absorptive cell to produce the beta-apoprotein required for final formation of the lipoprotein coat of chylomicrons.[156] The intestinal lamina propria in congenital β-lipoprotein deficiency appears to be normal except for the presence of an unusual number of macrophages containing bizarre inclusions (Figure 5-59).

Chylomicron retention disease is a malabsorption syndrome, recessively transmitted, in which there are normal fasting triglyceride levels unresponsive to a fat load, severe hypocholesterolemia, and epithelial cells filled with fat with the morphological appearance of chylomicrons but failure to release the chylomicrons even though epithelial cell apo-A and apo-B are present.[157] In a-β-lipoproteinemia there is a deficiency of apo-B and a failure to form chylomicrons, whereas in chylomicron retention disease apo-B is present and chylomicrons are formed but not released.

Familial hypo-β-lipoproteinemia is an inborn error of lipid metabolism in which serum β-lipoprotein content is moderately reduced and there is a milder degree of central nervous system involvement than seen in congenital a-β-lipoprotein deficiency.[119,158] It is likely that hypo-β-lipoproteinemia and chylomicron retention disease are the same disease process.[157]

Familial α-lipoprotein deficiency (Tangier disease)[155,159] is an autosomal recessive disease manifested by enlarged yellow-orange tonsils, hepatosplenomegaly, lymphadenopathy, and corneal infiltration. There is almost complete absence of α-lipoprotein from the serum, with reduced cholesterol and increased triglyceride levels. Storage of cholesterol ester has been demonstrated in many tissues, particularly in the reticuloendothelial system. The intestinal mucosa and the colonic mucosa and lamina propria may be infiltrated by numerous macrophages with a foamy cytoplasm. There is no retention of chylomicrons. Infiltration of the rectal mucosa with cholesterol-containing macrophages may be massive in extent, and may be associated with an ulcerative colitislike syndrome requiring colectomy. The other histological features of ulcerative colitis are not present, suggesting that the intense macrophage infiltrate results in "ischemia" leading to the bloody diarrhea. Not only is the lamina propria massively infiltrated with cholesterol ester-containing macrophages (Figure 5-60), but there is marked infiltration of smooth muscle with cholesterol esters and even presence of cholesterol esters in mast cells. Ganglion cells and Schwann cells of the myenteric plexus also contain accumulations of cholesterol esters.

Cholesterol ester storage disease[154] is characterized by enlargement of the liver and spleen, but in contrast to Tangier disease, serum cholesterol is moderately elevated. Further, in cholesterol ester storage disease, intestinal epithelial cells are morphologically normal by light and electron microscopy while the lamina propria is massively infiltrated with lipophages containing cholesterol

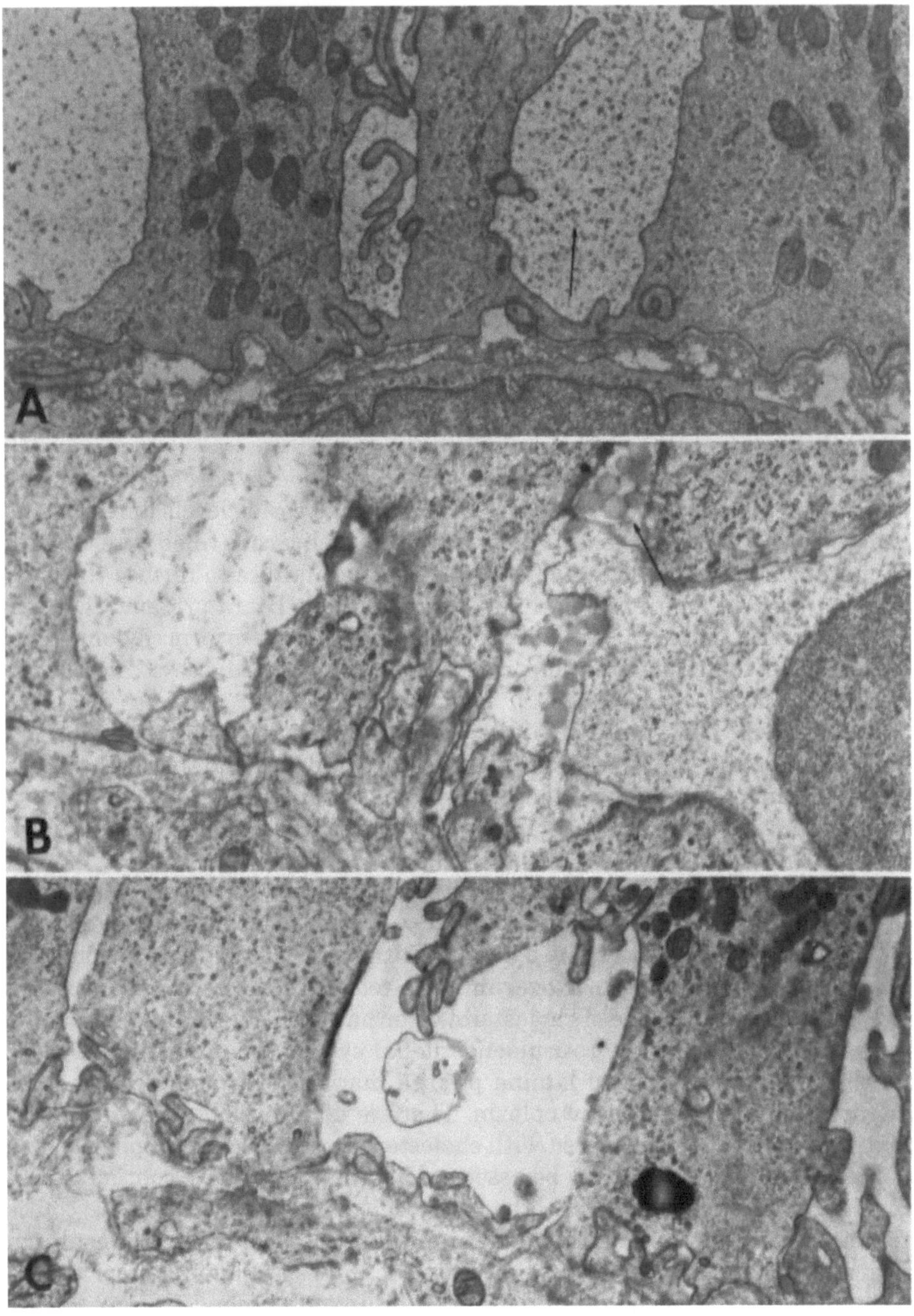

Figure 5-58. Electron micrographs of the base of intestinal absorptive cells illustrating appearance of basal intercellular spaces during normal fat absorption and in a-β-lipoproteinemia. (A) In normal control fasting state, showing very-low-density lipoprotein (arrow) particles (Tytgat GN, Rubin CE, Saunders DR. *J Clin Invest* **50**:2065, 1971). × 10,400. (B) In normal individual 45 minutes after a fat meal, showing numerous chylomicrons (arrow). × 10,400. (C) In a-β-lipoproteinemia 30 minutes after a fat meal, showing absence of lipoproteins and chylomicrons. × 10,400

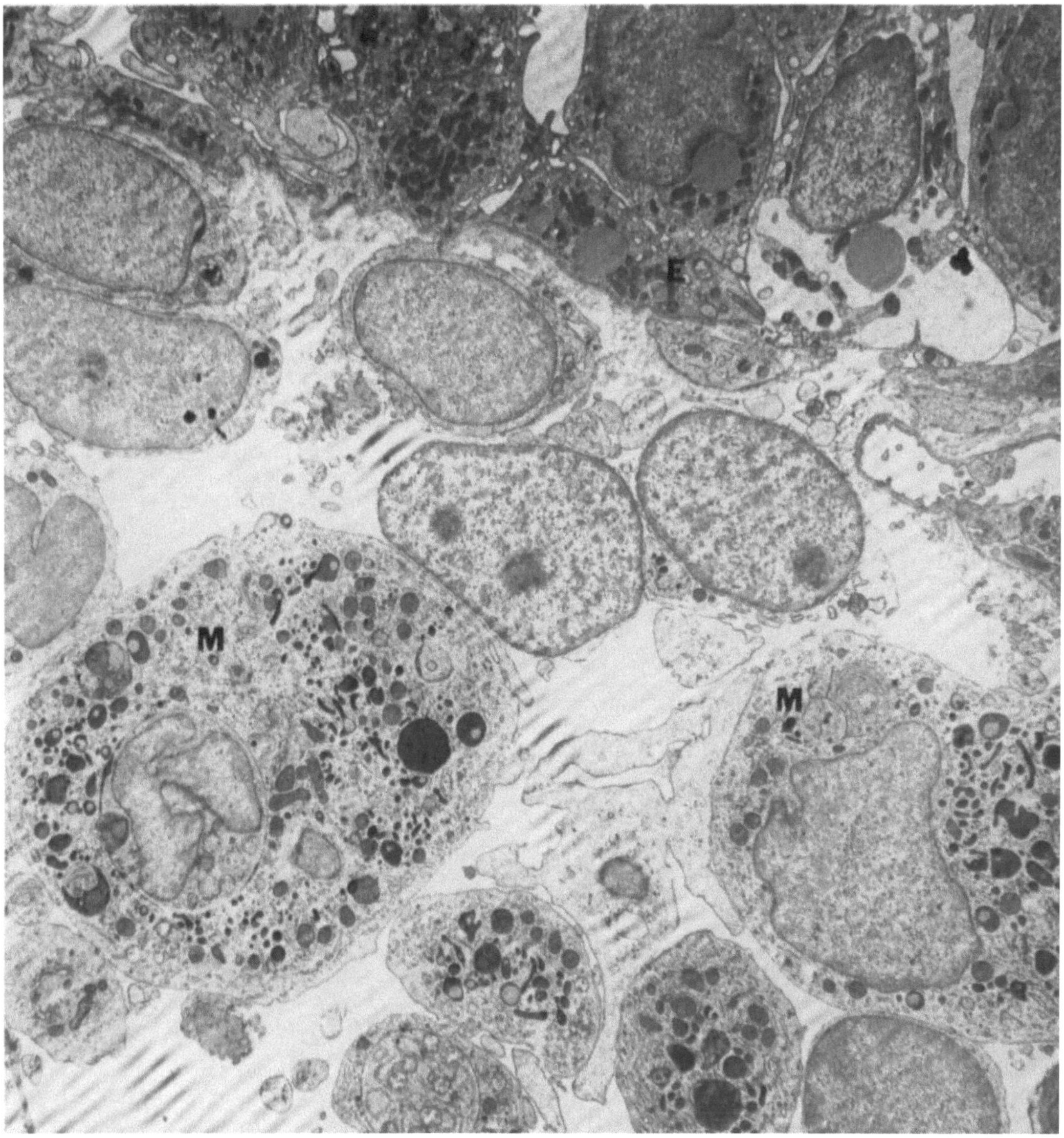

Figure 5-59. Electron micrograph in a-β-lipoproteinemia taken at the base of intestinal epithelial cells (E) and showing numerous lymphocytes and several macrophages (M) in the lamina propria. Note the macrophage inclusions, many of which are shaped like tennis rackets and like dumbbells. × 4,800

esters. There is marked retention of chylomicrons, possibly due to a block in transport of chylomicrons across lacteal endothelia.[154]

Fabry's disease (angiokeratoma corporis diffusum universale) is a disorder of glycolipid metabolism resulting from an inherited deficiency of a ceramide trihexoside-cleaving enzyme.[152,153] The enzyme defect results in the deposition of two closely related glycolipids in the endothelium of smooth muscle and in supporting cells of blood vessels of the kidney, skin, gastrointestinal tract, central

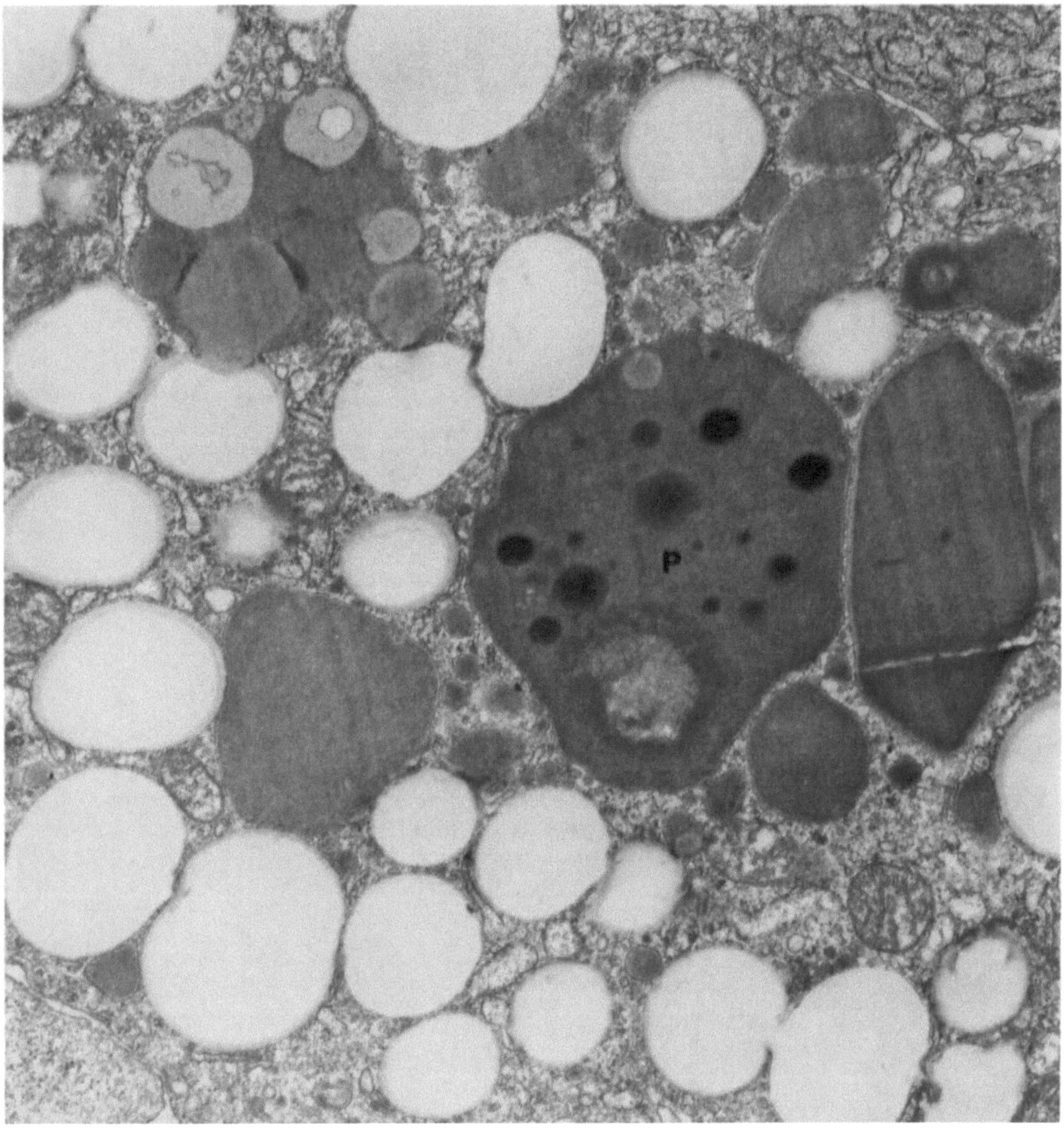

Figure 5-60. Electron micrograph of a portion of a macrophage in a rectal biopsy specimen from a patient with Tangier disease. The cell is filled with clear vacuoles representing cholesterol esters and with numerous phagosomes (P). × 25,000

nervous system, heart, and reticuloendothelial system. Intestinal and rectal biopsy specimens are normal at light microscopy. Only the ganglion cells of Meissner's plexus appear to be vacuolated. At electron microscopy, the vacuoles of the Meissner plexus have the appearance of myelin figures. Epithelial cells of Brunner's glands contain inclusions consisting of linear dense deposits with a periodicity of 5 to 6 nm. Similar inclusions are also present within macrophages of the lamina propria.[152,153]

Cystine storage disease, a metabolic disorder characterized by symptoms and signs of the Fanconi syndrome, is characterized pathologically by an abnormal accumulation of cystine crystals. Electron microscopy shows that these cystine crystals are confined to macrophages of the lamina propria.[160]

Amyloidosis

Rectal biopsy is preferable for the diagnosis of systemic amyloidosis because it is easy to perform and generally safe.[161] Both primary and secondary amyloidosis may involve the small intestine, and the diagnosis can then be established by peroral intestinal biopsy. Amyloid is demonstrated at light microscopy and its presence is confirmed using special stains or by electron microscopy.[138] The use of congo red staining and demonstration of dichroism by polarized light microscopy is the simplest approach to diagnosis. Small-bowel biopsy may provide the diagnosis when rectal biopsy examination is negative and amyloidosis is still suspected clinically.[162] Amyloid deposits consist of fibrillar proteins with a β-pleated sheet conformation.[163] Electron microscopy shows that most amyloid deposits consist of nodular masses of 8 to 10 nm-wide, randomly arrayed, non-branching fibrils (Figures 5-61, 5-62).

Melanosis and Lipofuscinosis

Melanosis of the gastrointestinal tract consists of a black pigmentation in the colonic mucosa, originating from pigment (iron and lipofuscin)-containing macrophages, that generally stops abruptly at the ileocecal valve. Extracolonic melanosis is a rare finding, although it has been noted in the ileum as an isolated manifestation. Seventeen cases of duodenal melanosis have been documented.[164,165] The villous architecture in duodenal biopsy specimens is normal but the lamina propria is filled with macrophages containing an intense brown-black pigment. The pigment is similar to that observed in melanosis coli and histochemically stains sometimes like melanin but usually like lipofuscin.[164,165] In melanosis coli, lipofuscinlike material is localized in macrophages (Figure 5-63), is related to laxative use, and is of no pathological significance.[166]

In brown bowel syndrome (lipofuscinosis), which is usually associated with chronic malabsorption due to a variety of causes, there is deposition of lipofuscin in smooth-muscle cells of the muscularis propria.[166] The lipofuscin aggregates contain myelin figures, round to irregular homogeneous bodies with varying density, and structures of possible mitochondrial origin. It is thought that vitamin E deficiency, secondary to the underlying malabsorption syndrome, results in mitochondrial damage and lipofuscin deposition.[166,167] Lipofuscin is a "wear and tear" pigment derived from the residues of lysosomal hydrolysis, is often called ceroid pigment, and may be found in the lamina propria in the absence of malabsorption.[167]

Mastocytosis

There may be a mucosal or submucosal infiltrate of mast cells in the intestine of individuals with systemic mast cell disease, but the villous architecture remains normal.[168] Several patients with systemic mast cell disease have also had celiac sprue, that is, they had a severe intestinal mucosal lesion that responded to a gluten-free diet.[1,169,170] The severe mucosal lesion in this setting has been either free of mast cell infiltration[170] or infiltrated with up to 30 mast cells per high power field on toluidine blue staining.[169] Mast cells, like eosinophils, are easily found when *looked for* in the intestinal mucosa, and thus

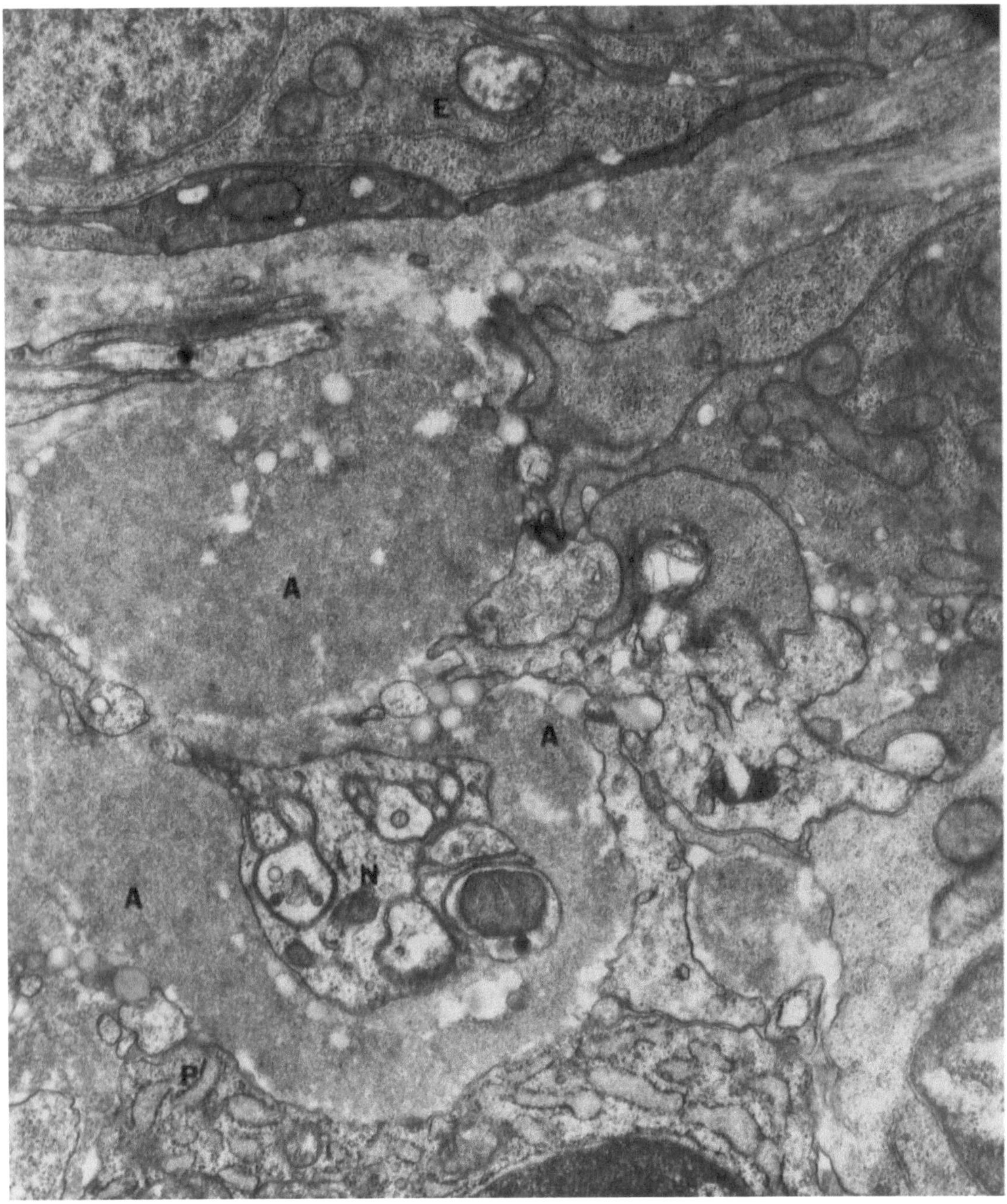

Figure 5-61. Electron micrograph of a portion of an intestinal biopsy specimen from a patient with systemic amyloidosis. Large deposits of amyloid fibrils (A) can be seen just below intestinal epithelial cells (E). Note that the amyloid surrounds a nerve fiber (N). Plasma cell (P) with dilated endoplasmic reticulum. × 26,900

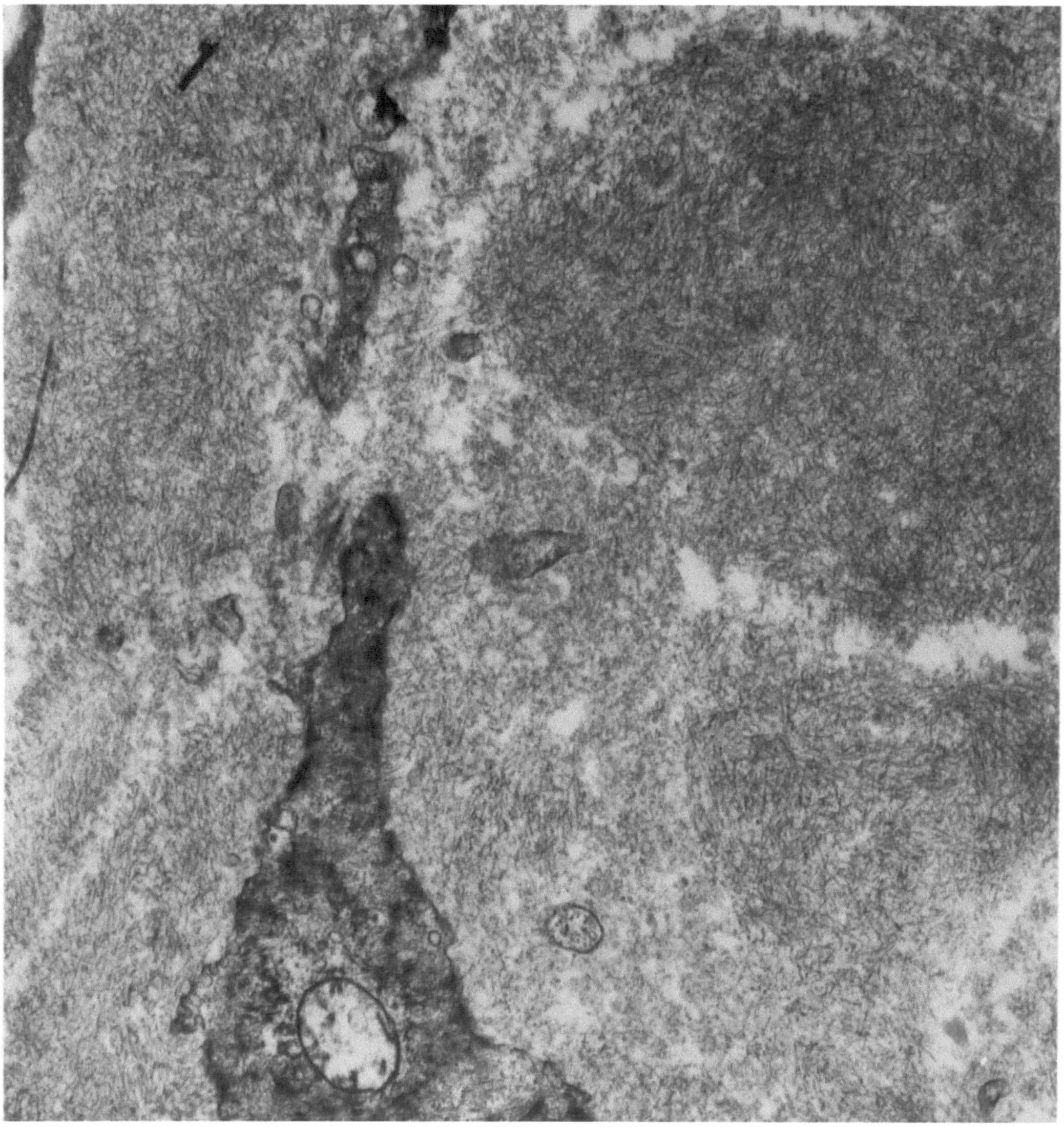

Figure 5-62. Electron micrograph from a different area of the specimen shown in Figure 5-61 illustrates the 8- to 10-nm wide, randomly arrayed, nonbranching fibrils that make up the amyloid deposits. × 33,600

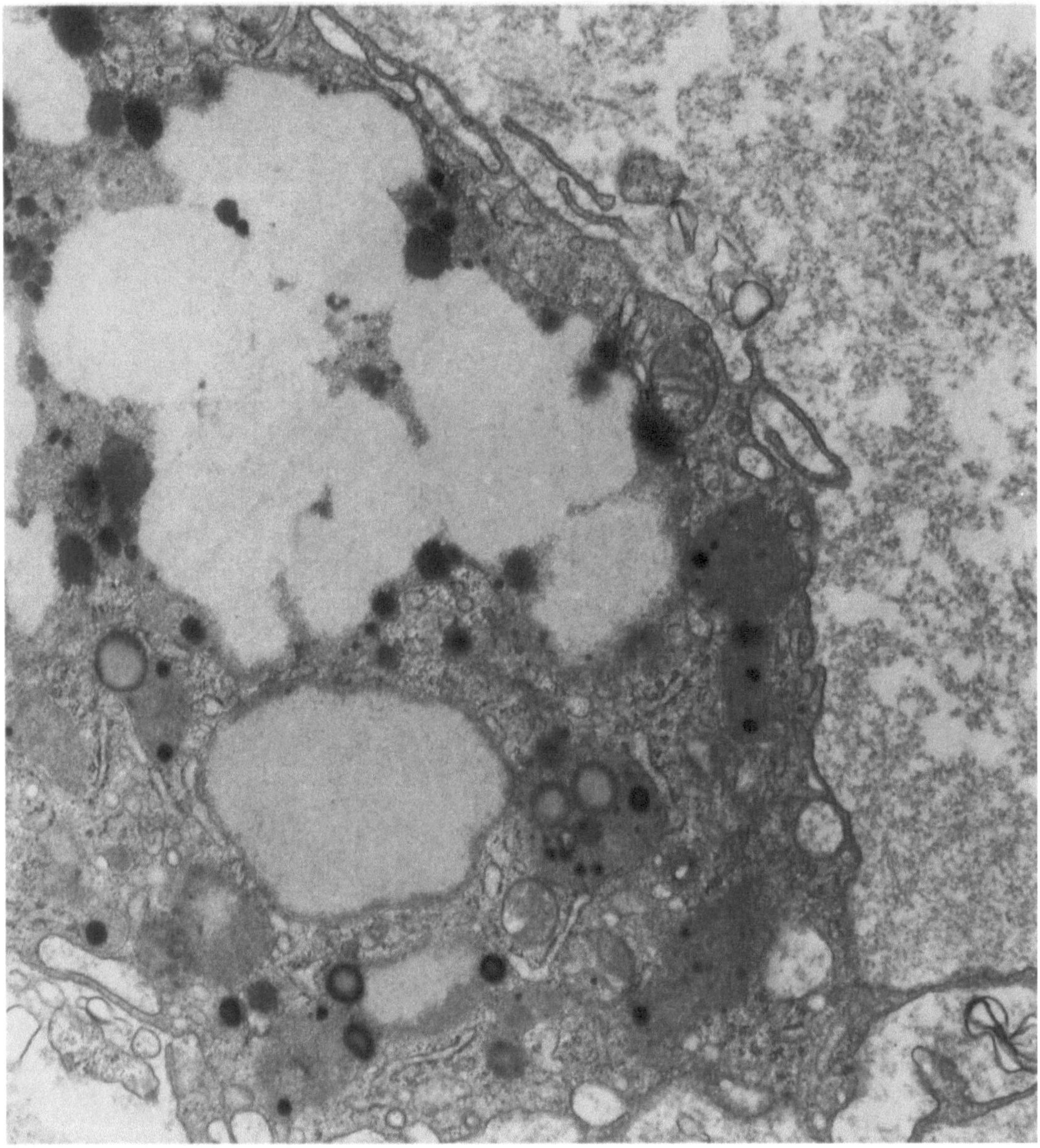

Figure 5-63. Electron micrograph of typical colonic macrophage, present in most normal rectal biopsy specimens. Note that the cytoplasm contains numerous lysosomes, many containing heterogeneous densities that are found in melanosis coli, and others (phagolysosomes) containing a clear mucuslike substance and which result in the macrophage being called a "muciphage." × 25,000

their presence must be quantitated before conclusions are drawn about their significance. Intraepithelial mast cells are often observed in normal intestine (Figure 5-34, Chapter 3).[171]

States Associated with Normal Proximal Jejunal Biopsy

There are conflicting reports that a variety of other diseases may result in abnormalities of the proximal jejunum. Most investigators have been unable to confirm that these diseases indeed result in abnormalities, and it is likely that many of the reported abnormalities are related to misinterpretation of artifacts. Table 5-7 lists the disease entities that reportedly result in abnormalities of the jejunum, a finding not confirmed by Rubin and the many individuals who have worked with him in his laboratory.[1,16]

Endoscopic Biopsy of the Duodenal Bulb

With the ready availability of fiberoptic duodenoscopes, it is difficult for the endoscopist to refrain from obtaining numerous tiny mucosal biopsy specimens of any abnormality that he or she might perceive in the duodenum. Perera et al.[1] have found that only about two thirds of the tiny biopsy specimens obtained by the endoscopist are interpretable. This has changed now that "jumbo" biopsy forceps that can be used through flexible endoscopes have become available.

It is important for the endoscopist to communicate clearly to the pathologist just why the biopsy specimen was taken. Generally, biopsy specimens are taken from the duodenal bulb to determine whether nonulcerative changes represent significant mucosal disease. If the endoscopist uses appropriate terms such as erythema, swelling, and deformity and avoids pathological terms such as inflammation, edema, and scar, the pathologist is more likely to be of help in interpretation of the specimens. Quite frequently the biopsy specimens are clearly normal, even though the endoscopist noted an apparent "abnormality." Often, histological abnormalities will be found in biopsy specimens obtained from apparently normal duodenal bulbs. Whether these histological abnormalities can account for the patient's symptoms has not yet been determined. It is hoped that prospective studies of duodenal bulb specimens and their comparison with endoscopic findings will clarify many of the unknowns of interpretation of duodenal bulb biopsy specimens.[172–177] Helpful discussions of the problem of duodenitis and its histological grading are available from the pathologist's viewpoint in the text by Whitehead[176] and from the gastroenterologists viewpoint in the paper by Weinstein.[177]

The histology of proximal duodenal biopsy specimens generally reflects the appearance of distal duodenal biopsy specimens.[172,178] Slight to moderate increases in lamina propria cell content and slight to moderate shortening of villi in proximal duodenal biopsy specimens should not be confused with specific pathological states. In normal individuals, mucosa overlying Brunner's glands may occasionally be flattened while biopsy specimens obtained from the ligament of Treitz are normal. Villi in the proximal duodenum tend to vary somewhat more in height than do villi in the distal duodenum, and the cellular content of the lamina propria tends to be slightly greater in the proximal duodenum. Histological findings in the proximal duodenum that are of questionable significance include the presence of hemorrhage, gastric surface epithelium, heterotopic fundic gland mucosa, and mild inflammatory changes. Hemorrhage is commonly observed in endoscopic biopsy specimens, and one can never be sure that it is not due to trauma of obtaining the biopsy. Villi in the proximal duodenum are often lined with cells similar to surface mucous cells of the stomach, and this finding appears to be more frequent in individuals with high gastric acid output than in individuals with low acid output.[172] It may be observed also in specimens obtained from normal individuals, and little significance can be attached to its finding (Figure 5-64), although on occasion I have been able to correctly suggest the diagnosis of Zollinger-Ellison syndrome

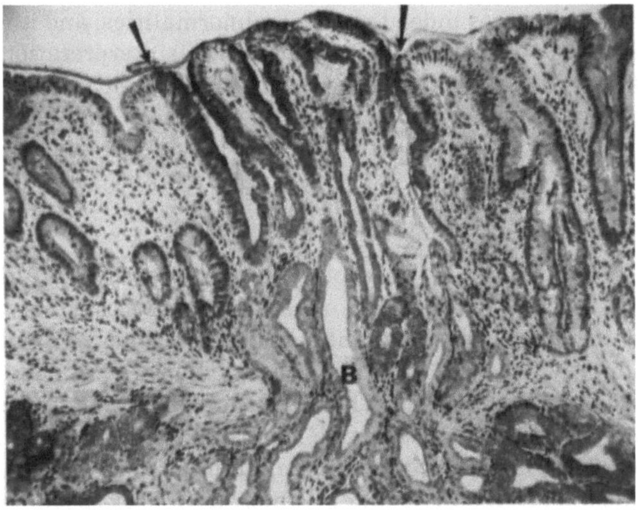

Figure 5-64. Duodenal bulb biopsy specimen showing gastric surface mucous cell metaplasia. The three central villi are covered with gastric-type epithelium rather than by absorptive cells. Note that a portion of the villus to the right and of the villus to the left of the three central villi are lined by gastric surface mucous cells, with a sharp demarcation between gastric and absorptive epithelium (arrows). Brunner (B) glands and their resulting artifact are also present. PAS and hematoxylin stain, × 150

when I observed extensive surface mucous cell metaplasia of the intestinal epithelium in jejunal biopsy specimens in patients with chronic diarrhea. Heterotopic fundic gland mucosa in the duodenum has been observed for many years and has no diagnostic significance.[1,16] The presence of mild increases in the round cell content of the lamina propria and of occasional polymorphonuclear leukocytes in the lamina propria of the proximal duodenum is of doubtful significance. Only if other pathological changes are present can any significance be placed on this observation.

The term *nonspecific duodenitis* should apply to changes seen in the mucosa obtained adjacent to duodenal ulcers as well as to the changes seen in celiac sprue and in Crohn's disease. In full-blown nonspecific duodenitis not due to specific inflammatory diseases such as celiac sprue, the villi are blunted and the surface epithelium (Figure 5-65) has an abnormal appearance.[177] The changes in the surface epithelium are often similar to those seen in reactive atypia. The surface epithelium in acute duodenitis resembles the surface epithelium seen in gastritis rather than the epithelial changes seen in celiac sprue.[177] The lamina propria contains a marked increase in number of round cells and in number of polymorphonuclear leukocytes. PMN leukocytes tend to invade the surface epithelium, and mitotic figures are increased within the crypt epithelium. The histological spectrum ranges from that of mild changes affecting the surface epithelium only to severe changes that on occasion may be difficult to distinguish from those of celiac sprue (Figures 5-12, 5-13, 5-65). The pathological significance of so-called nonspecific duodenitis has not been adequately defined and requires long-term prospective clinicopathological studies to determine its significance, if any.[177] Thus there is usually no need to biopsy duodenal erosions unless a specific cause (e.g., Crohn's disease, celiac sprue, Whipple's disease) is suspected.

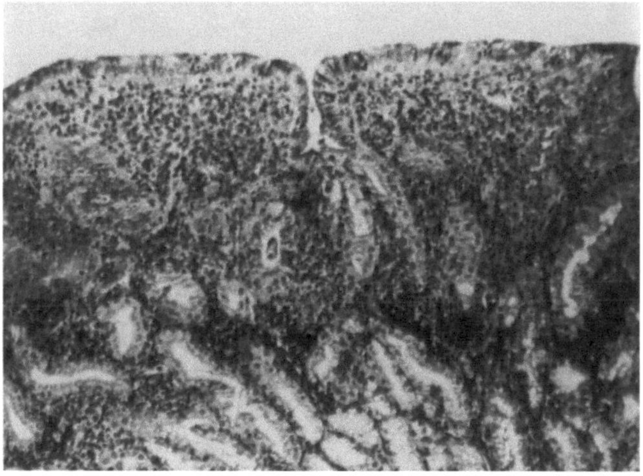

Figure 5-65. Severe changes of nonspecific duodenitis. The epithelial cells are cuboidal and there is a marked infiltrate of the lamina propria with polymorphonuclear leukocytes and plasma cells. Brunner glands are present. These changes are more like those seen in gastritis than those seen in celiac sprue. H & E stain, × 150. (Photograph kindly contributed by Dr. Wilfred M. Weinstein.)

Ileal Biopsy

With the ability to enter the ileum almost routinely during colonoscopy, the endoscopist faces the same temptations in the ileum as he or she does in the proximal duodenum and tends to obtain biopsy specimens from all "abnormalities," whether apparent or real. Many of the precautions in interpreting proximal duodenal biopsy specimens also apply to interpretation of ileal biopsy specimens.[179]

The normal structure of the distal ileum is similar to that of the rest of the intestine with a villus/crypt ratio of 3:1 to 4:1. Both villi and crypts are shorter than they are in the jejunum. There are more goblet cells in the epithelium than in the jejunum and there is increased cellularity of the lamina propria compared with the jejunum. Lymphoid follicles (Peyer's patches) may be prominent in the ileum and may distort the overlying mucosa, just as lymphoid follicles distort the duodenal mucosa (Figure 5-66).

Our knowledge of the pathology of the ileum is necessarily less extensive than our knowledge of duodenojejunal pathology. Villus shortening with inflammation in the lamina propria alone, especially when chronic, is nonspecific and must be interpreted with great caution. This finding is frequently observed and often overinterpreted. In celiac sprue, the ileal mucosa ranges from normal in appearance to that of a severe lesion similar to that of the jejunum.[180] When there is acute ischemic injury to the ileum, the lesion seen is a characteristic bland hemorrhagic necrosis with minimal inflammation. Histological features that support the diagnosis of Crohn's ileitis include a minimum of spotty necrosis and injury of surface epithelium and a PMN leukocyte infiltrate.[179] The presence of shortened villi, crypt abscesses, and frank ulceration provides further evidence for the diagnosis. Granulomas along with the other inflammatory changes described are strongly suggestive of Crohn's disease, but granulomas in

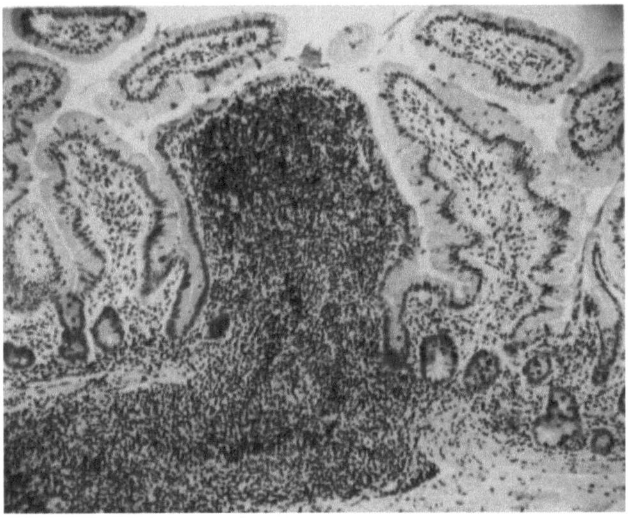

Figure 5-66. Section of proximal duodenal biopsy specimen in which a lymphoid follicle distorts a single villus in an otherwise normal specimen. Peyer's patches present similar and often more extensive artifacts in the ileum. H & E stain, × 132. (Reproduced from Dobbins,[16] by permission of Churchill Livingstone, Inc., New York.)

the ileum in the absence of other inflammatory changes may be nonspecific. "Backwash" ileitis seen in ulcerative colitis may mimic Crohn's ileitis.

When the ileum is utilized in place of the ureter, or of the urinary bladder, the mucosa generally becomes flat and presents a severe mucosal lesion.[181] This change is presumably due to stasis and bacterial overgrowth, but there are no direct data in this regard. When the ileum is utilized as an ileostomy reservoir (Koch pouch), there is shortening of villi and increased epithelial cell turnover in the mucosa. The lesion develops early, is only mild in degree, and is not progressive, at least over a 24-month observation period.[182] The bacterial flora within the pouch is intermediate between that of conventional ileostomy effluent and that of normal feces. At electron microscopy the epithelial cells show only mild shortening of microvilli. The mucosa of the ileostomy reservoir retains its ability to absorb d-xylose and phenylalanine.[182] An ileal pouch-anal anastomosis (IPAA) is constructed following subtotal colectomy and mucosal proctectomy, and utilized in the treatment of ulcerative colitis and of familial polyposis coli.[183] A syndrome called pouchitis develops in up to 20% of patients having this procedure for treatment of ulcerative colitis, but very rarely follows this procedure in patients with polyposis. Pouchitis is characterized by high stool frequency, watery stools, cramps, fat malabsorption, and occasionally by nocturnal leakage, incontinence, and rectal bleeding.[183] There is a poor correlation between the endoscopic appearance and biopsy findings. Biopsy specimens may show changes similar to those seen in the colon, including mild to moderate villous abnormalities and PMN leukocyte infiltrate of the mucosa including crypt abscesses.[174] Rapid clinical response to treatment with metronidazole suggests that anaerobic bacterial toxins and bile salt deconjugation play a role in the pathogenesis of the pouchitis.

In the syndrome of primary bile acid malabsorption, patients present with chronic watery diarrhea associated with excessive fecal bile acids, and respond to treatment with cholestyramine. Three patients with this syndrome were found to have normal fecal fat, normal Schilling test results, and normal barium studies of the intestine and colon.[184] Jejunal biopsies were normal but multiple biopsy specimens of the terminal ileum showed a flat mucosa with features of colonic metaplasia. None of the patients had a condition known to cause ileal mucosal injury, and all had evidence of autoimmunity. Thus this syndrome of primary bile acid malabsorption with colonic metaplasia of the ileum may be immune mediated.

Bile salt malabsorption secondary to chronic radiation injury is an occasional clinical problem.[185] Ileal biopsy specimens show variably abnormal villi, submucosal fibrosis and lymphangiectasia, and even secondary amyloidosis.

References

1. Perera DR, Weinstein WM, Rubin CE: Symposium on the pathology of the gastrointestinal tract. Part II. Small intestinal biopsy. *Hum Pathol* **6**:157–217, 1975.
2. Trier JS, Rubin CE: Electron microscopy of the small intestine: A review. *Gastroenterology* **49**:574–603, 1975.
3. Ginsel LA: Lysosomes and storage diseases. *A Morphological, Cytochemical, and Autoradiographical Study of the Function of Lysosomes in the Absorptive Cells of the Human Intestine in Relation to the Transport and Secretion of Cell-Coat Material.* Heidelberg, Springer-Verlag, 1979.
4. Potten CS, Allen TD: Ultrastructure of cell loss in intestinal mucosa. *J Ultrastruct Res* **60**:272–277, 1977.
5. Trier JS, Browning TH: Morphologic response of the mucosa of human small intestine to x-ray exposure. *J Clin Invest* **45**:194–204, 1966.
6. Hugon J, Borgers M: Ultrastructural and cytochemical studies on karyolytic bodies in the epithelium of the duodenal crypts of whole body x-irradiated mice. *Lab Invest* **15**:1528–1543, 1966.
7. Verbin RS, Liany H, Saez LM, et al: Effects of inhibitors of protein synthesis on structure and function of the crypts of the small intestine. *Exp Cell Res* **65**:81–93, 1971.
8. Mansbach CM II, Wilkins RM, Dobbins WO, et al: Intestinal mucosal function and structure in the steatorrhea of Zollinger-Ellison syndrome. *Arch Intern Med* **121**:487–494, 1968.
9. Portela-Gomes G, Martins MM, Correia JP: Ultrastructural changes of jejunal epithelial cells in liver cirrhosis. *Scand J Gastroenterol* **9**:657–663, 1974.
10. Rubin E, Rybak BJ, Lindenbaum J, et al: Ultrastructural changes in the small intestine induced by ethanol. *Gastroenterology* **63**:801–814, 1972.
11. Tomasini JT, Dobbins WO III: Intestinal mucosal morphology during water and electrolyte absorption: A light and electron microscopic study. *Am J Digest Dis* **15**:226–238, 1970.
12. Afzelius BA: Glycocalyx and glycocalyceal bodies in the respiratory epithelium of nose and bronchi. *Ultrastruct Pathol* **7**:1–18, 1984.
13. Boom A, Deams W Th, Luft JH: On the fixation of intestinal absorptive cells. *J Ultrastruct Res* **48**:350–360, 1974.
14. Hideshige I, Stein AA: Ultrastructure of adenocarcinoma of the colon. *Gastroenterology* **44**:410–418, 1963.

15. Rubin CE, Brandborg LL, Phelps PC, et al: Studies of celiac disease. I. The apparent identical and specific nature of the duodenal and proximal jejunal lesion in celiac disease and idiopathic sprue. *Gastroenterology* **38**:28–49, 1960.
16. Dobbins WO III: Small bowel biopsy in malabsorptive states. In HT Norris (Ed), *Contemporary Issues in Surgical Pathology*, Vol. 4, *Pathology of the Colon, Small Intestine and Anus*. Churchill Livingstone, 1983, pp 121–165.
17. MacDonald WC, Dobbins WO III, Rubin CE: Studies of the familial nature of celiac sprue using biopsy of the small intestine. *N Engl J Med* **272**:448–456, 1965.
18. Pena AS, Mann DL, Hague NE, et al: Genetic basis of gluten sensitive enteropathy. *Gastroenterology* **75**:230–235, 1978.
19. Trier JS, Moxey PC, Fordtran JS, et al: Ectopic gastric mucosa in celiac sprue. *Gastroenterology* **65**:712–727, 1973.
20. Ferguson A, MacDonald TT, McClure TP, et al: Cell-mediated immunity to gliadin within the small-intestinal mucosa in coeliac disease. *Lancet* **1**:895–897, 1975.
21. Brow J, Parker F, Weinstein W, et al: The small intestinal mucosa in dermatitis herpetiformis. I. Severity and distribution of the small intestinal lesion and associated malabsorption. *Gastroenterology* **60**:355–361, 1971.
22. Rubin W, Ross LL, Sleisenger MH, et al: An electron microscopic study of adult celiac disease. *Lab Invest* **15**:1720–1747, 1966.
23. Biempica L, Toccalino H, O'Donnell JC: Cytochemical and ultrastructural studies of the intestinal mucosa of children with celiac disease. *Am J Pathol* **52**:795–823, 1968.
24. Shiner M: Ultrastructure of jejunal surface epithelium in untreated idiopathic steatorrhoea. *Br Med Bull* **23**:223–225, 1967.
25. Laguens R, Schaposnik F, Echeverria R, et al: Fine structure of the small bowel in dermatitis herpetiformis. *Virch Arch Abt Path Anat* **352**:34–42, 1971.
26. Madara JL, Trier JS: Structural abnormalities of jejunal epithelial cell membranes in celiac sprue. *Lab Invest* **43**:254–261, 1980.
27. Baker AL, Rosenberg IH: Refractory sprue: Recovery after removal of nongluten dietary proteins. *Ann Intern Med* **89**:505–508, 1978.
28. Barnes G, Townley R: Duodenal mucosal damage in 31 infants with gastroenteritis. *Arch Dis Child* **48**:343–349, 1973.
29. Weinstein W, Brow J, Parker F, et al: The small intestinal mucosa in dermatitis herpetiformis. II. Relationship of the small intestinal lesion to gluten. *Gastroenterology* **60**:362–369, 1971.
30. Katz SI, Hall RP III, Lawley TJ, et al: Dermatitis herpetiformis: The skin and the gut. *Ann Intern Med* **93**:857–874, 1980.
31. Ament M, Rubin CE: Soy protein—another cause of the flat intestinal lesion. *Gastroenterology* **62**:227–234, 1972.
32. Walker-Smith JA, Harrison M, Kilby A, et al: Cow's milk-sensitive enteropathy. *Arch Dis Child* **53**:375–380, 1978.
33. Trier JS, Falchuk ZM, Carney MC, et al: Celiac sprue and refractory sprue. *Gastroenterology* **75**:307–316, 1978.
34. Baer AN, Bayless TM, Yardley JH: Intestinal ulceration and malabsorption syndromes. *Gastroenterology* **79**:754–765, 1980.
35. Weinstein W, Saunders D, Tytgat G, et al: Collagenous sprue—an unrecognized type of malabsorption. *N Engl J Med* **283**:1297–1301, 1970.
36. Bossart R, Henry K, Booth CC, et al: Subepithelial collagen in intestinal malabsorption. *Gut* **16**:18–22, 1975.
37. Dobbins WO III: Gut immunophysiology: A gastroenterologist's view with emphasis on pathophysiology. *Am J Physiol* **242**:G1–G8, 1982.
38. Ament M, Rubin CE: Relation of giardiasis to abnormal intestinal structure and function in gastrointestinal immunodeficiency syndromes. *Gastroenterology* **62**: 216–226, 1972.

39. Hartong WA, Gourley WK, Aravantakis C: Giardiasis: Clinical spectrum and functional-structural abnormalities of the small intestinal mucosa. *Gastroenterology* **77**:61–69, 1979.
40. Hughes WS, Cerda JJ, Holtzapple P, et al: Primary hypogammaglobulinemia and malabsorption. *Ann Intern Med* **74**:903–910, 1971.
41. Dobbins WO III, Sahba MM: A possible explanation for the gastrointestinal mucosal lesion in immunoglobulin deficiency state. *Am J Dig Dis* **17**:23–36, 1972.
42. Staley TE, Jones EW, Corley LD: Attachment and penetration of *Escherichia coli* into intestinal epithelium of the ileum of newborn pigs. *Am J Pathol* **56**:371–392, 1969.
43. Oxelius V-A, Laurell A-B, Lindquist B, et al: IgG subclasses in selective IgA deficiency: Importance of IgG2-IgA deficiency. *N Engl J Med* **304**:1476–1477, 1981.
44. McCarthy DM, Katz SI, Gazze L, et al: Selective IgA deficiency associated with total villous atrophy of the small intestine and an organ-specific antiepithelial cell antibody. *J Immunol* **120**:932–938, 1978.
45. Anderson KE, Finlayson NDC, Deschner EE: Intractable malabsorption with a flat jejunal mucosa and selective IgA deficiency. *Gastroenterology* **67**:709–716, 1974.
46. Ochs HD, Ament ME, Davis SD: Structure and function of the gastrointestinal tract in primary immunodeficiency syndromes (IDS) and in granulocyte dysfunction. *Birth Defects: Original Article Series* **11**:199–207, 1975.
47. Keren DF: *Immunology and Immunopathology of the Gastrointestinal Tract.* Chicago, American Society of Clinical Pathologists, Educational Products Division, 1980, pp 45–60.
48. Ament M, Ochs H, Davis S: Structure and function of the gastrointestinal tract in primary immunodeficiency syndromes: A study of 39 patients. *Medicine* **52**:227–241, 1973.
49. Brunser O, Eidelman S, Klipstein FA: Intestinal morphology of rural Haitians: A comparison between overt tropical sprue and asymptomatic subjects. *Gastroenterology* **58**:655–668, 1970.
50. Klipstein FA: Tropical sprue in travelers and expatriates living abroad. *Gastroenterology* **80**:590–600, 1981.
51. Mathan M, Mathan VI, Baker SJ: An electron-microscopic study of jejunal mucosal morphology in control subjects and in patients with tropical sprue in southern India. *Gastroenterology* **68**:17–32, 1975.
52. Brunser O, Reid A, Monkeberg F, et al: Jejunal biopsies in infant malnutrition: With special reference to mitotic index. *Pediatrics* **38**:605–612, 1966.
53. Brunser O, Catillo C, Araya M: Fine structure of the small intestinal mucosa in infantile marasmic malnutrition. *Gastroenterology* **70**:495–507, 1976.
54. Davidson GP, Cutz E, Hamilton JR, et al: Familial enteropathy: A syndrome of protracted diarrhea from birth, failure to thrive, and hypoplastic villus atrophy. *Gastroenterology* **75**:783–790, 1978.
55. Drucker MM, Polliade A, Yeivin R, et al: Immunofluorescent demonstration of enteropathic *E. Coli* in tissues of infants dying with enteritis. *Pediatrics* **46**:855–864, 1970.
56. Schreiber D, Blacklow N, Trier J: The intestinal lesion in acute infectious nonbacterial gastroenteritis. *N Engl J Med* **288**:1318–1323, 1973.
57. Rothbaum RJ, Partin JC, Saalfield K, et al: An ultrastructural study of enteropathogenic *Escherichia coli* infection in human infants. *Ultrastruct Pathol* **4**: 291–304, 1983.
58. Hampton JC, Rosario B: The attachment of microorganisms to epithelial cells in the distal ileum of the mouse. *Lab Invest* **14**:1464–1481, 1965.
59. Dobbins WO III, Weinstein WM: Electron microscopy of the intestine and rectum in acquired immuno-deficiency syndrome. *Gastroenterology* **88**:738–749, 1985.
60. Neutra MR: Ultrastructural studies of the interaction of bacteria with intestinal

cell surfaces. IN EC Boedeker (Ed), *Attachment of Organisms to the Gut Mucosa*, Vol I. Boca Raton, FL, CRC Press, 1984, pp 173–188.
61. Dobbins WO III: Alterations in intestinal structure associated with microorganisms. In EC Boedeker, (Ed), *Attachment of Organisms to the Gut Mucosa*, Vol I. Boca Raton, FL, CRC Press, 1984, pp 161–171.
62. Gebbers J-O, Ferguson DJP, Mason C, et al: Spirochaetosis of the human rectum associated with an intraepithelial mast cell and IgE plasma cell response. *Gut* **28**:588–593, 1987.
63. Surawicz CM, Goodell SE, Quinn TC, et al: Spectrum of rectal biopsy abnormalities in homosexual men with intestinal symptoms. *Gastroenterology* **91**:651–659, 1986.
64. Weinstein L, Edelstein SM, Madara JL, et al: Intestinal cryptosporidiosis complicated by disseminated cytomegalovirus infection. *Gastroenterology* **81**:584–591, 1981.
65. Current WL, Reese NC, Ernst JV, et al: Human cryptosporidiosis in immunocompetent and immunodeficient persons: Studies of an outbreak and experimental transmission. *N Engl J Med* **308**:1252–1257, 1983.
66. Marcial MA, Madara JL: *Cryptosporidium*: Cellular localization, structural analysis of absorptive cell-parasite-membrane interactions in guinea pigs, and suggestions of protozoan transport by M cells. *Gastroenterology* **90**:583–594, 1986.
67. Ament M, Shimoda S, Saunders D, et al: Pathogenesis of steatorrhea in three cases of small intestinal stasis syndrome. *Gastroenterology* **63**:728–746, 1972.
68. Fagundes-Neto V, Teichberg S, Bayne MA, et al: Bile salt-enhanced rat jejunal absorption of a macromolecular tracer. *Lab Invest* **44**:18–26, 1981.
69. Schuffler MD, Beegle RG: Progressive systemic sclerosis of the gastrointestinal tract and hereditary visceral myopathy: Two distinguishable disorders of intestinal smooth muscle. *Gastroenterology* **77**:664–671, 1979.
70. Mitros FA, Schuffler MD, Teja K, et al: Pathologic features of familial visceral myopathy. *Hum Pathol* **13**:825–833, 1982.
71. Schuffler MD, Kaplan LR, Johnson L: Small-intestinal mucosa in pseudoobstruction syndromes. *Am J Dig Dis* **23**:821–828, 1978.
72. Schuffler MD, Lowe MC, Bill AH: Studies of idiopathic intestinal pseudoobstruction. I. Hereditary hollow visceral myopathy: Clinical and pathological studies. *Gastroenterology* **73**:327–338, 1977.
73. Krishnamurthy S, Schuffler MD: Pathology of neuromuscular disorders of the small intestine and colon. *Gastroenterology* **93**:610–639, 1987.
74. Fang M, Ginsberg AL, Glassman L, et al: Clinical conference: Zollinger-Ellison syndrome with diarrhea as the predominant clinical feature. *Gastroenterology* **76**: 378–387, 1979.
75. Shimoda S, Saunders D, Rubin CE: The Zollinger-Ellison syndrome with steatorrhea. II. The mechanism of fat and vitamin B_{12} malabsorption. *Gastroenterology* **55**:705–723, 1968.
76. McDonald GB: Intestinal graft-versus-host disease. In Chadwick VS, Phillips SF (Eds), *The Small Intestine*. London, Butterworth, 1982.
77. Gallucci RB, McDonald GB, Epstein R, et al: The fine structure of human rectal epithelium in acute graft-versus-host disease. *Am J Surg Pathol* **6**:293–305, 1982.
78. Weber JR, Dobbins WO III: The intestinal and rectal epithelial lymphocyte in AIDS: An electron microscopic study. *Am J Surg Pathol* **10**:627–639, 1986.
79. Dobbins WO III: *Whipple's Disease*. Springfield, IL, CC Thomas, 1987.
80. Trier JS, Phelps PC, Eidelman E, et al: Whipple's disease: Light and electron microscopic correlation of jejunal mucosal histology with antibiotic treatment and clinical status. *Gastroenterology* **48**:684–707, 1965.
81. Dobbins WO III: Is there an immune deficit in Whipple's disease? *Dig Dis Sci* **26**:247–252, 1981.
82. Bank S, Trey C, Gaus I, et al: Histoplasmosis of the small bowel with "giant" intestinal villi and secondary protein-losing enteropathy. *Am J Med* **39**:492–501, 1965.

83. Gillin JS, Urmachen C, West R, et al: Disseminated *Mycobacterium avium intracellulare* infection in acquired immunodeficiency syndrome mimicking Whipple's disease. *Gastroenterology* **85**:1187–1191, 1983.
84. Waldmann T, Wochner R, Laster L, et al: Allergic gastroenteropathy. *N Engl J Med* **276**:761–769, 1967.
85. Leinbach G, Rubin CE: Eosinophilic gastroenteritis: A simple reaction to food allergens. *Gastroenterology* **59**:874–889, 1970.
86. Keshavarzian A, Saverymuttu SH, Tai P-C, et al: Activated eosinophils in familial eosinophilic gastroenteritis. *Gastroenterology* **88**:1041–1049, 1985.
87. Caldwell JH, Sharma HM, Hurtubise PE, et al: Eosinophilic gastroenteritis–extreme allergy. *Gastroenterology* **77**:560–564, 1979.
88. Katz AJ, Goldman H, Grand RJ: Gastric mucosal biopsy in eosinophilic (allergic) gastroenteritis. *Gastroenterology* **73**:705–709, 1977.
89. Dobbins JW, Sheahan DG, Behar J: Eosinophilic gastroenteritis with esophageal involvement. *Gastroenterology* **72**:1312–1316, 1977.
90. Eidelman S, Parkins RA, Rubin CE: Abdominal lymphoma presenting as malabsorption: A clinicopathologic study of nine cases in Israel, and a review of the literature. *Medicine* **45**:111–137, 1966.
91. Ramot B: Malabsorption due to lymphomatous disease. *Ann Rev Med* **22**:19–24, 1971.
92. Khojasteh A, Kaghsheness M, Haghighi P: Immunoproliferative small intestinal disease: A "third world lesion." *N Engl J Med* **308**:1401–1405, 1983.
93. Matuchansky C, Touchard G, Babin P, et al: Diffuse small intestinal lymphoid infiltration in nonimmunodeficient adults from Western Europe. *Gastroenterology* **95**:470–477, 1988.
94. Gray GM, Rosenberg SA, Cooper AD, et al: Lymphomas involving the gastrointestinal tract. *Gastroenterology* **82**:143–152, 1982.
95. Gudjonsson H, Jones M, Krawitt EL, et al: Pseudolymphoma of the jejunum. *Dig Dis Sci* **32**:1314–1318, 1987.
96. Levinson JD, Nastro LJ: Giardiasis with total villous atrophy. *Gastroenterology* **74**:271–275, 1978.
97. Saha TK, Ghosh TK: Invasion of small intestinal mucosa by *Giardia lamblia* in man. *Gastroenterology* **72**:402–405, 1977.
98. Liebman WM, Thaler MM, DeLorimier A, et al: Intractable diarrhea of infancy due to intestinal coccidiosis. *Gastroenterology* **78**:579–584, 1980.
99. Meisel JL, Perea DR, Meligro C, et al: Overwhelming watery diarrhea associated with a cryptosporidium in an immunosuppressed patient. *Gastroenterology* **70**:1156–1160, 1976.
100. Trier JS, Moxey PC, Schimmel EM, et al: Chronic intestinal coccidiosis in man: Intestinal morphology and response to treatment. *Gastroenterology* **66**:923–935, 1974.
101. Rijpstra AC, Canning EU, vanKetel RJ, et al: Use of light microscopy to diagnose small-intestinal microsporidiosis in patients with AIDS. *J Infect Dis* **157**:827–831, 1988.
102. Milner P, Irvine R, Barton C, et al: Intestinal malabsorption in *Strongyloides stercoralis* infestation. *Gut* **6**:574–581, 1965.
103. Weller PF: Case records of the Massachusetts General Hospital. *N Engl J Med* **314**:903–913, 1986.
104. Whalen G, Strickland G, Cross J, et al: Intestinal capillariasis. *Lancet* **1**:13–16, 1969.
105. Witham RR, Mossen RS: An unusual presentation of schistosomiasis duodenitis. *Gastroenterology* **77**:1316–1318, 1979.
106. Muigai R, Shaunak S, Gatei DG, et al: Jejunal function and pathology in visceral leishmaniasis. *Lancet* **2**:476–479, 1983.

107. Miller DP, Everett ED: Gastrointestinal histoplasmosis. *J Clin Gastroenterol* **1**:233–236, 1979.
108. Joshi SN, Garvin PJ, Sunwoo UC: Candidiasis of the duodenum and jejunum. *Gastroenterology* **80**:829–833, 1981.
109. Freeman HJ, Shnitka TK, Piercey JRA, et al: Cytomegalovirus infection of the gastrointestinal tract in a patient with late onset immunodeficiency syndrome. *Gastroenterology* **73**:1397–1403, 1977.
110. Franzin G, Muolo A, Griminelli T: Cytomegalovirus inclusions in the gastroduodenal mucosa of patients after renal transplantation. *Gut* **22**:698–701, 1981.
111. Knapp AB, Horst DA, Eliopoulos G, et al: Widespread cytomegalovirus gastroenterocolitis in a patient with acquired immunodeficiency syndrome. *Gastroenterology* **85**:1399–1402, 1983.
112. Blacklow NR, Cukor G: Viral gastroenteritis. *N Engl J Med* **304**:397–406, 1981.
113. Bishop RF, Davidson GP, Holmes IH, et al: Virus particles in epithelial cells of duodenal mucosa from children with acute non-bacterial gastroenteritis. *Lancet* **2**:1281–1283, 1973.
114. Brandt LJ, Davidoff A, Bernstein LH, et al: Small-intestinal involvement in Waldenstrom's macroglobulinemia: Case report and review of the literature. *Dig Dis Sci* **26**:174–180, 1981.
115. Harris M, Burton IE, Scarffe JH: Macroglobulinemia and intestinal lymphangiectasia: A rare association. *J Clin Pathol* **36**:30–36, 1983.
116. Waldmann TA, Steinfeld JL, Dutcher TF, et al: The role of the gastrointestinal system in "idiopathic hypoproteinemia." *Gastroenterology* **41**:197–207, 1961.
117. Popovic OS, Brkic S, Bojic P, et al: Sarcoidosis and protein losing enteropathy. *Gastroenterology* **78**:119–125, 1980.
118. Dobbins WO III, Rollins EL: Intestinal mucosal lymphatic permeability: An electron microscopic study of endothelial vesicles and cell junctions. *J Ultrastruct Res* **33**:29–59, 1970.
119. Dobbins WO III: Hypo-β-lipoproteinemia and intestinal lymphangiectasia: A new syndrome of malabsorption and protein losing enteropathy. *Arch Intern Med* **121**:31–38, 1968.
120. Maizel H, Ruffin JM, Dobbins WO III: Whipple's disease: A review of 19 patients from one hospital and a review of the literature since 1950. *Medicine* **49**:175–205, 1970.
121. Kalima T: The structure and function of intestinal lymphatics and the influence of impaired lymph flow on the ileum of rats. *Scand J Gastroenterol* **6**:Suppl 10:1–87, 1971.
122. Novi J, Duong HD, Hoang CT: Ultrastructure of intestinal lymphatics in Crohn's disease. *Am J Clin Pathol* **76**:385–394, 1981.
123. Foroozan P, Trier J: Mucosa of the small intestine in pernicious anemia. *N Engl J Med* **277**:553–559, 1967.
124. Bianchi A, Chipman D, Dreskin A, et al: Nutritional folic acid deficiency with megaloblastic changes in the small bowel epithelium. *N Engl J Med* **282**:859–861, 1970.
125. Trier J, Browning T: Morphologic response of mucosa of small intestine to x-ray exposure. *J Clin Invest* **45**:194–204, 1966.
126. Smith FP, Kisner DL, Widerlite L, et al: Chemotherapeutic alteration of small intestinal morphology and function: A progress report. *J Clin Gastroenterol* **1**:203–207, 1979.
127. Dobbins WO III: Drug-induced steatorrhea. *Gastroenterology* **54**:1193–1195, 1968.
128. Longstreth GF, Newcomer AD: Drug-induced malabsorption. *Mayo Clin Proc* **50**:284–293, 1975.
129. Freeman HJ: Sulindac-associated small bowel lesion. *J Clin Gastroenterol* **8**:569–571, 1986.

130. Race TF, Paes IC, Faloon WW: Intestinal malabsorption induced by oral colchicine: Comparison with neomycin and cathartic agents. *Am J Med Sci* **259**:32–41, 1970.
131. Dobbins WO III, Herrero BA, Mansbach CM: Morphologic alterations associated with neomycin induced malabsorption. *Am J Med Sci* **255**:63–77, 1968.
132. Baraona E, Pirola RC, Lieber CS: Small intestinal damage and changes in cell population produced by ethanol ingestion in the rat. *Gastroenterology* **66**:226–234, 1974.
133. Thomas FB, Baba N, Greenberger NJ, et al: Effect of phenobarbital on small intestinal structure and function in the rat. *J Lab Clin Med* **80**:548–558, 1972.
134. Trier JS: Morphologic alterations induced by methotrexate in the mucosa of human proximal intestine. II. Electron microscopic observations. *Gastroenterology* **43**:407–424, 1962.
135. Wiernik G: Changes in the villous pattern of the human jejunum associated with heavy radiation damage. *Gut* **7**:149–153, 1966.
135a. Bassen FA, Kornzweig AL: Malformation of the erythrocytes in a case of atypical retinitis pigmentosa. *Blood* **5**:381–387, 1950.
136. Gotto AM, Levy RI, John K, et al: On the protein defect in a-beta-lipoproteinemia. *N Engl J Med* **284**:813–818, 1971.
137. Dobbins WO III: An ultrastructural study of the intestinal mucosa in congenital β-lipoprotein deficiency with particular emphasis upon the intestinal absorptive cell. *Gastroenterology* **50**:195–210, 1966.
138. Dobbins WO III: Diagnostic pathology of the intestine and colon. In Trump BF, Jones RT (eds), *Diagnostic Electron Microscopy*, Vol. I. New York, John Wiley, 1978, pp 253–339.
139. Hermos J, Cooper H, Kramer P, et al: Histological diagnosis by peroral biopsy of Crohn's disease of the proximal intestine. *Gastroenterology* **59**:868–873, 1970.
140. Schuffler MD, Chaffee RG: Small inestinal biopsy in a patient with Crohn's disease of the duodenum: The spectrum of abnormal findings in the absence of granulomas. *Gastroenterology* **76**:1009–1014, 1979.
141. Sprague R, Harper P, McClain S, et al: Disseminated gastrointestinal sarcoidosis: Case report and review of the literature. *Gastroenterology* **87**:421–425, 1984.
142. Bjorneklett A, Fausa O, Refsum SB, et al: Jejunal villous atrophy and granulomatous inflammation responding to a gluten-free diet. *Gut* **18**:814–816, 1977.
143. Douglas JG, Logan RFA, Gillon J, et al: Sarcoidosis and coeliac disease: An association? *Lancet* **2**:13–14, 1984.
144. Dvoark AM, Connell AB, Dickersin GR: Crohn's disease: A scanning electron microscopic study. *Hum Pathol* **10**:165–177, 1979.
145. Rickert RR, Carter HW: The early ulcerative lesion of Crohn's disease: Correlative light and scanning electron microscopic studies. *J Clin Gastroenterol* **2**:11–19, 1980.
146. Dvorak AM, Dickersin GR: Crohn's disease: Transmission electron microscopic studies. I. Barrier function. Possible changes related to alterations of cell coat, mucous coat, epithelial cells, and Paneth cells. *Hum Pathol* **11**:561–571, 1980.
147. Dvorak AM, Monahan RA, Osage JE, et al: Crohn's disease: Transmission electron microscopic studies. II. Immunologic inflammatory response. Alternations of mast cells, basophils, eosinophils, and the microvasculature. *Hum Pathol* **11**:606–619, 1980.
148. Dvorak AM, Osage GE, Monahan RA, et al: Crohn's disease: Transmission electron microscopic studies. III. Target tissue. Proliferation and injury to smooth muscle and the autonomic nervous system. *Hum Pathol* **11**:620–634, 1980.
149. Bishop AE, Polak JM, Bryant MG: Abnormalities of vasoactive intestinal polypeptide-containing nerves in Crohn's disease. *Gastroenterology* **79**:853–860, 1980.
150. Marin ML, Greenstein AJ, Geller SA, et al: Freeze-fracture analysis of epithelial cell lysosomal inclusions in Crohn's disease. *Ultrastruct Pathol* **6**:39–44, 1984.

151. Dourmashkin RR, Davies J, Wells C, et al: Epithelial patchy necrosis in Crohn's disease. *Hum Pathol* **14**:643–648, 1984.
152. Bagdade JD, Parker F, Ways PO, et al: Fabry's disease: A correlative clinical, morphological and biochemical study. *Lab Invest* **18**:681–688, 1968.
153. O'Brien BD, Shnitka TK, McDougall R, et al: Pathophysiologic and ultrastructural basis for intestinal systems in Fabry's disease. *Gastroenterology* **82**:957–962, 1982.
154. Partin J, Schubert W: Small intestinal mucosa in cholesterol ester storage disease: A light and electron microscope study. *Gastroenterology* **57**:542–558, 1969.
155. Ferrans VJ, Fredrickson DS: The pathology of Tangier disease: A light and electron microscopic study. *Am J Pathol* **78**:101–158, 1975.
156. Kayden HJ: Abetalipoproteinemia. *Ann Rev Med* **23**:285–296, 1972.
157. Roy CC, Levy E, Green PHR, et al: Malabsorption, hypocholesterolemia, and fat-filled enterocytes with increased intestinal apoprotein B: Chylomicron retention disease. *Gastroenterology* **92**:390–399, 1987.
158. Mars H, Lewis LA, Robertson AL, et al: Familial hypo-β-lipoproteinemia: A genetic disorder of lipid metabolism with nervous system involvement. *Am J Med* **46**:886–900, 1969.
159. Kocen RS, Lloyd JK, Lascelles PT, et al: Familial α-lipoprotein deficiency (Tangier disease) with neurological abnormalities. *Lancet* **1**:1341–1345, 1967.
160. Morecki R, Paunier, Hamilton JR: Intestinal mucosa in cystinosis: A fine structure study. *Arch Pathol* **86**:297–307, 1968.
161. Gear EV Jr, Dobbins WO III: Rectal biopsy: A review of its diagnostic usefulness. *Gastroenterology* **55**:522–544, 1968.
162. Mordechai R, Sohar E: Intestinal malabsorption: First manifestation of amyloidosis in familial Mediterranean fever. *Gastroenterology* **66**:446–449, 1974.
163. Glenner GG: Amyloid deposits and amyloidosis: The β-fibrilloses. *N Engl J Med* **302**:1283–1292, 1333–1343, 1980.
164. Rex DK, Jersild RA Jr: Further characterization of the pigment in pseudomelanosis duodeni in three patients. *Gastroenterology* **95**:177–182, 1988.
165. Lin H-J, Isay S-H, Chiang H, et al: Pseudomelanosis duodeni: Case report and review of the literature. *J Clin Gastroenterol* **10**:155–159, 1988.
166. Horn T, Svendensen LB, Aage J, et al: Brown bowel syndrome. *Ultrastruct Pathol* **8**:357–361, 1985.
167. Troncale FJ, Miller LH, Keusch GT, et al: Occurrence of ceroid pigment without malabsorption in Thailand. *Gastroenterology* **54**:1122–1128, 1968.
168. Jarnum S, Zachariae H: Mastocytosis (urticaria pigmentosa) of skin, stomach, and gut with malabsorption. *Gut* **8**:64–68, 1967.
169. Scott BB, Hardy GJ, Losowsky MS: Involvement of the small intestine in systemic mast cell disease. *Gut* **16**:918–924, 1975.
170. Fishman RS, Fleming CR, Li C-Y: Systemic mastocytosis with review of gastrointestinal manifestations. *Mayo Clin Proc* **4**:51–54, 1979.
171. Dobbins WO III, Tomasini JT, Rollins EL: Electron and light microscopic identification of the mast cell of the gastrointestinal tract. *Gastroenterology* **56**:268–279, 1969.
172. Gear EV Jr, Dobbins WO III: The histologic spectrum of proximal duodenal biopsy in man. *Am J Med Sci* **257**:90–99, 1969.
173. Korn ER, Foroozan P: Endoscopic biopsies of normal duodenal mucosa. *Gastrointest Endosc* **21**:51–54, 1974.
174. Cotton PB, Prince AB, Tighe JR, et al: Preliminary evaluation of "duodenitis" by endoscopy and biopsy. *Br Med J* **3**:430–433, 1973.
175. Scott BB, Jenkins D: Endoscopic small intestinal biopsy. *Gastrointest Endosc* **27**:162–167, 1981.
176. Whitehead R: *Mucosal Biopsy of the Gastrointestinal Tract* (3rd ed). In *Major Problems in Pathology*, Vol. 3. Philadelphia, WB Saunders, 1985.

177. Weinstein WM: The diagnosis and classification of gastritis and duodenitis. *J Clin Gastroenterol* **3**(Suppl 2):7–16, 1981.
178. Scott BB, Losowsky MS: Patchiness and duodenal-jejunal variation of the mucosal abnormality in coeliac disease and dermatitis herpetiformis. *Gut* **17**:984–992, 1976.
179. Goldman H, Antonioli DA: Mucosal biopsy of the rectum, colon, and distal ileum. *Hum Pathol* **13**:981–1012, 1982.
180. McDonald WC, Brandborg LL, Flick AL, et al: Studies of celiac sprue. IV. The response of the whole length of the small bowel to a gluten free diet. *Gastroenterology* **47**:573–589, 1964.
181. Goldstein MJ, Melamed MR, Grabstald H, et al: Progressive villous atrophy of the ileum used as a urinary conduit. *Gastroenterology* **52**:859–864, 1967.
182. Philipson B, Brandberg A, Jagenberg R, et al: Mucosal morphology, bacteriology, and absorption in intra-abdominal ileostomy reservoir. *Scand J Gastroenterol* **10**:145–153, 1975.
183. Mortensen N: Progress with the pouch-restorative proctocolectomy for ulcerative colitis. *Gut* **29**:561–565, 1988.
184. Popovic OS, Kostic KM, Milovic VB, et al: Primary bile acid malabsorption: Histologic and immunologic study in three patients. *Gastroenterology* **92**:1851–1858, 1987.
185. Austad WI, Lack L, Tyor MP: Importance of bile acids and of an intact distal small intestine for fat absorption. *Gastroenterology* **52**:638–646, 1967.

Index

MIX
Papier aus verantwortungsvollen Quellen
Paper from responsible sources
FSC® C105338

If you have any concerns about our products,
you can contact us on
ProductSafety@springernature.com

In case Publisher is established outside the EU,
the EU authorized representative is:
Springer Nature Customer Service Center GmbH
Europaplatz 3, 69115 Heidelberg, Germany

Printed by Libri Plureos GmbH
in Hamburg, Germany